Mathematical
Surveys
and
Monographs

Volume 182

Introduction to Heat Potential Theory

Neil A. Watson

American Mathematical Society
Providence, Rhode Island

EDITORIAL COMMITTEE

Ralph L. Cohen, Chair
Michael A. Singer

Benjamin Sudakov
Michael I. Weinstein

2010 *Mathematics Subject Classification.* Primary 31-02, 31B05, 31B20, 31B25, 31C05, 31C15, 35-02, 35K05, 31B15.

For additional information and updates on this book, visit
www.ams.org/bookpages/surv-182

Library of Congress Cataloging-in-Publication Data
Watson, N. A., 1948–
  Introduction to heat potential theory / Neil A. Watson.
    p. cm. – (Mathematical surveys and monographs ; v. 182)
  Includes bibliographical references and index.
  ISBN 978-0-8218-4998-9 (alk. paper)
  1. Potential theory (Mathematics) I. Title.

QA404.7.W38 2012
515′.96–dc23
2012004904

Copying and reprinting. Individual readers of this publication, and nonprofit libraries acting for them, are permitted to make fair use of the material, such as to copy a chapter for use in teaching or research. Permission is granted to quote brief passages from this publication in reviews, provided the customary acknowledgment of the source is given.

Republication, systematic copying, or multiple reproduction of any material in this publication is permitted only under license from the American Mathematical Society. Requests for such permission should be addressed to the Acquisitions Department, American Mathematical Society, 201 Charles Street, Providence, Rhode Island 02904-2294 USA. Requests can also be made by e-mail to reprint-permission@ams.org.

© 2012 by the American Mathematical Society. All rights reserved.
The American Mathematical Society retains all rights
except those granted to the United States Government.
Printed in the United States of America.

∞ The paper used in this book is acid-free and falls within the guidelines
established to ensure permanence and durability.
Visit the AMS home page at http://www.ams.org/

10 9 8 7 6 5 4 3 2 1   17 16 15 14 13 12

Contents

Preface	vii
Notation and Terminology	xi

Chapter 1. The Heat Operator, Temperatures and Mean Values	1
1.1. Temperatures and Heat Balls	1
1.2. Mean Values of Smooth Functions over Heat Spheres	3
1.3. Mean Values of Smooth Subtemperatures over Heat Spheres	7
1.4. Mean Values of Smooth Subtemperatures over Heat Balls	13
1.5. The Boundary Maximum Principle on Circular Cylinders	17
1.6. Modified Heat Balls	19
1.7. Harnack Theorems	25
1.8. Equicontinuous Families of Temperatures	29
1.9. Notes and Comments	31

Chapter 2. The Poisson Integral for a Circular Cylinder	35
2.1. The Cauchy Problem on a Half-Space	35
2.2. The Dirichlet Problem on a Circular Cylinder	37
2.3. Double Layer Heat Potentials	39
2.4. The Poisson Integral and the Caloric Measure	44
2.5. Characterizations of Temperatures	47
2.6. Extensions of some Harnack Theorems	51
2.7. Notes and Comments	52

Chapter 3. Subtemperatures and the Dirichlet Problem on Convex Domains of Revolution	53
3.1. Semicontinuous Functions	53
3.2. Subtemperatures	55
3.3. The Dirichlet Problem on Convex Domains of Revolution	64
3.4. Boundary Behaviour of the PWB Solution	69
3.5. Characterizations of Hypotemperatures and Subtemperatures	71
3.6. Properties of Hypotemperatures	80
3.7. Thermic Majorants	82
3.8. Notes and Comments	83

Chapter 4. Temperatures on an Infinite Strip	85
4.1. An Extension of the Maximum Principle on an Infinite Strip	85
4.2. Gauss-Weierstrass Integrals	87
4.3. Nonnegative Temperatures	95
4.4. Minimality of the Fundamental Temperature	101
4.5. Notes and Comments	103

Chapter 5. Classes of Subtemperatures on an Infinite Strip	105
5.1. Hyperplane Mean Values and Classes of Subtemperatures	105
5.2. Behaviour of the Hyperplane Mean Values of Subtemperatures	114
5.3. Classes of Subtemperatures and Nonnegative Thermic Majorants	119
5.4. Characterizations of the Gauss-Weierstrass Integrals of Functions	123
5.5. Notes and Comments	126
Chapter 6. Green Functions and Heat Potentials	127
6.1. Green Functions	127
6.2. Green Functions and the Adjoint Heat Equation	131
6.3. Heat Potentials	134
6.4. The Distributional Heat Operator	140
6.5. The Riesz Decomposition Theorem	146
6.6. Monotone Approximation by Smooth Supertemperatures	150
6.7. Further Characterizations of Subtemperatures	151
6.8. Supertemperatures on an Infinite Strip or Half-Space	152
6.9. Notes and Comments	157
Chapter 7. Polar Sets and Thermal Capacity	159
7.1. Polar Sets	159
7.2. Families of Supertemperatures	162
7.3. The Natural Order Decomposition	166
7.4. Reductions and Smoothed Reductions	170
7.5. The Thermal Capacity of Compact Sets	175
7.6. The Thermal Capacity of More General Sets	178
7.7. Thermal and Cothermal Capacities	183
7.8. Capacitable Sets	183
7.9. Polar Sets and Heat Potentials	187
7.10. Thermal Capacity and Lebesgue Measure	188
7.11. Notes and Comments	192
Chapter 8. The Dirichlet Problem on Arbitrary Open Sets	195
8.1. Classification of Boundary Points	196
8.2. Upper and Lower PWB Solutions	199
8.3. Resolutivity and PWB Solutions	205
8.4. The Caloric Measure on the Essential Boundary	207
8.5. Boundary Behaviour of PWB Solutions	214
8.6. Geometric Tests for Regularity	222
8.7. Green Functions, Heat Potentials, and Thermal Capacity	225
8.8. Notes and Comments	228
Chapter 9. The Thermal Fine Topology	231
9.1. Definitions and Basic Properties	231
9.2. Further Properties of Reductions	237
9.3. The Fundamental Convergence Theorem	240
9.4. Applications of the Fundamental Convergence Theorem to Reductions	244
9.5. Thermal Thinness and the Regularity of Normal Boundary Points	249
9.6. Thermal Fine Limits and Euclidean Limits	252
9.7. Thermal Thinness and the Quasi-Lindelöf Property	253

9.8. Notes and Comments 257

Bibliography 259

Index 263

Preface

This book is the first to be devoted entirely to the potential theory of the heat (or diffusion) equation

$$\sum_{i=1}^{n} \frac{\partial^2 u}{\partial x_i^2} = \frac{\partial u}{\partial t}$$

in Euclidean space $\mathbb{R}^{n+1} = \mathbb{R}^n \times \mathbb{R}$. It thus deals with time dependent potential theory. Its purpose is to give a logical, coherent introduction to a subject that has been approached in several conflicting ways.

The subject has had an unusual history. Some of the elementary results have appeared only in the twenty-first century, but some other results were included in an axiomatic theory in the nineteen sixties. I make no reference to the axiomatic theory in the text, but the definition and treatment of subtemperatures in Chapter 3 is designed to reconcile the harmonic spaces definition with the one that I have used in my researches. This approach is very recent, having first appeared in 2008.

Most results in the heat potential theory have been modelled on the classical results for Laplace's equation. However, after a great deal of thought, I decided to make no mention of the classical theory in the text. I wanted to write a book in which heat potential theory stands as a subject in its own right, free from the clutter of perpetual references to the classical case. Many of the proofs in heat potential theory are similar to those in classical potential theory, so if the classical case is covered first, then either the classical proofs have to be duplicated, or the proofs for the case of the heat equation have to be replaced by a claim that the proofs are similar to the classical case. Many times I have seen such claims in print, but in a substantial number of cases the claim has proved to be false. There is no substitute for writing out all the details of a proof. One could, of course, deduce the results of classical potential theory from those of heat potential theory, as the former is the special case of the latter in which nothing depends on time. But I don't think that would be of much interest. Such an approach to the classical case would be unnecessarily tortuous, and there are already some excellent texts that deal solely with classical potential theory.

The reader already familiar with the heat equation, may be surprised to find that the explicit Poisson integral representation of solutions of the heat equation on a rectangular domain, is not mentioned in the text. I have found it unnecessary, and so have been delighted to omit it because of its complication, which may even have deterred some mathematicians from researching on the heat equation. Its place has been taken by a caloric measure interpretation of the representation of solutions on a circular cylinder.

I have reworked the entire content of this book, including not only most of the individual proofs but also the overall approach. Despite this, the only essentially new material here is that on caloric measure in Chapter 8, a few minor theorems, and a few examples. Otherwise, there is just a great deal of fine-tuning, including a different approach to the Riesz decomposition theorem in Chapter 6. As befits an introduction, I have treated the subject in as elementary a way as I could. I have not attempted a complete coverage, and in particular have made no mention of the probabilistic approach. The prerequisites for understanding all the proofs are a good background in the calculus of functions of several variables, in the limiting processes and inequalities of analysis, in measure theory, and in general topology for Chapter 9. Some general analytic results that are not easy to find elsewhere, or are not necessarily presented in the most suitable form elsewhere, have been included where they are needed.

The material is presented in logical order, which differs from the chronological order in which the results were first discovered. Chapter 1 deals with elementary issues, although the results are more recent than might be expected, and those in Section 1.6 first appeared in 2002. Chapter 2 presents the classical existence theory for temperatures on a circular cylinder, which is much older than the results in Chapter 1. Subtemperatures are introduced in Chapter 3, using a definition based on the representation theorem in Chapter 2. However, heat balls and modified heat balls are essential to our treatment, and Chapter 3 contains some necessary existence theory for temperatures on such, and other, domains. This introduces the PWB approach to the Dirichlet problem without the added complications of arbitrary open sets and arbitrary boundary functions. Chapters 4 and 5 deal with temperatures and subtemperatures, respectively, on domains of the form $\mathbb{R}^n \times\,]0, a[$, where $0 < a \leq +\infty$. Although potentials are used as early as Chapter 2, the general theory of heat potentials begins in earnest in Chapter 6, where Green functions and heat potentials are introduced, and a distributional approach to the Riesz decomposition theorem is taken. Chapter 7 deals with polar sets, reductions, and thermal capacity. In Chapter 8 we consider the generalized Dirichlet problem, where the open sets and boundary functions are arbitrary. This chapter includes a new treatment of caloric measure for such sets. Finally, in Chapter 9 we discuss the thermal fine topology, which gives us an insight into the continuity properties of subtenperatures, and thus enables us to improve upon some earlier results. Each chapter concludes with bibliographical notes and comments, which include mention of matters not covered in the text, and of open questions even in Chapter 1. They do not provide a detailed historical account of the theory, except for the more recent results. They contain very few references to the axiomatic approach, and none to the probabilistic approach, mainly because those approaches have very different starting points and linking them with the main text would take too much space.

It is a pleasure to acknowledge my debt to those who have written earlier books on potential theory. In particular, L. L. Helms' book *Introduction to Potential Theory* [**33**] first inspired me to take up the subject, with the eventual goal of writing a book such as this one. J. L. Doob's *Classical Potential Theory and its Probabilistic Counterpart* [**14**] has been a source of ideas for clever proofs, and for most of the material in Chapter 9, but is definitely not for the novice. D. H. Armitage & S. J. Gardiner's *Classical Potential Theory* [**3**] has provided many ideas for clever proofs, which I have been able to adapt to the case of the heat equation. The material in Chapter 2, on the existence of solutions to the Dirichlet problem on a circular cylinder, is based on the treatment in E. M. Landis' *Second Order Equations of*

Elliptic and Parabolic Type [**49**].

The reader who wants to look at an axiomatic approach to potential theory that includes heat potential theory, could consult H. Bauer's *Harmonische Räume und ihre Potentialtheorie* [**5**], C. Constantinescu & A. Cornea's *Potential Theory on Harmonic Spaces* [**12**], or J. Bliedtner & W. Hansen's *Potential Theory: An Analytic and Probabilistic Approach to Balayage* [**7**].

Notation and Terminology

We summarize here our basic notations and conventions. Most notation will be explained as it is introduced during the course of the book, and is indexed.

We say that a number or function f is *positive* if $f > 0$, *negative* if $f < 0$, *nonnegative* if $f \geq 0$, and *nonpositive* if $f \leq 0$. However, we say that a real function f is *increasing* if $f(a) \leq f(b)$ whenever $a \leq b$, and that it is *decreasing* if $f(a) \geq f(b)$ whenever $a \leq b$. Similarly, for sequences of numbers or functions, the terms *increasing* and *decreasing* are used in the wide sense.

We use $\mathbb{R}^n$ to denote real Euclidean space of dimension n, with $n \geq 1$, but often omit the superscript if $n = 1$. We also denote the set of positive integers by $\mathbb{N}$, and the set of rational numbers by $\mathbb{Q}$. A typical point of $\mathbb{R}^n$ is denoted by $x = (x_1, ..., x_n)$, and we write $|x|$ for the Euclidean norm $\left(\sum_{i=1}^{n} x_i^2\right)^{1/2}$ of x. The inner product $\sum_{i=1}^{n} x_i y_i$ of two points x and y in $\mathbb{R}^n$ is written $\langle x, y \rangle$. Most of the material is presented in the context of $\mathbb{R}^{n+1} = \mathbb{R}^n \times \mathbb{R} = \{(x,t) : x \in \mathbb{R}^n, t \in \mathbb{R}\}$, where the variables $x = (x_1, ..., x_n)$ are called the spatial variables and t is called the temporal variable. Where there is no need to specify these variables separately, we use p or q to denote a typical point of $\mathbb{R}^{n+1}$, reserving x and y for points of $\mathbb{R}^n$. We denote the Euclidean norm of a point p in $\mathbb{R}^{n+1}$ by $|p|$, leaving the notation for a point to distinguish between $|p|$ and $|x|$. Similarly, the open balls of radius r in $\mathbb{R}^n$ and $\mathbb{R}^{n+1}$ are denoted by $B(x,r)$ and $B(p,r)$, respectively; thus $B(x,r) = \{y \in \mathbb{R}^n : |x-y| < r\}$ and $B(p,r) = \{q \in \mathbb{R}^{n+1} : |p-q| < r\}$. The points x and p are called the centres of the respective balls. A unit ball is a ball of radius 1, and a unit sphere is its boundary. The origin of Euclidean space is denoted by 0, regardless of the dimension of the space.

All topological concepts are relative to the Euclidean topology of $\mathbb{R}^{n+1}$, unless otherwise stated. The symbol E denotes an open set in $\mathbb{R}^{n+1}$, which is always assumed to be nonempty. For any set S in $\mathbb{R}^{n+1}$, we denote its boundary by ∂S and its closure by $\overline{S}$, although we denote the closure of a ball by $\overline{B}(p,r)$ rather than $\overline{B(p,r)}$, and similarly for other sets that depend on listed parameters. The boundary of a set is taken with respect to the one-point compactification of $\mathbb{R}^{n+1}$, so that the point at infinity is included if the set is unbounded. The interior of S is denoted by S°. The connected components of a set are referred to simply as its *components*, and a nonempty connected open set is called a *domain*. By a *hyperplane*, we mean a set of the form $\{p \in \mathbb{R}^{n+1} : \langle p, q \rangle = a\}$ for some $q \in \mathbb{R}^{n+1}$ and $a \in \mathbb{R}$. If A and B are two sets, we put $A \backslash B = \{p \in A : p \notin B\}$. A set is called a G_δ set if it can be expressed as a countable intersection of open sets, and an F_σ set if it can be expressed as a countable union of closed sets.

All of our functions are extended real-valued, that is, their values are in $\mathbb{R}$ or are $\pm\infty$. This necessitates a limited arithmetic with $\pm\infty$ when we add or multiply functions, so we adopt the following conventions, in which $t \in \mathbb{R}$:

$$(\pm\infty) + (\pm\infty) = \pm\infty = t + (\pm\infty) = (\pm\infty) + t,$$

$$(\pm\infty).(\pm\infty) = +\infty, \qquad (\pm\infty).(\mp\infty) = -\infty,$$

$$t.(\pm\infty) = (\pm\infty).t = \begin{cases} \pm\infty & \text{if } t > 0, \\ 0 & \text{if } t = 0, \\ \mp\infty & \text{if } t < 0. \end{cases}$$

Other expressions, such as $(\pm\infty) + (\mp\infty)$, are left undefined. We put $\inf \emptyset = +\infty$ and $\sup \emptyset = -\infty$.

If f is an extended real-valued function defined on a set $S \subset \mathbb{R}^{n+1}$, q is a limit point of S in some topology, and $\mathcal{N}_q$ is the collection of neighbourhoods of q in that topology, then we define

$$\liminf_{p \to q,\, p \in S} f(p) = \sup_{N \in \mathcal{N}_q} \left(\inf_{p \in N \cap S \setminus \{q\}} f(p) \right)$$

and

$$\limsup_{p \to q,\, p \in S} f(p) = \inf_{N \in \mathcal{N}_q} \left(\sup_{p \in N \cap S \setminus \{q\}} f(p) \right).$$

We say that $\lim_{p \to q,\, p \in S} f(p)$ exists if $\liminf_{p \to q,\, p \in S} f(p) = \limsup_{p \to q,\, p \in S} f(p)$, and if that common value is l we write $\lim_{p \to q,\, p \in S} f(p) = l$. Here l may be a real number or $\pm\infty$. If S is the domain of definition of f, or if $S \in \mathcal{N}_q$, then we may omit the qualification "$p \in S$". We say that f is continuous at q if f is defined at q and $\lim_{p \to q} f(p) = f(q)$, regardless of whether $f(q) \in \mathbb{R}$.

If u and v are extended real-valued functions defined on the same set, we use $u \vee v$ to denote $\max\{u, v\}$ and $u \wedge v$ to denote $\min\{u, v\}$. We also put $u^+ = u \vee 0$ and $u^- = -(u \wedge 0)$, thus obtaining the identities $u = u^+ - u^-$ and $|u| = u^+ + u^-$. If S is a subset of the domain of definition of u, and $u(p) \leq M$ for all $p \in S$ and some real number M, then we say that u is *upper bounded* on S. Similarly, if $u(p) \geq m$ for all $p \in S$ and some $m \in \mathbb{R}$, then we say that u is *lower bounded* on S. If u is both upper bounded on S and lower bounded on S, we say that u is *bounded* on S. If u is bounded on K for each compact subset K of S, then we say that u is *locally bounded* on S. We define *locally upper bounded* and *locally lower bounded* analogously. If D is the domain of definition of u, we define the *support* of u to be the set $D \setminus \{p \in D : u = 0 \text{ on } D \cap B(p, r) \text{ for some } r > 0\}$.

A family $\mathcal{F}$ of functions defined on a set S is said to be *uniformly bounded* on S if there is a real number M such that $|u(p)| \leq M$ for all $u \in \mathcal{F}$ and all $p \in S$. The family $\mathcal{F}$ is said to be *locally uniformly bounded* on S if it is uniformly bounded on K for each compact subset K of S. We define the phrases *uniformly upper bounded*, *uniformly lower bounded*, *locally uniformly upper bounded*, and *locally uniformly lower bounded*, analogously. A sequence $\{u_j\}$ is said to converge *locally uniformly* on S if it converges uniformly on each compact subset of S.

Let X be a subset of the one-point compactification of $\mathbb{R}^{n+1}$. The class $\mathcal{B}$ of *Borel* subsets of X is the smallest σ-algebra to contain the open subsets of X. We say that an extended real-valued function u on X is *Borel measurable* if the set $\{p \in X : u(p) > a\}$ belongs to $\mathcal{B}$ for every real number a. Continuous functions are Borel measurable. A *nonnegative (Borel) measure* on X is a countably additive set function μ, defined on a σ-algebra that contains $\mathcal{B}$, taking nonnegative extended-real values, such that $\mu(\emptyset) = 0$ and $\mu(K) < +\infty$ for every compact subset K of X. Such a measure is regular, in the sense that

$$\mu(S) = \inf\{\mu(E) : S \subseteq E, \ E \text{ is open}\} = \sup\{\mu(K) : K \subseteq S, \ K \text{ is compact}\}.$$

The *support* of a nonnegative measure μ is the set of points $p \in X$ such that $\mu(N) > 0$ for every open neighbourhood N of p. It is the smallest closed set F such that $\mu(X \backslash F) = 0$. If S belongs to the σ-algebra upon which μ is defined, we say that S is μ-*measurable*, and define the *restriction* of μ to S by $\mu_S(T) = \mu(T \cap S)$ for all μ-measurable sets T. If $X \subseteq Y$ and $X \neq Y$, we define the restriction of μ to S as a nonnegative measure on Y by adding the condition $\mu_S(Y \backslash X) = 0$. We say that an extended real-valued function u on X is μ-*measurable* if the set $\{p \in X : u(p) > a\}$ is μ-measurable for every $a \in \mathbb{R}$. In the case where μ is Lebesgue measure, we omit the prefix μ-. A μ-measurable function u on X is said to be μ-*integrable* on X if $\int_X |u| \, d\mu < +\infty$, and *locally μ-integrable* on X if $\int_K |u| \, d\mu < +\infty$ for every compact subset K of X. The prefix μ- is omitted if μ is Lebesgue measure. When writing integrals with respect to Lebesgue measure, we usually use the traditional notation $\int_X u(p) \, dp$. A relation which holds on a μ-measurable set Y such that $\mu(X \backslash Y) = 0$, is said to hold μ-*almost everywhere* on X, and again the prefix μ- is omitted if μ is Lebesgue measure.

A *signed measure* on X is a countably additive set function ν, defined on a σ-algebra that contains $\mathcal{B}$, taking only real values, such that $\nu(\emptyset) = 0$. (Some relaxation of the finiteness is described in Chapter 4.) In view of the Hahn-Jordan decomposition theorem, there are disjoint ν-measurable sets P and N such that $P \cup N = X$, and nonnegative finite measures ν^+ and ν^- on X, such that for all ν-measurable subsets S of X we have $\nu^+(S) = \nu(S \cap P)$ and $\nu^-(S) = \nu(S \cap N)$. Then ν has the decomposition $\nu = \nu^+ - \nu^-$, and the nonnegative measure $|\nu| = \nu^+ + \nu^-$ is called the *total variation* of ν.

CHAPTER 1

The Heat Operator, Temperatures and Mean Values

In this chapter we introduce some basic tools of heat potential theory, using mostly calculus. These tools include the fundamental solution of the heat equation, the heat balls and heat spheres, the fundamental mean values over heat balls and heat spheres, the modified heat balls and mean values over them, the maximum principle, the boundary uniqueness principle, and theorems about the convergence of sequences of solutions of the heat equation.

1.1. Temperatures and Heat Balls

We work throughout in Euclidean space of dimension $n+1$, namely
$$\mathbb{R}^{n+1} = \{(x,t) : x = (x_1, ..., x_n) \in \mathbb{R}^n, t \in \mathbb{R}\}.$$
The x variables represent coordinates in $\mathbb{R}^n$, and may be referred to as the *spatial variables*, while the t variable represents time and may be called the *temporal variable*.

We let Θ denote the *Heat Operator* and Θ^* its *adjoint*, defined by
$$\Theta = \sum_{i=1}^n \frac{\partial^2}{\partial x_i^2} - \frac{\partial}{\partial t} \quad \text{and} \quad \Theta^* = \sum_{i=1}^n \frac{\partial^2}{\partial x_i^2} + \frac{\partial}{\partial t}.$$
The symbol Θ is used because theta is an anagram of heatt! If Θ is operating on a function of other variables as well as x and t, we shall use $\Theta_{x,t}$ to signify that the operation is relative to the coordinates (x, t).

If E is an arbitrary open set in $\mathbb{R}^{n+1}$, we denote by $C^{2,1}(E)$ the set of real-valued functions u on E such that the partial derivatives $\partial^2 u/\partial x_i \partial x_j$ ($i, j \in \{1, ..., n\}$) and $\partial u/\partial t$ all exist and are continuous on E. The *Heat Equation* is $\Theta u = 0$, and its *adjoint* is $\Theta^* v = 0$. A solution of the heat equation on E is a function $u \in C^{2,1}(E)$ that satisfies $\Theta u = 0$; it is called a *temperature*. Similarly, a solution of its adjoint equation belongs to $C^{2,1}(E)$ and satisfies $\Theta^* v = 0$; it is called a *cotemperature*. If $u(x,t)$ is a temperature, and $v(x,t) = u(x,-t)$, then trivially v is a cotemperature. So results relative to Θ^* can easily be deduced from those for Θ. However, in some instances the theories of the two operators interact in a nontrivial way.

Note that both equations are *linear*. That is, if u is a temperature and $\alpha \in \mathbb{R}$, then αu is a temperature; and if v is another temperature then $u + v$ is also a temperature.

If u is a temperature and $v(x,t) = u(x+x_0, t+t_0)$ for some fixed $(x_0, t_0) \in \mathbb{R}^{n+1}$, then v is also a temperature. We describe this by saying that temperatures are *translation invariant*. Furthermore, if w is defined by $w(x,t) = u(ax, a^2 t)$ for some

1

fixed real number a then w is a temperature, and we say that temperatures are *invariant under parabolic dilation*.

Polynomial temperatures are easy to find, the simplest to depend on all the variables being $u(x,t) = |x|^2 + 2nt$.

A rather more interesting temperature v is defined as follows. Given any fixed $\alpha \in \mathbb{R}^n$, put $v(x,t) = \exp(\langle \alpha, x \rangle + |\alpha|^2 t)$, where $\langle \alpha, x \rangle$ denotes the usual inner product between α and x. Then v is a *positive* temperature on the whole of $\mathbb{R}^{n+1}$.

One particular temperature is the most important of all. Let

$$W(x,t) = \begin{cases} (4\pi t)^{-\frac{n}{2}} \exp\left(-\frac{|x|^2}{4t}\right) & \text{if } t > 0, \\ 0 & \text{if } t \leq 0. \end{cases}$$

Then W satisfies the heat equation on $\mathbb{R}^{n+1} \setminus \{0\}$, and is called the *Fundamental Temperature*. The constant $(4\pi)^{-\frac{n}{2}}$ is chosen to give the following result.

LEMMA 1.1. *Whenever $x \in \mathbb{R}^n$ and $a, t \in \mathbb{R}$ with $t > a$, we have*

$$\int_{\mathbb{R}^n} W(x-y, t-a)\, dy = 1.$$

PROOF. Put $r = |x - y|$. Then, if

$$\omega_n = \frac{2\pi^{\frac{n}{2}}}{\Gamma(\frac{n}{2})}$$

denotes the $(n-1)$-dimensional surface area of the unit sphere in $\mathbb{R}^n$, a change to polar coordinates gives

$$\int_{\mathbb{R}^n} W(x-y, t-a)\, dy = (4\pi(t-a))^{-\frac{n}{2}} \int_0^\infty \exp\left(-\frac{r^2}{4(t-a)}\right) \omega_n r^{n-1}\, dr.$$

Putting $z = r^2/[4(t-a)]$, we deduce that

$$\int_{\mathbb{R}^n} W(x-y, t-a)\, dy = \left(\frac{\omega_n}{2\pi^{\frac{n}{2}}}\right) \int_0^\infty e^{-z} z^{\frac{n}{2}-1}\, dz = 1.$$

□

One of the many uses of W is in the definition of the heat ball, which to some extent takes the place of the Euclidean ball when dealing with the heat equation. For any point $p_0 = (x_0, t_0) \in \mathbb{R}^{n+1}$ and any positive number c, the set

$$\Omega(p_0; c) = \Omega(x_0, t_0; c) = \{(y, s) \in \mathbb{R}^{n+1} : W(x_0 - y, t_0 - s) > (4\pi c)^{-\frac{n}{2}}\}$$

is called the *Heat Ball* with *centre* (x_0, t_0) and *radius* c. In the sequel, we shall usually write $\tau(c)$ for $(4\pi c)^{-\frac{n}{2}}$. Note that the centre of the heat ball is actually on the boundary. The heat ball is a convex domain (that is, convex connected open set), axially symmetric about the line $\{x_0\} \times \mathbb{R}$, and contained in the circular cylinder

$$\left\{(y, s) : |x_0 - y| < \sqrt{\frac{2nc}{e}},\ t_0 - c < s < t_0\right\},$$

as the equivalent expression

$$\Omega(x_0, t_0; c) = \left\{(y, s) : |x_0 - y| < \sqrt{2n(t_0 - s)\log\left(\frac{c}{t_0 - s}\right)},\ t_0 - c < s < t_0\right\}$$

shows. As $c \to 0$, the heat ball $\Omega(x_0, t_0; c)$ shrinks to the point (x_0, t_0). If $0 < d < c$, then $\overline{\Omega}(x_0, t_0; d) \backslash \{(x_0, t_0)\} \subseteq \Omega(x_0, t_0; c)$.

The boundary of the heat ball $\Omega(x_0, t_0; c)$ is called the *Heat Sphere* with centre (x_0, t_0) and radius c.

1.2. Mean Values of Smooth Functions over Heat Spheres

In this section we collect three results which connect the heat operator on a heat ball to a fundamental mean value over heat spheres. In later sections, we will explore the consequences of these results for temperatures and, more generally, for functions u which satisfy $\Theta u \geq 0$.

Throughout this chapter, we use E to denote an arbitrary open set in $\mathbb{R}^{n+1}$.

Our starting point is Green's formula for the heat equation. If $v, w \in C^{2,1}(E)$, then

$$v\Theta w - w\Theta^* v = \sum_{i=1}^n \left(v \frac{\partial^2 w}{\partial x_i^2} - w \frac{\partial^2 v}{\partial x_i^2} \right) - \left(v \frac{\partial w}{\partial t} + w \frac{\partial v}{\partial t} \right)$$

$$= \sum_{i=1}^n \frac{\partial}{\partial x_i} \left(v \frac{\partial w}{\partial x_i} - w \frac{\partial v}{\partial x_i} \right) - \frac{\partial}{\partial t}(vw).$$

The last expression is the divergence of the continuously differentiable function $\phi : E \to \mathbb{R}^{n+1}$ whose i-th component is $v \partial w / \partial x_i - w \partial v / \partial x_i$ if $i \leq n$, and is $-vw$ if $i = n + 1$. It therefore follows from Gauss's divergence theorem that, for any bounded open set D with $\overline{D} \subseteq E$ whose boundary is piecewise smooth,

$$\iint_D (v\Theta w - w\Theta^* v)\, dx\, dt = \int_{\partial D} \left(\sum_{i=1}^n \left(v \frac{\partial w}{\partial x_i} - w \frac{\partial v}{\partial x_i} \right) \nu_i - vw\nu_t \right) d\sigma$$

(1.1)
$$= \int_{\partial D} \left(\langle v \nabla_x w - w \nabla_x v, \nu_x \rangle - vw\nu_t \right) d\sigma,$$

where $\langle \cdot, \cdot \rangle$ denotes the inner product in $\mathbb{R}^n$,

$$\nabla_x w = \left(\frac{\partial w}{\partial x_1}, ..., \frac{\partial w}{\partial x_n} \right),$$

denotes the gradient in the spatial variables, $\nu = (\nu_x, \nu_t)$ with $\nu_x = (\nu_1, ..., \nu_n)$ is the outward unit normal to ∂D, and σ denotes the surface area measure on ∂D. Formula (1.1) is called *Green's Formula for the Heat Equation*. The special case where $v \equiv 1$,

(1.2)
$$\iint_D \Theta w\, dx\, dt = \int_{\partial D} \left(\langle \nabla_x w, \nu_x \rangle - w\nu_t \right) d\sigma,$$

is also important.

Given $(x_0, t_0) \in \mathbb{R}^{n+1}$ and $c > 0$, we would like to apply formula (1.1) with $D = \Omega(x_0, t_0; c)$ and $v(x, t) = W(x_0 - x, t_0 - t)$, in order to obtain a fundamental mean value formula. It is not possible to do this directly because of the singularity at (x_0, t_0), so we will have to truncate the heat ball. In preparation, we consider the term $-\langle \nabla_x v, \nu_x \rangle$ for this choice of v. If $(x, t) \in \partial \Omega(x_0, t_0; c)$, then a routine calculation gives

$$-\langle \nabla_x v(x, t), \nu_x(x, t) \rangle = \tau(c) Q(x_0 - x, t_0 - t)$$

where
$$Q(x_0 - x, t_0 - t) = \frac{|x_0 - x|^2}{\left(4|x_0-x|^2(t_0-t)^2 + \left(|x_0-x|^2 - 2n(t_0-t)\right)^2\right)^{1/2}}.$$

Putting $y = x_0 - x$ and $s = t_0 - t$ we see that, as $(x,t) \to (x_0, t_0)$ from within $\partial\Omega(x_0, t_0; c)$,

$$Q(y,s) = \left(\frac{4s^2}{|y|^2} + \left(1 - \frac{2ns}{|y|^2}\right)^2\right)^{-\frac{1}{2}}$$

$$= \left(\frac{2s}{n\log\frac{c}{s}} + \left(1 - \frac{1}{\log\frac{c}{s}}\right)^2\right)^{-\frac{1}{2}}$$

$$\to 1.$$

Therefore, if we define $Q(0,0)$ to be 1, then $Q(x_0 - x, t_0 - t)$ is a continuous function for $(x,t) \in \partial\Omega(x_0, t_0; c)$. It is also positive, except for a zero at $(x_0, t_0 - c)$.

DEFINITION 1.2. The fundamental *mean value over heat spheres* is defined by

$$\mathcal{M}(u; x_0, t_0; c) = \tau(c) \int_{\partial\Omega(x_0, t_0; c)} Q(x_0 - x, t_0 - t) u(x,t) \, d\sigma$$

for any function u such that the integral exists.

The following general result shows the connection between $\mathcal{M}$ and the heat operator.

THEOREM 1.3. *If $\overline{\Omega}(x_0, t_0; c) \subseteq E$, $u \in C^{2,1}(E)$, and*

$$W_0(x,t) = W(x_0 - x, t_0 - t),$$

then

$$\mathcal{M}(u; x_0, t_0; c) - u(x_0, t_0) = \int\int_{\Omega(x_0, t_0; c)} (W_0(x,t) - \tau(c)) \Theta u(x,t) \, dx \, dt.$$

PROOF. We apply Green's formula (1.1), with D the following truncation of $\Omega(x_0, t_0; c)$. For any s such that $t_0 - c < s < t_0$, we put

$$\Psi(s) = \{(x,t) \in \Omega(x_0, t_0; c) : t < s\}.$$

We divide $\partial\Psi(s)$ into two parts:

$$A(s) = \partial\Psi(s) \cap (\mathbb{R}^n \times \{s\}), \qquad B(s) = \partial\Psi(s) \backslash A(s).$$

Taking $D = \Psi(s)$, $v = W_0$, and $w = u$ in (1.1), we obtain

$$(1.3) \quad \int\int_{\Psi(s)} W_0 \Theta u \, dx \, dt = \int_{\partial\Psi(s)} \left(\langle W_0 \nabla_x u - u \nabla_x W_0, \nu_x\rangle - u W_0 \nu_t\right) d\sigma,$$

because $\Theta^* W_0 = 0$ on $\Psi(s)$. The right-hand side of (1.3) can be written as

$$(1.4) \quad -\int_{A(s)} u W_0 \, d\sigma + \tau(c) \int_{B(s)} \left(\langle \nabla_x u, \nu_x\rangle - u \nu_t\right) d\sigma - \int_{B(s)} u \langle \nabla_x W_0, \nu_x\rangle \, d\sigma,$$

1.2. MEAN VALUES OF SMOOTH FUNCTIONS OVER HEAT SPHERES

because $\nu_x = 0$ and $\nu_t = 1$ on $A(s)$, and $W_0 = \tau(c)$ on $B(s)$.

As $s \to t_0-$, we have

$$-\int_{B(s)} u \langle \nabla_x W_0, \nu_x \rangle \, d\sigma \to \mathcal{M}(u; x_0, t_0; c),$$

and

$$\int_{B(s)} \left(\langle \nabla_x u, \nu_x \rangle - u\nu_t \right) d\sigma \to \int_{\partial \Omega(x_0, t_0; c)} \left(\langle \nabla_x u, \nu_x \rangle - u\nu_t \right) d\sigma$$
$$= \int\int_{\Omega(x_0, t_0; c)} \Theta u \, dx \, dt$$

by formula (1.2) with $w = u$ and $D = \Omega(x_0, t_0; c)$.

Considering the integral over $A(s)$ in (1.4), we first note that, writing $r = t_0 - s$ and $\rho(r) = 2nr\log(c/r)$ for $r \in]0, c[$, we have

$$\int_{A(s)} W_0 \, d\sigma = \int_{|x_0 - x| < \sqrt{\rho(r)}} (4\pi r)^{-\frac{n}{2}} \exp\left(-\frac{|x_0 - x|^2}{4r}\right) dx$$
$$= \frac{1}{\Gamma(\frac{n}{2})} \int_0^{\rho(r)/(4r)} z^{\frac{n}{2}-1} e^{-z} \, dz$$
$$\to 1-$$

as $r \to 0+$. Therefore, as $s \to t_0-$, Lemma 1.1 shows that

$$\left| \int_{A(s)} u(x, s) W_0(x, s) \, dx - u(x_0, t_0) \right|$$
$$\leq \left| \int_{A(s)} \left(u(x, s) - u(x_0, t_0) \right) W_0(x, s) \, dx \right| + \left| u(x_0, t_0) \left(\int_{A(s)} W_0(x, s) \, dx - 1 \right) \right|$$
$$\leq \sup \left\{ |u(x, s) - u(x_0, t_0)| : |x_0 - x| < \sqrt{\rho(r)} \right\} + o(1)$$
$$= o(1).$$

Taking the limit as $s \to t_0-$ in (1.3), we therefore obtain

$$\int\int_{\Omega(x_0, t_0; c)} W_0 \Theta u \, dx \, dt$$
$$= -u(x_0, t_0) + \tau(c) \int\int_{\Omega(x_0, t_0; c)} \Theta u \, dx \, dt + \mathcal{M}(u; x_0, t_0; c),$$

and the result follows. $\square$

Given any point $(x_0, t_0) \in \mathbb{R}^{n+1}$, and numbers b, c such that $0 < b < c$, we put

$$A(x_0, t_0; b, c) = \{(y, s) : \tau(c) < W(x_0 - y, t_0 - s) < \tau(b)\}$$
$$= \Omega(x_0, t_0; c) \backslash \overline{\Omega}(x_0, t_0; b),$$

and call $A(x_0, t_0; b, c)$ the *Heat Annulus* with *centre* (x_0, t_0), *inner radius b* and *outer radius c*.

THEOREM 1.4. *If $u \in C^{2,1}(E)$, the closed heat ball $\overline{\Omega}(x_0, t_0; c)$ is contained in E, and $W_0(x,t) = W(x_0 - x, t_0 - t)$, then whenever $0 < b < c$,*

$$\mathcal{M}(u; x_0, t_0; c) - \mathcal{M}(u; x_0, t_0; b)$$
$$(1.5) \qquad = \int\!\!\int_{\Omega(x_0, t_0; c)} \big((W_0(x,t) \wedge \tau(b)) - \tau(c)\big) \Theta u(x,t) \, dx \, dt.$$

PROOF. Since (x_0, t_0) is fixed, we shall abbreviate $\Omega(x_0, t_0; b)$ to $\Omega(b)$, etc. If $0 < b < c$, then by Theorem 1.3,

$$\mathcal{M}(u; x_0, t_0; c) - \mathcal{M}(u; x_0, t_0; b)$$
$$= \int\!\!\int_{\Omega(c)} (W_0(x,t) - \tau(c)) \Theta u(x,t) \, dx \, dt - \int\!\!\int_{\Omega(b)} (W_0(x,t) - \tau(b)) \Theta u(x,t) \, dx \, dt$$
$$= \int\!\!\int_{A(b,c)} (W_0(x,t) - \tau(c)) \Theta u(x,t) \, dx \, dt + \int\!\!\int_{\Omega(b)} (\tau(b) - \tau(c)) \Theta u(x,t) \, dx \, dt$$
$$= \int\!\!\int_{\Omega(c)} \big((W_0(x,t) \wedge \tau(b)) - \tau(c)\big) \Theta u(x,t) \, dx \, dt,$$

which proves (1.5). $\square$

Reading the statement of Theorem 1.4, it is natural to ask if anything can be said if we assume only that $\overline{A}(x_0, t_0; b, c) \subseteq E$, rather than that $\overline{\Omega}(x_0, t_0; c) \subseteq E$. Theorem 1.5 below answers this question.

THEOREM 1.5. *Suppose that $u \in C^{2,1}(E)$, and that $\overline{A}(x_0, t_0; b, c) \subseteq E$. If $W_0(x,t) = W(x_0 - x, t_0 - t)$, then*

$$\mathcal{M}(u; x_0, t_0; c) - \mathcal{M}(u; x_0, t_0; b)$$
$$= \int\!\!\int_{A(x_0, t_0; b, c)} W_0(x,t) \Theta u(x,t) \, dx \, dt - \tau(c) \kappa(c) + \tau(b) \kappa(b),$$

where

$$\kappa(a) = \int_{\partial \Omega(x_0, t_0; a)} \big(\langle \nabla_x u, \nu_x \rangle - u \nu_t\big) \, d\sigma$$

for $a \in \{b, c\}$.

PROOF. The proof is essentially similar to that of Theorem 1.3. We apply Green's formula (1.1), with D the following truncation of $A(x_0, t_0; b, c)$. For any s such that $t_0 - b < s < t_0$, we put

$$\Psi(s) = \{(x,t) \in A(x_0, t_0; b, c) : t < s\}.$$

We divide $\partial \Psi(s)$ into three parts:

$$T(s) = \partial \Psi(s) \cap (\mathbb{R}^n \times \{s\}),$$

and

$$U(a, s) = \{(x,t) \in \partial \Omega(x_0, t_0; a) : t < s\} \qquad \text{for} \qquad a \in \{b, c\}.$$

Taking $D = \Psi(s)$, $v = W_0$, and $w = u$ in (1.1), we obtain

$$(1.6) \qquad \int\!\!\int_{\Psi(s)} W_0 \Theta u \, dx \, dt = \int_{\partial \Psi(s)} \big(\langle W_0 \nabla_x u - u \nabla_x W_0, \nu_x \rangle - u W_0 \nu_t\big) \, d\sigma,$$

because $\Theta^* W_0 = 0$ on $\Psi(s)$. The right-hand side of (1.6) can be written as

$$-\int_{T(s)} u W_0 \, d\sigma$$

(1.7)
$$+ \tau(c) \int_{U(c,s)} \left(\langle \nabla_x u, \nu_x \rangle - u\nu_t \right) d\sigma - \tau(b) \int_{U(b,s)} \left(\langle \nabla_x u, \nu_x \rangle - u\nu_t \right) d\sigma$$
$$- \int_{U(c,s)} u \langle \nabla_x W_0, \nu_x \rangle \, d\sigma + \int_{U(b,s)} u \langle \nabla_x W_0, \nu_x \rangle \, d\sigma$$

because $\nu_x = 0$ and $\nu_t = 1$ on $T(s)$, and $W_0 = \tau(a)$ on $U(a,s)$ for $a \in \{b,c\}$.

As $s \to t_0-$, we have

$$-\int_{U(a,s)} u \langle \nabla_x W_0, \nu_x \rangle \, d\sigma \to \mathcal{M}(u; x_0, t_0; a),$$

and

$$\int_{U(a,s)} \left(\langle \nabla_x u, \nu_x \rangle - u\nu_t \right) d\sigma \to \int_{\partial \Omega(x_0, t_0; a)} \left(\langle \nabla_x u, \nu_x \rangle - u\nu_t \right) d\sigma = \kappa(a).$$

We now consider the integral over $T(s)$ in (1.7). We write $r = t_0 - s$, and put $\beta(r) = 2nr\log(b/r)$ for $0 < r < b$, and $\gamma(r) = 2nr\log(c/r)$ for $0 < r < c$. Then

$$\int_{T(s)} W_0 \, d\sigma = \int_{\sqrt{\beta(r)} < |x_0 - x| < \sqrt{\gamma(r)}} (4\pi r)^{-\frac{n}{2}} \exp\left(-\frac{|x_0 - x|^2}{4r}\right) dx$$
$$= \int_{\sqrt{\beta(r)}}^{\sqrt{\gamma(r)}} (4\pi r)^{-\frac{n}{2}} \exp\left(-\frac{\rho^2}{4r}\right) \omega_n \rho^{n-1} \, d\rho$$
$$= \frac{1}{\Gamma(\frac{n}{2})} \int_{\beta(r)/(4r)}^{\gamma(r)/(4r)} e^{-z} z^{\frac{n}{2}-1} \, dz$$
$$\to 0$$

as $r \to 0+$. Since u is bounded on $\overline{A}(x_0, t_0; b, c)$, it follows that

$$\int_{T(s)} u W_0 \, d\sigma \to 0$$

as $s \to t_0-$. Taking the limit as $s \to t_0-$ in (1.6), we therefore obtain

$$\int\!\!\int_{A(x_0,t_0;b,c)} W_0(x,t) \Theta u(x,t) \, dx \, dt$$
$$= \tau(c)\kappa(c) - \tau(b)\kappa(b) + \mathcal{M}(u; x_0, t_0; c) - \mathcal{M}(u; x_0, t_0; b),$$

and the result follows. $\square$

1.3. Mean Values of Smooth Subtemperatures over Heat Spheres

Functions $u \in C^{2,1}(E)$ such that $\Theta u \geq 0$, and their generalizations, are very important. The generalizations will be called 'subtemperatures' in the sequel, so now we shall define a *smooth subtemperature* u on E to be any function $u \in C^{2,1}(E)$ such that $\Theta u \geq 0$. The term comes from the fact that, if D is a bounded open set such that $\overline{D} \subseteq E$, and v is a temperature on D which is continuous on $\overline{D}$ and satisfies $v = u$ on ∂D, then $u \leq v$ on D. We shall be able to prove this for D a circular cylinder in Section 1.5.

In this section, we look at the implications that the results in Section 1.2 have for smooth subtemperatures.

Theorem 1.3 leads to the following mean value theorem about temperatures and smooth subtemperatures.

THEOREM 1.6. *Let $u \in C^{2,1}(E)$. If u is a temperature on E, then the equality*

(1.8) $$u(x_0, t_0) = \mathcal{M}(u; x_0, t_0; c)$$

holds whenever $\overline{\Omega}(x_0, t_0; c) \subseteq E$. If, more generally, u is a smooth subtemperature on E, then the inequality

(1.9) $$u(x_0, t_0) \leq \mathcal{M}(u; x_0, t_0; c)$$

holds whenever $\overline{\Omega}(x_0, t_0; c) \subseteq E$.

Conversely if, given any point $(x_0, t_0) \in E$ and $\epsilon > 0$, we can find $c < \epsilon$ such that (1.9) holds, then u is a smooth subtemperature on E.

PROOF. If $\overline{\Omega}(x_0, t_0; c) \subseteq E$, then by Theorem 1.3,

$$\mathcal{M}(u; x_0, t_0; c) - u(x_0, t_0) = \int\!\!\int_{\Omega(x_0, t_0; c)} \bigl(W_0(x,t) - \tau(c)\bigr) \Theta u(x,t)\, dx\, dt.$$

So if u is a temperature on E, then (1.8) holds. If, more generally, $\Theta u \geq 0$ on E, then since

$$\Omega(x_0, t_0; c) = \{(y, s) : W_0(y, s) > \tau(c)\}$$

we have $\mathcal{M}(u; x_0, t_0; c) - u(x_0, t_0) \geq 0$, so that (1.9) holds. On the other hand, if $\Theta u < 0$ at some point of E, then the continuity of Θu implies that $\Theta u < 0$ on some open neighbourhood D of (x_0, t_0). Then $\mathcal{M}(u; x_0, t_0; c) - u(x_0, t_0) < 0$ for all c so small that $\overline{\Omega}(x_0, t_0; c) \subseteq D$, so that the conditions for the converse fail to hold. □

REMARK 1.7. Since constant functions are temperatures, we can take $u = 1$ in (1.8) to get $1 = \mathcal{M}(1; x_0, t_0; c)$, or

$$1 = \tau(c) \int_{\partial \Omega(x_0, t_0; c)} Q(x_0 - x, t_0 - t)\, d\sigma,$$

for all $c > 0$.

Theorem 1.6 shows that the fundamental mean values $\mathcal{M}$ characterize smooth subtemperatures, and so their behaviour for smooth subtemperatures is of interest. Theorem 1.4 easily implies the following result.

THEOREM 1.8. *If $u \in C^{2,1}(E)$, the closed heat ball $\overline{\Omega}(x_0, t_0; c) \subseteq E$, and u is a smooth subtemperature on $\Omega(x_0, t_0; c)$, then the function $\mathcal{M}(u; x_0, t_0; \cdot)$ is increasing on $]0, c]$.*

PROOF. If $0 < a < b \leq c$, then by Theorem 1.4,

$$\mathcal{M}(u; x_0, t_0; b) - \mathcal{M}(u; x_0, t_0; a)$$
$$= \int\!\!\int_{\Omega(x_0, t_0; b)} \bigl((W_0(x,t) \wedge \tau(a)) - \tau(b)\bigr) \Theta u(x,t)\, dx\, dt$$
$$\geq 0.$$

□

1.3. MEAN VALUES OF SMOOTH SUBTEMPERATURES OVER HEAT SPHERES

The next property of the fundamental means $\mathcal{M}$ for smooth subtemperatures is a consequence of Theorem 1.5. It requires some preliminary explanation.

Recall that a real-valued function ϕ, defined on an interval $J \subseteq \mathbb{R}$, is called *convex* if the inequality

$$\phi\big((1-\lambda)a + \lambda b\big) \leq (1-\lambda)\phi(a) + \lambda\phi(b) \tag{1.10}$$

holds whenever $a, b \in J$ and $0 \leq \lambda \leq 1$. Graphically, this condition means that the line segment joining the points $(a, \phi(a))$ and $(b, \phi(b))$ lies above or on the graph of ϕ. A function ϕ is called *concave* if $-\phi$ is convex, and *affine* if it is both convex and concave. Any affine function ϕ has the form $\phi(r) = \alpha r + \beta$ for some real numbers α and β.

We need to re-write (1.10) in an equivalent form. Let $r, s, t \in J$, with $r < s < t$. If we take

$$a = t, \qquad b = r, \qquad \lambda = \frac{t-s}{t-r}$$

in (1.10), then it becomes

$$\phi(s) \leq \frac{s-r}{t-r}\phi(t) + \frac{t-s}{t-r}\phi(r), \tag{1.11}$$

which re-arranges to

$$\frac{\phi(s) - \phi(r)}{s - r} \leq \frac{\phi(t) - \phi(s)}{t - s}.$$

Graphically, this condition means that the slope of the line joining the points $(r, \phi(r))$ and $(s, \phi(s))$ is no more than that of the line joining $(s, \phi(s))$ and $(t, \phi(t))$.

More generally, if ψ is a continuous, strictly monotone function on J, then ϕ is said to be a *convex function of ψ* if the inequality

$$\frac{\phi(s) - \phi(r)}{\psi(s) - \psi(r)} \leq \frac{\phi(t) - \phi(s)}{\psi(t) - \psi(s)} \tag{1.12}$$

holds whenever $r, s, t \in J$ and $\psi(r) < \psi(s) < \psi(t)$. This means that $\phi = \chi \circ \psi$ for some convex function χ defined on the interval $\psi(J)$.

We shall prove that, if $u \in C^{2,1}(E)$, the closed heat annulus $\overline{A}(x_0, t_0; b, c) \subseteq E$, and u is a smooth subtemperature on $A(x_0, t_0; b, c)$, then $\mathcal{M}(u; x_0, t_0; \cdot) = \chi \circ \tau$ on $[b, c]$, for some convex function χ. But first we present some of the properies of convex functions that we shall find useful in the sequel.

LEMMA 1.9. *Let J be an interval in $\mathbb{R}$, let $r, s, t \in J$ with $r < s < t$, and let ϕ be a convex function on J. Then*

$$\frac{\phi(s) - \phi(r)}{s - r} \leq \frac{\phi(t) - \phi(r)}{t - r} \leq \frac{\phi(t) - \phi(s)}{t - s}. \tag{1.13}$$

If J is open, then the left and right derivatives ϕ'_- and ϕ'_+ exist and are finite at every point of J, are increasing functions on J, and satisfy $\phi'_- \leq \phi'_+$ on J. Furthermore, ϕ is continuous on J, and for each point $s_0 \in J$ there is an affine function χ such that $\chi(s_0) = \phi(s_0)$ and $\chi \leq \phi$ on J.

PROOF. The inequalities (1.13) both follow from (1.11).

Now suppose that J is open. Given any point $s \in J$, we define the function f on a deleted neighbourhood N of 0 by putting

$$f(h) = \frac{\phi(s+h) - \phi(s)}{h}.$$

It follows from (1.13) that, if $h_1 < h_2$ in N, we have
$$\frac{\phi(s+h_1) - \phi(s)}{h_1} \leq \frac{\phi(s+h_2) - \phi(s)}{h_2},$$
so that f is increasing. Therefore
$$\lim_{h \to 0-} f(h) = \sup\{f(h) : h < 0\} \leq \inf\{f(h) : h > 0\} = \lim_{h \to 0+} f(h),$$
which implies that ϕ'_- and ϕ'_+ exist and are finite at s, and that $\phi'_-(s) \leq \phi'_+(s)$.

The inequalities (1.13) also imply that, if $r < s$ in J,
$$\phi'_+(r) = \lim_{h \to 0+} \frac{\phi(r+h) - \phi(r)}{h} \leq \frac{\phi(s) - \phi(r)}{s - r} \leq \lim_{t \to s+} \frac{\phi(t) - \phi(s)}{t - s} \leq \phi'_+(s),$$
so that ϕ'_+ is increasing. Similarly, ϕ'_- is increasing.

To show that ϕ is continuous, we take two points $r_0 < t_0$ in J, and put $M = |\phi'_+(r_0)| \vee |\phi'_-(t_0)|$, where $a \vee b = \max\{a, b\}$. Then, whenever $r_0 < r < t < t_0$, we have
$$-M \leq \phi'_+(r_0) \leq \phi'_+(r) \leq \frac{\phi(t) - \phi(r)}{t - r} \leq \phi'_-(t) \leq \phi'_-(t_0) \leq M,$$
so that $|\phi(t) - \phi(r)| \leq M(t - r)$. Interchanging r and t, we deduce that
$$|\phi(t) - \phi(r)| \leq M|t - r|$$
for any points $r, t \in [r_0, t_0]$. Hence ϕ is continuous on J.

Finally, given any point $s_0 \in J$, we put
$$\chi(t) = \phi(s_0) + \phi'_+(s_0)(t - s_0).$$
Then χ is an affine function such that $\chi(s_0) = \phi(s_0)$. Furthermore, if $t > s_0$ then (1.13) implies that
$$\phi'_+(s_0) \leq \frac{\phi(t) - \phi(s_0)}{t - s_0},$$
so that $\chi(t) \leq \phi(t)$; and if $t > s_0$ then
$$\frac{\phi(t) - \phi(s_0)}{t - s_0} \leq \phi'_-(s_0) \leq \phi'_+(s_0),$$
so that again $\chi(t) \leq \phi(t)$. □

COROLLARY 1.10. *Let ϕ be a nonnegative, convex function on $[0, +\infty[$ such that $\phi(0) = 0$. Then the function $r \mapsto r^{-1}\phi(r)$ is nonnegative and increasing on $]0, +\infty[$.*

PROOF. Taking $J = [0, +\infty[$ and $r = 0$ in Lemma 1.9, we find that
$$\frac{\phi(s)}{s} = \frac{\phi(s) - \phi(0)}{s - 0} \leq \frac{\phi(t) - \phi(0)}{t - 0} = \frac{\phi(t)}{t}$$
whenever $0 < s < t$. This proves the result. □

THEOREM 1.11. *Let $u \in C^{2,1}(E)$, let $\overline{A}(x_0, t_0; b, c) \subseteq E$, and suppose that u is a smooth subtemperature on $A(x_0, t_0; b, c)$. Then $\mathcal{M}(u; x_0, t_0; \cdot)$ is a convex function of τ on $[b, c]$. In particular, if u is a temperature on $A(x_0, t_0; b, c)$, then there are real numbers α and β such that*
$$\mathcal{M}(u; x_0, t_0; a) = \alpha\tau(a) + \beta$$
whenever $b \leq a \leq c$.

PROOF. Let δ, η be numbers such that $b \leq \delta < \eta \leq c$. Applying Theorem 1.5 to u on $A(\delta, \eta) = A(x_0, t_0; \delta, \eta)$, we obtain

$$\mathcal{M}(\eta) - \mathcal{M}(\delta) = \iint_{A(\delta,\eta)} W_0(x,t) \Theta u(x,t) \, dx \, dt - \tau(\eta) \kappa(\eta) + \tau(\delta) \kappa(\delta),$$

where $\mathcal{M}(\gamma) = \mathcal{M}(u; x_0, t_0; \gamma)$ for $\gamma \in \{\delta, \eta\}$. By Green's formula (1.2), with $D = A(\delta, \eta)$ and $w = u$, we have

$$\kappa(\eta) - \kappa(\delta) = \int_{\partial A(\delta,\eta)} (\langle \nabla_x u, \nu_x \rangle - u \nu_t) \, d\sigma$$

$$= \iint_{A(\delta,\eta)} \Theta u(x,t) \, dx \, dt,$$

so that

$$\mathcal{M}(\eta) - \mathcal{M}(\delta)$$
$$= \iint_{A(\delta,\eta)} W_0(x,t) \Theta u(x,t) \, dx \, dt - \tau(\delta)\big(\kappa(\eta) - \kappa(\delta)\big) - \kappa(\eta)\big(\tau(\eta) - \tau(\delta)\big)$$
$$= \iint_{A(\delta,\eta)} \big(W_0(x,t) - \tau(\delta)\big) \Theta u(x,t) \, dx \, dt - \kappa(\eta)\big(\tau(\eta) - \tau(\delta)\big).$$

It follows that, if $b \leq p < q < r \leq c$, then

$$\frac{\mathcal{M}(r) - \mathcal{M}(q)}{\tau(r) - \tau(q)} - \frac{\mathcal{M}(q) - \mathcal{M}(p)}{\tau(q) - \tau(p)}$$
$$= \iint_{A(q,r)} \left(\frac{W_0(x,t) - \tau(q)}{\tau(r) - \tau(q)}\right) \Theta u(x,t) \, dx \, dt - \kappa(r)$$
$$\quad - \iint_{A(p,q)} \left(\frac{W_0(x,t) - \tau(p)}{\tau(q) - \tau(p)}\right) \Theta u(x,t) \, dx \, dt - \kappa(q)$$
$$= \iint_{A(q,r)} \left(\frac{W_0(x,t) - \tau(q)}{\tau(r) - \tau(q)} - 1\right) \Theta u(x,t) \, dx \, dt$$
$$\quad - \iint_{A(p,q)} \left(\frac{W_0(x,t) - \tau(p)}{\tau(q) - \tau(p)}\right) \Theta u(x,t) \, dx \, dt$$
$$= \iint_{A(q,r)} \left(\frac{W_0(x,t) - \tau(r)}{\tau(r) - \tau(q)}\right) \Theta u(x,t) \, dx \, dt$$
$$\quad + \iint_{A(p,q)} \left(\frac{\tau(p) - W_0(x,t)}{\tau(q) - \tau(p)}\right) \Theta u(x,t) \, dx \, dt.$$

By definition of the heat annulus, $\tau(\eta) < W_0 < \tau(\delta)$ on $A(\delta, \eta)$, and so the last two integrands are nonpositive. Hence

$$\frac{\mathcal{M}(r) - \mathcal{M}(q)}{\tau(r) - \tau(q)} \leq \frac{\mathcal{M}(q) - \mathcal{M}(p)}{\tau(q) - \tau(p)}$$

whenever $b \leq p < q < r \leq c$. Since τ is strictly decreasing, this shows that $\mathcal{M}$ is a convex function of τ.

If u is a temperature on $A(x_0, t_0; b, c)$, then both of the means $\mathcal{M}(u; x_0, t_0; \cdot)$ and $\mathcal{M}(-u; x_0, t_0; \cdot) = -\mathcal{M}(u; x_0, t_0; \cdot)$ are convex functions of τ, which implies that $\mathcal{M}(u; x_0, t_0; \cdot)$ is an affine function of τ. $\square$

EXAMPLE 1.12. Given two points $(x_0, t_0), (x^*, t^*) \in \mathbb{R}^{n+1}$ with $t^* < t_0$, we evaluate the mean values $\mathcal{M}(W^*; x_0, t_0; c)$ of the function
$$W^*(x,t) = W(x - x^*, t - t^*)$$
for every $c > 0$. This will be used several times in the sequel.

Let c_0 be the positive number such that $W(x_0 - x^*, t_0 - t^*) = \tau(c_0)$; that is, such that $(x^*, t^*) \in \partial\Omega(x_0, t_0; c_0)$. If $0 < c < c_0$, then W^* is a temperature on an open superset of $\overline{\Omega}(x_0, t_0; c)$, so that
$$\mathcal{M}(W^*; x_0, t_0; c) = W^*(x_0, t_0) = \tau(c_0),$$
by Theorem 1.6.

Consider the case where $c > c_0$. Theorem 1.11 shows that there are real numbers α and β such that $\mathcal{M}(W^*; x_0, t_0; c) = \alpha \tau(c) + \beta$. Since $W^* \geq 0$, and $\tau(c) \to 0$ as $c \to \infty$, we have $\beta \geq 0$. We use Theorem 1.5 to show that $\alpha = 1$, a process which involves the evaluation of
$$\kappa(a) = \int_{\partial\Omega(x_0,t_0;a)} (\langle \nabla_x W^*, \nu_x \rangle - W^* \nu_t) \, d\sigma$$
for all $a > c_0$. To achieve this, we go back to Green's formula. Given $a > c_0$, we choose $r, \rho > 0$ such that the closed cylinder
$$C = \{(y,s) : |x^* - y| \leq r, \, t^* \leq s \leq t^* + \rho\}$$
is a subset of $\Omega(x_0, t_0; a)$. We apply Green's formula (1.2) with $D = \Omega(a) \backslash C$, where $\Omega(a) = \Omega(x_0, t_0; a)$, and $w = W^*$. This gives
$$0 = \int_{\partial\Omega(a)} (\langle \nabla_x W^*, \nu_x \rangle - W^* \nu_t) \, d\sigma - \int_{\partial C} (\langle \nabla_x W^*, \nu_x \rangle - W^* \nu_t) \, d\sigma.$$
We evaluate the limit of the latter integral as $\rho \to 0+$. Let
$$T = \{(y,s) : |x^* - y| \leq r, \, s = t^* + \rho\}$$
and
$$L = \{(y,s) : |x^* - y| = r, \, t^* \leq s \leq t^* + \rho\},$$
so that the integral can be written as
$$-\int_T W^* \, d\sigma + \int_L \langle \nabla_x W^*, \nu_x \rangle \, d\sigma.$$
Now
$$\lim_{\rho \to 0+} \int_L \langle \nabla_x W^*, \nu_x \rangle \, d\sigma = 0$$
because the integrand is bounded on L. Furthermore,
$$\int_T W^* \, d\sigma = \int_{|x - x^*| \leq r} W(x - x^*, \rho) \, dx$$
$$= \frac{1}{\Gamma(\frac{n}{2})} \int_0^{r^2/(4\rho)} e^{-z} z^{\frac{n}{2}-1} \, dz$$
$$\to 1$$
as $\rho \to 0+$. It follows that
$$\kappa(a) = \lim_{\rho \to 0+} \int_{\partial C} (\langle \nabla_x W^*, \nu_x \rangle - W^* \nu_t) \, d\sigma = -1.$$

We now take $E = \mathbb{R}^{n+1}\backslash\{(x^*,t^*)\}$ and $u = W^*$ in Theorem 1.5, with $c > b > c_0$. This gives
$$\mathcal{M}(W^*;x_0,t_0;c) - \mathcal{M}(W^*;x_0,t_0;b) = -\tau(c)\kappa(c) + \tau(b)\kappa(b) = \tau(c) - \tau(b),$$
which implies that $\alpha = 1$. Thus
$$\mathcal{M}(W^*;x_0,t_0;c) = \begin{cases} \tau(c_0) & \text{if } 0 < c < c_0, \\ \tau(c) + \beta & \text{if } c > c_0, \end{cases}$$
where $\beta \geq 0$.

We now prove that $\mathcal{M}(W^*;x_0,t_0;\cdot)$ is a decreasing function on $]0,\infty[$, which implies that $\beta = 0$ and that $\mathcal{M}(W^*;x_0,t_0;c_0) = \tau(c_0)$. To do this, we approximate W^* with an increasing sequence $\{w_k\}$ of functions such that $-w_k$ is a smooth subtemperature on $\mathbb{R}^{n+1}$ for every k. This sequence is constructed in the following way. Let ψ be a continuously differentiable function on $\mathbb{R}$ such that $\psi(t) = 0$ for all $t \leq \frac{1}{2}$, $\psi(t) = 1$ for all $t \geq 1$, and $\psi'(t) \geq 0$ for all $t \in \mathbb{R}$. Put
$$w_k(x,t) = W^*(x,t)\psi(k(t-t^*))$$
whenever $(x,t) \in \mathbb{R}^{n+1}$ and $k \in \mathbb{N}$. Since $\psi(k(t-t^*)) = 0$ for all $t \leq t^* + (1/(2k))$, each function w_k belongs to $C^{2,1}(\mathbb{R}^{n+1})$. Furthermore,
$$\Theta w_k(x,t) = \Theta W^*(x,t)\psi(k(t-t^*)) - W^*(x,t)k\psi'(k(t-t^*)) \leq 0$$
for all $(x,t) \in \mathbb{R}^{n+1}$. Theorem 1.8 shows that each function $\mathcal{M}(w_k;x_0,t_0;\cdot)$ is decreasing. Furthermore, since ψ is an increasing function, the sequence $\{w_k\}$ is increasing; and since $\psi(k(t-t^*)) = 1$ for all $t \geq t^* + (1/k)$, we have that $w_k \to W^*$ as $k \to \infty$. Hence, by the Lebesgue monotone convergence theorem,
$$\mathcal{M}(W^*;x_0,t_0;\cdot) = \lim_{k\to\infty}\mathcal{M}(w_k;x_0,t_0;\cdot),$$
and the latter function is decreasing. Hence $\mathcal{M}(W^*;x_0,t_0;\cdot)$ is decreasing, which implies that $\beta = 0$ and that $\mathcal{M}(W^*;x_0,t_0;c_0) = \tau(c_0) = W^*(x_0,t_0)$. Therefore
(1.14) $$\mathcal{M}(W^*;x_0,t_0;c) = W^*(x_0,t_0) \wedge \tau(c)$$
for all $c > 0$.

EXAMPLE 1.13. The increasing property of the function $\mathcal{M}(u;x_0,t_0;\cdot)$, given in Theorem 1.8, fails to hold if we assume only that $\overline{A}(x_0,t_0;b,c) \subseteq E$ and $\Theta u \geq 0$ on $A(x_0,t_0;b,c)$. Let (x_0,t_0) and (x^*,t^*) be points such that $t^* < t_0$, and let $W^*(x,t) = W(x-x^*,t-t^*)$ for all $(x,t) \in E = \mathbb{R}^{n+1}\backslash\{(x^*,t^*)\}$. Then W^* is a temperature on E, and $\mathcal{M}(W^*;x_0,t_0;c) = W^*(x_0,t_0) \wedge \tau(c)$ for all $c > 0$, by Example 1.12. If b is chosen so that $(x^*,t^*) \in \Omega(x_0,t_0;b)$ then $W^*(x_0,t_0) = \tau(b)$, so that for all $c > b$ we have $\mathcal{M}(W^*;x_0,t_0;c) = \tau(c) < \tau(b) = \mathcal{M}(W^*;x_0,t_0;b)$.

1.4. Mean Values of Smooth Subtemperatures over Heat Balls

In this section, we use the integral mean values over heat spheres to develop an integral mean over heat balls that also characterizes temperatures and smooth subtemperatures. We get from one to the other by integration, and there are uncountably many different possibilities for the kernel in the heat ball case. We choose the kernel that appears to be the simplest, but unfortunately there is no choice that leads to a bounded kernel. In the next section, we show how to modify the heat ball to get a bounded, although complicated, kernel.

In order to pass from an integral over heat spheres to one over heat balls, we need to calculate the Jacobian. For all $(x,t) \in \mathbb{R}^n \times]0,\infty[$, we define the function J by

(1.15) $$J(x,t) = 2nt \exp\left(-\frac{|x|^2}{2nt}\right) \left(4|x|^2 t^2 + (|x|^2 - 2nt)^2\right)^{-1/2}.$$

LEMMA 1.14. *Let $(x_0, t_0) \in \mathbb{R}^{n+1}$, let $c > 0$, and let f be a measurable function on $\Omega(x_0, t_0; c)$. If one of the integrals*

$$\iint_{\Omega(x_0,t_0;c)} f(x,t) \, dx \, dt,$$

$$\int_0^c \int_{\partial\Omega(x_0,t_0;l)} f(x,t) J(x_0 - x, t_0 - t) \, d\sigma \, dl$$

exists, then the other exists and the two are equal.

PROOF. Put $y = x_0 - x$ and $s = t_0 - t$. Take hyperspherical coordinates $\theta_1, ..., \theta_{n-1}, r$ in $\mathbb{R}^n$. We consider the following three transformations. The first is from hyperspherical to rectangular coordinates in $\mathbb{R}^n$:

$$\Psi : (\theta_1, ..., \theta_{n-1}, r) \mapsto (y_1, ..., y_n).$$

The second is obtained by putting $r = \sqrt{2ns \log(c/s)}$ in Ψ. We then consider

$$\Xi : (\theta_1, ..., \theta_{n-1}, c, s) \mapsto (y_1, ..., y_n, s).$$

The third is derived from Ξ by fixing c:

$$\Phi : (\theta_1, ..., \theta_{n-1}, s) \mapsto (y_1, ..., y_n, s).$$

Note that, as $\theta_1, ..., \theta_{n-1}, s$ vary with c fixed, the surface $\partial\Omega(x_0, t_0; c)$ is described.

Let Ψ', Ξ' and Φ' denote the Jacobians of the transformations. Using the standard formula for the Jacobian, we obtain the formula

(1.16) $$|\Xi'| = \left(\frac{\partial r}{\partial c}\right) |\Psi'|.$$

If Ψ_j denotes the vector whose components are the partial derivates of $(y_1, ..., y_n)$ with respect to the jth coordinate (that is, θ_j if $1 \leq j \leq n-1$, and r if $j = n$), then $|\Psi'|$ is given by

$$|\Psi'|^2 = \begin{vmatrix} \Psi_1\Psi_1 & \cdots & \Psi_1\Psi_n \\ \cdot & \cdots & \cdot \\ \Psi_n\Psi_1 & \cdots & \Psi_n\Psi_n \end{vmatrix}.$$

If Φ_j ($1 \leq j \leq n$) is defined analogously to Ψ_j, it is easily verified that

$$\Psi_j \Psi_i = \Psi_i \Psi_j = \Phi_i \Phi_j = \Phi_j \Phi_i$$

if $1 \leq i, j \leq n-1$, that

$$\Phi_j \Phi_n = \Phi_n \Phi_j = 0 = \Psi_n \Psi_j = \Psi_j \Psi_n$$

if $1 \leq j \leq n-1$, and that

$$\Phi_n \Phi_n = \left(\frac{\partial r}{\partial s}\right)^2 + 1, \qquad \Psi_n \Psi_n = 1.$$

Therefore, because

$$|\Phi'|^2 = \begin{vmatrix} \Phi_1\Phi_1 & \cdots & \Phi_1\Phi_n \\ \cdot & \cdots & \cdot \\ \Phi_n\Phi_1 & \cdots & \Phi_n\Phi_n \end{vmatrix}$$

we have

(1.17) $$|\Phi'|^2 = \left[1 + \left(\frac{\partial r}{\partial s}\right)^2\right]|\Psi'|^2.$$

Hence, by (1.16),

$$|\Xi'| = \left(\frac{\partial r}{\partial c}\right)\left[1 + \left(\frac{\partial r}{\partial s}\right)^2\right]^{-\frac{1}{2}}|\Phi'|.$$

A routine calculation now gives

$$|\Xi'| = J|\Phi'|,$$

where J is defined by (1.15), and the result follows. $\square$

DEFINITION 1.15. Given a function u on the heat ball $\Omega(x_0, t_0; c)$ for which the integral exists, we define the *volume mean value* of u by

(1.18) $$\mathcal{V}(u; x_0, t_0; c) = \frac{n}{2} c^{-\frac{n}{2}} \int_0^c l^{\frac{n}{2}-1} \mathcal{M}(u; x_0, t_0; l) \, dl.$$

Since the kernel Q for the surface mean value over the heat sphere satisfies

$$Q(x_0 - x, t_0 - t) = \frac{|x_0 - x|^2}{2n(t_0 - t)} \exp\left(\frac{|x_0 - x|^2}{2n(t_0 - t)}\right) J(x_0 - x, t_0 - t),$$

and

$$l = (t_0 - t)\exp\left(\frac{|x_0 - x|^2}{2n(t_0 - t)}\right)$$

whenever $(x, t) \in \partial\Omega(x_0, t_0; l)\setminus\{(x_0, t_0)\}$, we obtain

$$\mathcal{V}(u; x_0, t_0; c) = \frac{n}{2} c^{-\frac{n}{2}} \int_0^c l^{\frac{n}{2}-1} \tau(l) \int_{\partial\Omega(x_0, t_0; l)} Q(x_0 - x, t_0 - t) u(x, t) \, d\sigma$$

$$= \frac{n}{2} \tau(c) \int_0^c l^{-1} \int_{\partial\Omega(x_0, t_0; l)} \frac{|x_0 - x|^2}{2n(t_0 - t)} \exp\left(\frac{|x_0 - x|^2}{2n(t_0 - t)}\right)$$
$$\times J(x_0 - x, t_0 - t) u(x, t) \, d\sigma$$

$$= \tau(c) \int_0^c \int_{\partial\Omega(x_0, t_0; l)} \frac{|x_0 - x|^2}{4(t_0 - t)^2} J(x_0 - x, t_0 - t) u(x, t) \, d\sigma$$

(1.19) $$= \tau(c) \int\int_{\Omega(x_0, t_0; c)} \frac{|x_0 - x|^2}{4(t_0 - t)^2} u(x, t) \, dx \, dt.$$

We now present variants of Theorems 1.6, 1.8, and 1.11, for the volume means.

THEOREM 1.16. *Let $u \in C^{2,1}(E)$. If u is a temperature on E, then the equality*

(1.20) $$u(x_0, t_0) = \mathcal{V}(u; x_0, t_0; c)$$

holds whenever $\overline{\Omega}(x_0, t_0; c) \subseteq E$. If, more generally, u is a smooth subtemperature on E, then the inequality

(1.21) $$u(x_0, t_0) \leq \mathcal{V}(u; x_0, t_0; c)$$

holds whenever $\overline{\Omega}(x_0, t_0; c) \subseteq E$.

Conversely if, given any point $(x_0, t_0) \in E$ and $\epsilon > 0$, we can find $c < \epsilon$ such that (1.21) holds, then u is a smooth subtemperature on E.

PROOF. Suppose that $\overline{\Omega}(x_0, t_0; c) \subseteq E$. If u is a smooth subtemperature on E, then $u(x_0, t_0) \leq \mathcal{M}(u; x_0, t_0; l)$ whenever $0 < l \leq c$, by Theorem 1.6. It therefore follows from (1.18) that

$$\mathcal{V}(u; x_0, t_0; c) \geq \frac{n}{2} c^{-\frac{n}{2}} \int_0^c l^{\frac{n}{2}-1} u(x_0, t_0) \, dl = u(x_0, t_0),$$

so that (1.21) holds. If u is a temperature, then (1.20) follows.

For the converse, if $\Theta u < 0$ at some point of E, then the continuity of Θu implies that $\Theta u < 0$ on an open subset D of E. If $\overline{\Omega}(x_0, t_0; c) \subseteq D$, then Theorem 1.6 shows that $u(x_0, t_0) > \mathcal{M}(u; x_0, t_0; l)$ whenever $0 < l \leq c$. It now follows from (1.18) that $\mathcal{V}(u; x_0, t_0; b) < u(x_0, t_0)$ whenever $0 < b \leq c$, so that the condition in the converse fails to hold. □

REMARK 1.17. Since constant functions are temperatures, we can take $u = 1$ in (1.20) to get $1 = \mathcal{V}(1; x_0, t_0; c)$, or

$$1 = \tau(c) \int\!\!\int_{\Omega(x_0, t_0; c)} \frac{|x_0 - x|^2}{4(t_0 - t)^2} \, dx \, dt,$$

for all $c > 0$.

THEOREM 1.18. *If u is a smooth subtemperature on E, and $\overline{\Omega}(x_0, t_0; c) \subseteq E$, then the function $\mathcal{V}(u; x_0, t_0; \cdot)$ is increasing on $]0, c]$.*

PROOF. If $0 < a < b \leq c$, then by (1.18)

$$\mathcal{V}(u; x_0, t_0; b) - \mathcal{V}(u; x_0, t_0; a)$$
$$= \frac{n}{2} b^{-\frac{n}{2}} \int_0^b l^{\frac{n}{2}-1} (\mathcal{M}(u; x_0, t_0; l) - \mathcal{M}(u; x_0, t_0; al/b)) \, dl,$$

and the integrand is non-negative by Theorem 1.8. □

THEOREM 1.19. *Let $u \in C^{2,1}(E)$, let $\overline{\Omega}(x_0, t_0; c_0) \subseteq E$, and suppose that u is a smooth subtemperature on $\Omega(x_0, t_0; c_0)$. Then $\mathcal{V}(u; x_0, t_0; \cdot)$ is a convex function of τ on $]0, c_0]$. In particular, if u is a temperature on $\Omega(x_0, t_0; c_0)$, then there are real numbers α and β such that*

$$\mathcal{V}(u; x_0, t_0; a) = \alpha \tau(a) + \beta$$

whenever $0 < a \leq c_0$.

PROOF. The result is a consequence of Theorem 1.11 and (1.18). First observe that, if $a, b, c, r \in {]0, \infty[}$ then

$$\tau(b) - \tau(a) = \left(\frac{r}{c}\right)^{\frac{n}{2}} \left(\tau\left(\frac{br}{c}\right) - \tau\left(\frac{ar}{c}\right)\right).$$

It follows that, if $0 < a < b < c \leq c_0$, and we write $\mathcal{V}(l) = \mathcal{V}(u; x_0, t_0; l)$ and $\mathcal{M}(l) = \mathcal{M}(u; x_0, t_0; l)$, then by (1.18),

$$\frac{\mathcal{V}(c) - \mathcal{V}(b)}{\tau(c) - \tau(b)} - \frac{\mathcal{V}(b) - \mathcal{V}(a)}{\tau(b) - \tau(a)}$$
$$= \frac{n}{2} c^{-\frac{n}{2}} \int_0^c \left(\frac{\mathcal{M}(r) - \mathcal{M}(br/c)}{\tau(c) - \tau(b)} - \frac{\mathcal{M}(br/c) - \mathcal{M}(ar/c)}{\tau(b) - \tau(a)}\right) r^{\frac{n}{2}-1} \, dr$$
$$= \frac{n}{2} \int_0^c \left(\frac{\mathcal{M}(r) - \mathcal{M}(br/c)}{\tau(r) - \tau(br/c)} - \frac{\mathcal{M}(br/c) - \mathcal{M}(ar/c)}{\tau(br/c) - \tau(ar/c)}\right) r^{-1} \, dr.$$

By Theorem 1.11, the last integrand is nonpositive. Therefore

$$\frac{\mathcal{V}(c) - \mathcal{V}(b)}{\tau(c) - \tau(b)} \leq \frac{\mathcal{V}(b) - \mathcal{V}(a)}{\tau(b) - \tau(a)}$$

whenever $0 < a < b < c \leq c_0$, which means that $\mathcal{V}$ is a convex function of τ.

If u is a temperature on $\Omega(x_0, t_0; c_0)$, then the above can be applied to both u and $-u$ to deduce that $\mathcal{V}$ is an affine function of τ on ${]0, c_0]}$. $\square$

1.5. The Boundary Maximum Principle on Circular Cylinders

As an application of our mean value theorems, we prove a boundary maximum principle for smooth subtemperatures on circular cylinders which we shall need in Chapter 2. In fact, we prove the principle for more general functions, because the proof does not require the $C^{2,1}$ smoothness of the functions.

Henceforth, given any set $S \in \mathbb{R}^{n+1}$, we denote the class of all continuous, real-valued functions on S by $C(S)$.

DEFINITION 1.20. If E is an open set, and u is a function in $C(E)$ such that

$$u(x, t) \leq \mathcal{V}(u; x, t; c)$$

whenever $\overline{\Omega}(x, t; c) \subseteq E$, then u is called a *real continuous subtemperature* on E.

Theorem 1.16 shows that every smooth subtemperature is a real continuous subtemperature. Furthermore, if u and v are smooth subtemperatures on the same open set E, then

$$(u \vee v)(p) \leq \mathcal{V}(u; x, t; c) \vee \mathcal{V}(v; x, t; c) \leq \mathcal{V}(u \vee v; x, t; c)$$

whenever $\overline{\Omega}(x, t; c) \subseteq E$, so that $u \vee v$ is a real continuous subtemperature on E, but not usually a smooth subtemperature. In particular, $u^+ = u \vee 0$ is a real continuous subtemperature on E. If u is a temperature on E, then $u^- = (-u) \vee 0$ and $|u| = u^+ \vee u^-$ are also real continuous subtemperatures on E.

Observe that, if u is a function in $C(E)$ such that

$$u(x, t) \leq \mathcal{M}(u; x, t; c)$$

whenever $\overline{\Omega}(x,t;c) \subseteq E$, then

$$u(x,t) = \frac{n}{2}c^{-\frac{n}{2}}\int_0^c l^{\frac{n}{2}-1}u(x,t)\,dl \leq \frac{n}{2}c^{-\frac{n}{2}}\int_0^c l^{\frac{n}{2}-1}\mathcal{M}(u;x,t;l)\,dl = \mathcal{V}(u;x,t;c)$$

whenever $\overline{\Omega}(x,t;c) \subseteq E$, so that u is a real continuous subtemperature on E.

Consider an open ball B in $\mathbb{R}^n$, and a bounded time interval $]a,b[$. We denote by D the *circular cylinder* $D = B \times]a,b[\subseteq \mathbb{R}^{n+1}$. We denote by $\partial_n D$ the *normal boundary* of D, which consists of the union of the *lateral surface* $\partial B \times]a,b]$ and the *initial surface* $\overline{B} \times \{a\}$.

The next theorem gives the *boundary maximum principle* for real continuous subtemperatures on D.

THEOREM 1.21. *Let u belong to $C(D \cup \partial_n D)$ and the class of real continuous subtemperatures on D. If there is a point $(x_0, t_0) \in D$ such that $u(x_0, t_0) \geq u(x,t)$ whenever $(x,t) \in D$ and $t < t_0$, then $u(x_0, t_0) = u(x,t)$ for all such points (x,t). Consequently,*

$$\sup_{D \cup \partial_n D} u = \max_{\partial_n D} u.$$

PROOF. Put $M = u(x_0, t_0)$, choose an arbitrary point $q_0 = (y_0, s_0) \in D$ with $s_0 < t_0$, and join the point $p_0 = (x_0, t_0)$ to q_0 with a line segment λ. We need to prove that $u(q_0) = M$. Since the distance between λ and $\mathbb{R}^{n+1}\setminus \overline{D}$ is positive, we can find $c_0 > 0$ such that $\overline{\Omega}(p;c_0) \equiv \overline{\Omega}(x,t;c_0) \subseteq D$ for every point $p = (x,t) \in \lambda$.

The proof proceeds inductively. We know that $u(p_0) = M$. Suppose that, for some integer $k \geq 0$, there is a point $p_k \in \lambda$ such that $u(p_k) = M$. Then

$$M = u(p_k) \leq \mathcal{V}(u;p_k;c_0) \leq \mathcal{V}(M;p_k;c_0) = M,$$

so the continuity of u implies that $u(p) = M$ for all $p \in \overline{\Omega}(p_k;c_0)$. If $q_0 \in \overline{\Omega}(p_k;c_0)$, then $u(q_0) = M$ as required. If not, take p_{k+1} to be the sole point of $\partial\Omega(p_k;c_0) \cap \lambda$. Then $p_{k+1} \in \lambda$ and $u(p_{k+1}) = M$. Since λ has finite length, and the length of the segment between p_k and p_{k+1} does not depend on k, after finitely many iterations we obtain a point, p_m say, such that $u(p_m) = M$ and $q_0 \in \overline{\Omega}(p_m;c_0)$. So $u(q_0) = M$, as required.

For the last part, given any α such that $a < \alpha < b$, we put $D_\alpha = B \times]a,\alpha[$ and $M_\alpha = \max\{u(p) : p \in \overline{D}_\alpha\}$. Choose a point $(x',t') \in \overline{D}_\alpha$ such that $u(x',t') = M_\alpha$. If $(x',t') \in D$, then the first part of the theorem shows that $u(x,t) = M_\alpha$ for all $(x,t) \in \overline{D}_\alpha$ such that $t < t'$, so we can assume that $(x',t') \in \overline{D}_\alpha \setminus D$. It follows that

$$\sup_{D \cup \partial_n D} u = \sup_{\alpha \in]a,b[} M_\alpha$$

is attained at some point of $\partial_n D$. $\square$

REMARK 1.22. If $u \in C(D \cup \partial_n D)$ and is a temperature on D, then we can also apply Theorem 1.21 to $-u$ to obtain a *minimum principle* for u. In particular, we get

$$\inf_{D \cup \partial_n D} u = \min_{\partial_n D} u.$$

As a consequence of the maximum principle, we can prove the statement we made in Section 1.3 motivating the term 'subtemperature'.

THEOREM 1.23. *Let u be a real continuous subtemperature on an open set E, and let D be a circular cylinder such that $\overline{D} \subseteq E$. If $v \in C(D \cup \partial_n D)$, is a temperature on D, and satisfies $u \leq v$ on $\partial_n D$, then $u \leq v$ on D.*

PROOF. The function $w = u - v$ belongs to $C(D \cup \partial_n D)$, is a real continuous subtemperature on D, and takes nonpositive values at all points of $\partial_n D$. Therefore, by Theorem 1.21,
$$\sup_{D \cup \partial_n D} w = \max_{\partial_n D} w \leq 0,$$
so that $w \leq 0$ on D. Hence $u \leq v$ on D, as asserted. □

Theorem 1.23 does not address the *existence* of such a temperature v. We shall prove in Chapter 2 that, given any function $f \in C(\partial_n D)$, there is a temperature on D which is continuous on $\overline{D}$ and coincides with f on $\partial_n D$.

The boundary maximum principle enables us to prove the important *boundary uniqueness principle for temperatures* on a circular cylinder.

THEOREM 1.24. *Let u and v be temperatures on the circular cylinder D, and in the class $C(D \cup \partial_n D)$. If $u(p) = v(p)$ for all $p \in \partial_n D$, then $u(p) = v(p)$ for all $p \in D$.*

PROOF. The function $w = u - v$ is in the class $C(D \cup \partial_n D)$, is a temperature on D, and takes the value 0 at all points of $\partial_n D$. Therefore, by Theorem 1.21,
$$\sup_{D \cup \partial_n D} w = \max_{\partial_n D} w = 0,$$
so that $w \leq 0$ on D. Since $-w$ has the same properties as w, Theorem 1.21 also shows that $-w \leq 0$ on D. Hence $w = 0$ on D, as asserted. □

1.6. Modified Heat Balls

The kernel for the heat ball, given in formula (1.19), is unbounded near the centre of the ball. For most purposes this does not cause problems, but for some it causes substantial ones. For example, in Theorem 1.16 we gave a characterization of temperatures amongst the class $C^{2,1}(E)$; but if we want a similar characterization of temperatures amongst the class of *locally integrable* functions on E, then we need a bounded kernel. We also need a bounded kernel to prove the Harnack Inequality.

We now consider a family of modified heat balls indexed by an integer $m \geq 1$. As m increases, the kernel's behaviour improves. For $m \geq 3$ the kernel is bounded, while for $m \geq 5$ it has a smooth extension by zero to the whole of $\mathbb{R}^{n+1}$.

Let m be an integer, $m \geq 1$. Given $(x_0, t_0) \in \mathbb{R}^{n+1}$ and $c > 0$, we put

$$\Omega_m(x_0, t_0; c)$$
$$= \left\{ (y,s) : (t_0 - s)^{-\frac{m+n}{2}} \exp\left(-\frac{|x_0 - y|^2}{4(t_0 - s)}\right) > c^{-\frac{m+n}{2}} \right\}$$
$$= \left\{ (y,s) : |x_0 - y|^2 < 2(m+n)(t_0 - s) \log\left(\frac{c}{t_0 - s}\right), t_0 - c < s < t_0 \right\}.$$

Thus $\Omega_m(x_0, t_0; c)$ is the projection onto $\mathbb{R}^{n+1}$ of a heat ball in $\mathbb{R}^{m+n+1}$. Let $u \in C^{2,1}(E)$, and put

(1.22) $\quad \widehat{u}(\xi, x, t) = u(x, t) \quad$ for all $\quad \xi \in \mathbb{R}^m \quad$ and $\quad (x, t) \in E$.

Then $\widehat{u} \in C^{2,1}(\mathbb{R}^m \times E)$, and we can apply Theorem 1.16 to $\widehat{u}$. Note that the volume mean formula (1.19), when applied to $\widehat{u}$, becomes

$$\mathcal{V}(\widehat{u}; \xi_0, x_0, t_0)$$
$$= (4\pi c)^{-\frac{m+n}{2}} \int\int\int_{\Omega(\xi_0, x_0, t_0; c)} \frac{|\xi_0 - \xi|^2 + |x_0 - x|^2}{4(t_0 - t)^2} \widehat{u}(\xi, x, t) \, d\xi \, dx \, dt.$$

Since $\widehat{u}(\xi, x, t) = u(x, t)$ does not depend on ξ, we can integrate out ξ and obtain a new volume mean for u, which we denote by $\mathcal{V}_m(u; x_0, t_0; c)$. Thus

$$\mathcal{V}_m(u; x_0, t_0; c)$$
$$= (4\pi c)^{-\frac{m+n}{2}} \int\int_{\Omega_m(x_0, t_0; c)} \left(\int_{|\xi_0 - \xi| < R} \frac{|\xi_0 - \xi|^2 + |x_0 - x|^2}{4(t_0 - t)^2} \, d\xi \right) u(x, t) \, dx \, dt,$$

where

$$R = R(x_0 - x, t_0 - t) = \sqrt{2(m+n)(t_0 - t) \log\left(\frac{c}{t_0 - t}\right) - |x_0 - x|^2}.$$

The innermost integral can be evaluated explicitly. We have

$$\int_{|\xi_0 - \xi| < R} \frac{|\xi_0 - \xi|^2 + |x_0 - x|^2}{4(t_0 - t)^2} \, d\xi = \omega_m \int_0^R \frac{r^2 + |x_0 - x|^2}{4(t_0 - t)^2} r^{m-1} \, dr$$
$$= \frac{\omega_m R^m}{4m(t_0 - t)^2} \left(\frac{m}{m+2} R^2 + |x_0 - x|^2 \right),$$

where ω_m is the surface area of the unit sphere in $\mathbb{R}^m$. Hence

$$(1.23) \qquad \mathcal{V}_m(u; x_0, t_0; c) = \int\int_{\Omega_m(x_0, t_0; c)} K_{m,c}(x_0 - x, t_0 - t) u(x, t) \, dx \, dt,$$

where

$$K_{m,c}(x_0 - x, t_0 - t)$$
$$= \frac{\omega_m (4\pi c)^{-\frac{m+n}{2}}}{2m(m+2)} R(x_0 - x, t_0 - t)^m \left(\frac{m(m+n)}{t_0 - t} \log\left(\frac{c}{t_0 - t}\right) + \frac{|x_0 - x|^2}{(t_0 - t)^2} \right)$$

is a continuous function of (x, t) on $\overline{\Omega}_m(x_0, t_0; c) \setminus \{(x_0, t_0)\}$, positive on $\Omega_m(x_0, t_0; c)$, and zero on $\partial\Omega_m(x_0, t_0; c) \setminus \{(x_0, t_0)\}$.

If m is sufficiently large, then the function $(x, t) \mapsto K_{m,c}(x_0 - x, t_0 - t)$ has a continuous extension by zero at (x_0, t_0). For if

$$|x_0 - x|^2 < 2(m+n)(t_0 - t) \log \frac{c}{t_0 - t},$$

then

$$0 < R(x_0 - x, t_0 - t)^2 < 2(m+n)(t_0 - t) \log \frac{c}{t_0 - t},$$

so that

$$K_{m,c}(x_0 - x, t_0 - t) \leq A(t_0 - t)^{\frac{m-2}{2}} \left(\log \frac{c}{t_0 - t} \right)^{\frac{m+2}{2}}$$

for some positive constant A which depends only on c, m, and n. It follows that, if $m \geq 3$, then $K_{m,c}(x_0 - x, t_0 - t) \to 0$ as $t \to t_0$. In particular, the kernel is bounded if $m \geq 3$.

For any $m \geq 1$, we can give a characterization of smooth subtemperatures in the class $C^{2,1}(E)$ using the means $\mathcal{V}_m$.

THEOREM 1.25. *Let $u \in C^{2,1}(E)$, and let m be an integer, $m \geq 1$. If u is a temperature on E, then the equality*

(1.24) $$u(x_0, t_0) = \mathcal{V}_m(u; x_0, t_0; c)$$

holds whenever $\overline{\Omega}_m(x_0, t_0; c) \subseteq E$. If, more generally, u is a smooth subtemperature on E, then the inequality

(1.25) $$u(x_0, t_0) \leq \mathcal{V}_m(u; x_0, t_0; c)$$

holds whenever $\overline{\Omega}_m(x_0, t_0; c) \subseteq E$.

Conversely if, given any point $(x_0, t_0) \in E$ and $\epsilon > 0$, we can find $c < \epsilon$ such that (1.25) holds, then u is a smooth subtemperature on E.

PROOF. Given $u \in C^{2,1}(E)$, we define $\hat{u}$ by (1.22), and denote by $\hat{\Theta}$ the heat operator in $\mathbb{R}^{m+n+1}$.

Suppose that u is a smooth subtemperature on E, and that $\overline{\Omega}_m(x_0, t_0; c) \subseteq E$. Then $\hat{\Theta}\hat{u} = \Theta u \geq 0$ on $\mathbb{R}^m \times E$, and $\overline{\Omega}(\xi_0, x_0, t_0; c) \subseteq \mathbb{R}^m \times E$ for every $\xi_0 \in \mathbb{R}^m$. Applying Theorem 1.16 to $\hat{u} \in C^{2,1}(\mathbb{R}^m \times E)$, we obtain

$$u(x_0, t_0) = \hat{u}(\xi_0, x_0, t_0) \leq \mathcal{V}(\hat{u}; \xi_0, x_0, t_0; c) = \mathcal{V}_m(u; x_0, t_0; c).$$

This proves (1.25), and (1.24) follows if u is a temperature.

Conversely, suppose that we are given $(x_0, t_0) \in E$ and $\epsilon > 0$, and that $c < \epsilon$ is chosen so that (1.25) holds. Then, whenever $\xi_0 \in \mathbb{R}^m$, we have

$$\hat{u}(\xi_0, x_0, t_0) = u(x_0, t_0) \leq \mathcal{V}_m(u; x_0, t_0; c) = \mathcal{V}(\hat{u}; \xi_0, x_0, t_0; c),$$

and another application of Theorem 1.16 to $\hat{u}$ shows that $\hat{\Theta}\hat{u} \geq 0$ on $\mathbb{R}^m \times E$. Hence $\Theta u \geq 0$ on E. $\square$

COROLLARY 1.26. *Let u be a temperature on E, let m be an integer, $m \geq 3$, and put $l = 2(m+n)/e$. If the closed circular cylinder*

$$\overline{B}(x_0, l\sqrt{c}) \times [t_0 - c, t_0] = \{(y, s) : |y - x_0| \leq l\sqrt{c},\ t_0 - c \leq s \leq t_0\}$$

is contained in E, then there is a positive number κ, which depends only on c, m and n, such that

$$|u(x_0, t_0)| \leq \kappa \int_{t_0-c}^{t_0} \int_{B(x_0, l\sqrt{c})} |u(y, s)|\, dy\, ds.$$

PROOF. The closed modified heat ball $\overline{\Omega}_m(x_0, t_0; c)$ is contained in the cylinder $\overline{B}(x_0, l\sqrt{c}) \times [t_0 - c, t_0]$, and therefore is a subset of E. Hence, by Theorem 1.25,

$$u(x_0, t_0) = \int\!\!\int_{\Omega_m(x_0, t_0; c)} K_{m,c}(x_0 - y, t_0 - s) u(y, s)\, dy\, ds.$$

The kernel $K_{m,c}$ is nonnegative, and bounded by a number κ which depends only on c, m and n. It follows that

$$|u(x_0, t_0)| \leq \kappa \int\!\!\int_{\Omega_m(x_0, t_0; c)} |u(y, s)|\, dy\, ds \leq \kappa \int_{t_0-c}^{t_0} \int_{B(x_0, l\sqrt{c})} |u(y, s)|\, dy\, ds.$$

$\square$

The key to our characterization of temperatures in the class of locally integrable functions lies in parts (a) and (c) of the next theorem. Part (b) is essential for the proofs of theorems in Chapter 6.

We extract part of the proof as a lemma.

LEMMA 1.27. *Let m be an integer with $m \geq 5$, and let $c > 0$. If*
$$\rho(r,t) = \sqrt{2(m+n)t \log \frac{c}{t} - r^2}$$
whenever $r \in \mathbb{R}$ and $0 < t < c$, and if
$$\lambda_{m,c}(r,t) = \begin{cases} \alpha_{m,c}\rho(r,t)^m \left(m(m+n)\frac{1}{t}\log\frac{c}{t} + \frac{r^2}{t^2} \right) & \text{if } r^2 < 2(m+n)t\log\frac{c}{t}, \\ 0 & \text{otherwise,} \end{cases}$$
where
$$\alpha_{m,c} = \frac{\omega_m (4\pi c)^{-(m+n)/2}}{2m(m+2)},$$
then $\lambda_{m,c} \in C^{2,1}(\mathbb{R}^2)$, and
$$\lambda_{m,c}(|x-y|, t-s) = K_{m,c}(x-y, t-s)$$
whenever $(y,s) \in \Omega_m(x,t;c)$.

PROOF. Routine calculations show that there are constants $a_1, ..., a_9$ such that
$$\frac{\partial \lambda_{m,c}}{\partial r}(r,t) = \alpha_{m,c}\rho(r,t)^{m-2}\left(a_1 \frac{r}{t}\log\frac{c}{t} + a_2\frac{r^3}{t^2} \right),$$

$$\frac{\partial^2 \lambda_{m,c}}{\partial r^2}(r,t) = \alpha_{m,c}\rho(r,t)^{m-4}\left(a_3 \frac{r^2}{t}\log\frac{c}{t} + a_4\left(\log\frac{c}{t}\right)^2 + a_5\frac{r^4}{t^2} \right),$$

$$\frac{\partial \lambda_{m,c}}{\partial t}(r,t) = \alpha_{m,c}\rho(r,t)^{m-2}\left(a_6 \frac{1}{t}\left(\log\frac{c}{t}\right)^2 + a_7\frac{1}{t}\log\frac{c}{t} + a_8\frac{r^2}{t^2}\log\frac{c}{t} + a_9\frac{r^4}{t^3} \right)$$

whenever $r^2 < 2(m+n)t\log(c/t)$. Therefore, because $m \geq 5$, all these derivatives tend to zero as (r,t) approaches any point (R,T) where $R^2 = 2(m+n)T \log(c/T)$ and $T > 0$. Furthermore, whenever $r^2 < 2(m+n)t\log(c/t)$ we have

$$\lambda_{m,c}(r,t) \leq At^{\frac{m-2}{2}}\left(\log\frac{c}{t}\right)^{\frac{m+2}{2}},$$

$$\left|\frac{\partial \lambda_{m,c}}{\partial r}(r,t)\right| \leq At^{\frac{m-3}{2}}\left(\log\frac{c}{t}\right)^{\frac{m+1}{2}},$$

$$\left|\frac{\partial^2 \lambda_{m,c}}{\partial r^2}(r,t)\right| \leq At^{\frac{m-4}{2}}\left(\log\frac{c}{t}\right)^{\frac{m}{2}},$$

$$\left|\frac{\partial \lambda_{m,c}}{\partial t}(r,t)\right| \leq At^{\frac{m-4}{2}}\left(\log\frac{c}{t}\right)^{\frac{m}{2}}\left(1 + \log\frac{c}{t}\right),$$

for some constant A. Since $m \geq 5$, it follows that all these functions tend to zero as $t \to 0+$, so that $\lambda_{m,c} \in C^{2,1}(\mathbb{R}^2)$. The rest of the statement is trivial. □

THEOREM 1.28. *Let u be a locally integrable function on E, let m be an integer with $m \geq 5$, let $c > 0$, and let*

$$E_{m,c} = \{(y,s) : \overline{\Omega}_m(y,s;c) \subseteq E\}.$$

For all $(x,t) \in E_{m,c}$, we define

$$u_{m,c}(x,t) = \mathcal{V}_m(u; x, t; c).$$

Then:

(a) $u_{m,c} \in C^{2,1}(E_{m,c})$.
(b) If K is any compact subset of $E_{m,c}$, we have

$$\lim_{c \to 0+} \int \int_K |u_{m,c}(x,t) - u(x,t)| \, dx \, dt = 0.$$

(c) If there is an integer $j \geq 0$ such that the inequality

(1.26) $$u(x,t) \leq \mathcal{V}_j(u; x, t; b)$$

holds whenever $\overline{\Omega}_j(x,t;b) \subseteq E$ (where $\mathcal{V}_0 = \mathcal{V}$ and $\Omega_0 = \Omega$), then $u_{m,c}$ is a smooth subtemperature on $E_{m,c}$.

PROOF. Let D be a bounded open set with $\overline{D} \subseteq E$, and put

$$D_{m,c} = \{(y,s) : \overline{\Omega}_m(y,s;c) \subseteq D\}.$$

Then u is integrable on D, and if we prove that $u_{m,c} \in C^{2,1}(D_{m,c})$ then part (a) will follow. Since

$$\mathcal{V}_m(u; x, t; c) = \int \int_{\Omega_m(x,t;c)} K_{m,c}(x-y, t-s) u(y,s) \, dy \, ds$$
$$= \int \int_{\mathbb{R}^{n+1}} \lambda_{m,c}(|x-y|, t-s) u(y,s) \, dy \, ds,$$

the smoothness follows from Lemma 1.27.

To prove part (b), we need only show that the family $\{\phi_c : c > 0\}$ of functions given by

$$\phi_c(y,s) = \lambda_c(|y|, s)$$

is an approximate identity. Since $\phi_c \geq 0$, and

$$\int \int_{\mathbb{R}^{n+1}} \phi_c(y,s) \, dy \, ds = \mathcal{V}_m(1; 0, 0; c) = 1$$

by Theorem 1.25, and the support of ϕ_c is contained in the set

$$\{(y,s) : |y| \leq \sqrt{2(m+n)c/e}, \; 0 \leq s \leq c\},$$

the result of part (b) follows.

Now suppose that there is an integer $j \geq 0$ such that (1.26) holds whenever $\overline{\Omega}_j(x,t;b) \subseteq E$. To prove that $u_{m,c}$ is a smooth subtemperature on $E_{m,c}$, it suffices to show that, given any point $(x,t) \in E_{m,c}$ and any $\epsilon > 0$, we can find $b < \epsilon$ such that $u_{m,c}(x,t) \leq \mathcal{V}_j(u_{m,c}; x, t; b)$, in view of Theorem 1.16 if $j = 0$, Theorem 1.25

if $j \geq 1$. Given D as above, suppose that $\overline{\Omega}_j(x,t;a) \subseteq D_{m,c}$ and that $0 < b < a$. Then

$$\mathcal{V}_j(u_{m,c};x,t;b)$$
$$= \int\!\!\int_{\Omega_j(x,t;b)} K_{j,b}(x-y,t-s)$$
$$\times \left(\int\!\!\int_{\Omega_m(y,s;c)} K_{m,c}(y-z,s-r)u(z,r)\,dz\,dr\right)dy\,ds$$
$$= \int\!\!\int_{\Omega_j(x,t;b)} K_{j,b}(x-y,t-s)$$
$$\times \left(\int\!\!\int_{\Omega_m(0,0;c)} K_{m,c}(-z,-r)u(z+y,r+s)\,dz\,dr\right)dy\,ds$$
$$= \int\!\!\int_{\Omega_m(0,0;c)} K_{m,c}(-z,-r)$$
$$\times \left(\int\!\!\int_{\Omega_j(x,t;b)} K_{j,b}(x-y,t-s)u(y+z,s+r)\,dy\,ds\right)dz\,dr$$
$$= \int\!\!\int_{\Omega_m(0,0;c)} K_{m,c}(-z,-r)$$
$$\times \left(\int\!\!\int_{\Omega_j(x+z,t+r;b)} K_{j,b}(x+z-y,t+r-s)u(y,s)\,dy\,ds\right)dz\,dr$$
$$\geq \int\!\!\int_{\Omega_m(0,0;c)} K_{m,c}(-z,-r)u(x+z,t+r)\,dz\,dr$$
$$= \int\!\!\int_{\Omega_m(x,t;c)} K_{m,c}(x-z,t-r)u(z,r)\,dz\,dr$$
$$= u_{m,c}(x,t).$$

The change in the order of the integrals is justified by Fubini's Theorem. For, if $M = \max K_{m,c}$, then

$$\int\!\!\int_{\Omega_j(x,t;b)} K_{j,b}(x-y,t-s)\left(\int\!\!\int_{\Omega_m(y,s;c)} K_{m,c}(y-z,s-r)|u(z,r)|\,dz\,dr\right)dy\,ds$$
$$\leq M \int\!\!\int_{\Omega_j(x,t;b)} K_{j,b}(x-y,t-s)\left(\int\!\!\int_{\Omega_m(y,s;c)} |u(z,r)|\,dz\,dr\right)dy\,ds$$
$$\leq M \int\!\!\int_D |u(z,r)|\,dz\,dr\,\mathcal{V}_j(1;x,t;b)$$
$$< +\infty.$$

This proves part (c), and completes the proof of the theorem. $\square$

Now we can characterize temperatures within the class of locally integrable functions.

THEOREM 1.29. *Let u be a locally integrable function on E, and let m be an integer with $m \geq 5$. If there is a positive number c_0 such that the identity*

(1.27) $$u(x,t) = \mathcal{V}_m(u;x,t;b)$$

holds whenever $\overline{\Omega}_m(x,t;b) \subseteq E$ and $b \leq c_0$, then u is a temperature on E.

PROOF. Let B be an open euclidean ball such that $B \subseteq E$ and the diameter of B is no greater than c_0. Then whenever $\overline{\Omega}_m(x,t;b) \subseteq B$, we have $b \leq c_0$. If we prove that u is a temperature on B, then it will follow from the arbitrary nature of B that u is a temperature on E.

For each $c > 0$, we put $B_{m,c} = \{(y,s) : \overline{\Omega}_m(y,s;c) \subseteq B\}$, and we define a function $u_{m,c}$ on $B_{m,c}$ by putting

$$u_{m,c}(x,t) = \mathcal{V}_m(u;x,t;c).$$

Then Theorem 1.28 and (1.27) imply that $u_{m,c}$ is a smooth subtemperature on $B_{m,c}$. Furthermore, because $-u$ satisfies the same conditions as u, the function $-u_{m,c} = (-u)_{m,c} = -(u_{m,c})$ is also a smooth subtemperature on $B_{m,c}$. Hence $u_{m,c}$ is a temperature on $B_{m,c}$. Since $u_{m,c} = u$ on $B_{m,c}$, by (1.27), it follows that u is a temperature on $\bigcup_{c>0} B_{m,c} = B$, and therefore on E. □

1.7. Harnack Theorems

The characterization of temperatures in Theorem 1.29 enables us to prove results about convergence of sequences of temperatures, the simplest of which is this one.

THEOREM 1.30. *If $\{u_k\}$ is a sequence of temperatures which converges to a function u locally uniformly on E, then u is a temperature on E.*

PROOF. Let m be an integer, $m \geq 5$. Whenever k is a positive integer and the modified heat ball $\overline{\Omega}_m(p;c) \subseteq E$, we have $u_k(p) = \mathcal{V}_m(u_k;p;c)$, by Theorem 1.25. Since $u_k \to u$ locally uniformly on E, we see that $u \in C(E)$ and $u(p) = \mathcal{V}_m(u;p;c)$ whenever $\overline{\Omega}_m(p;c) \subseteq E$. Therefore u is a temperature on E, by Theorem 1.29. □

For the other results in this section, we need to introduce some notation. Given an open set E, and a point $p_0 = (x_0, t_0) \in E$, we denote by $\Lambda(p_0; E) \equiv \Lambda(x_0, t_0; E)$ the set of points p that are lower than p_0 relative to E, in the sense that there is a polygonal path $\gamma \subseteq E$ joining p_0 to p, along which the temporal variable is strictly decreasing. By a polygonal path, we mean a path which is a union of finitely many line segments.

As a simple example, if D is a circular cylinder and $p_0 \in D$, then $\Lambda(p_0; D)$ is just the set $\{(x,t) \in D : t < t_0\}$. For arbitrary E, the set $\Lambda(p_0; E)$ plays a similar role.

The next result is known as the *Harnack Monotone Convergence Theorem* for temperatures.

THEOREM 1.31. *Let $\{u_k\}$ be an increasing sequence of temperatures on E, and let $u = \lim_{k \to \infty} u_k$. If there is a point $(x_0, t_0) \in E$ such that $u(x_0, t_0) < +\infty$, then u is a temperature on $\Lambda(x_0, t_0; E)$.*

PROOF. We show that u is locally integrable on $\Lambda(x_0, t_0; E)$, with a view to applying Theorem 1.29. It is immediate that $-\infty < u(x,t) \leq +\infty$ for all $(x,t) \in E$. Let m be an integer, $m \geq 5$. If $\overline{\Omega}_m(x,t;c) \subseteq E$, then

$$u_k(x,t) = \mathcal{V}_m(u_k; x, t; c)$$

for all k, by Theorem 1.25, so that Lebesgue's Monotone Convergence Theorem yields

(1.28) $$u(x,t) = \mathcal{V}_m(u; x, t; c).$$

We prove the contrapositive. If u is not locally integrable on $\Lambda = \Lambda(x_0, t_0; E)$, we can find a point $(x_1, t_1) \in \Lambda$ such that u is not integrable on any neighbourhood of (x_1, t_1). Join (x_0, t_0) to (x_1, t_1) by a polygonal path $\gamma \subseteq E$ along which the temporal variable is strictly decreasing. Since γ is compact, its distance from $\mathbb{R}^{n+1} \backslash E$ is positive, and so we can find $c_0 > 0$ such that $\overline{\Omega}_m(x, t; c_0) \subseteq E$ for all $(x,t) \in \gamma$. Given $(x,t) \in \gamma$, we put

$$P(x,t) = \{(y,s) : |y-x|^2 < 2(m+n)(s-t),\ s-t < c_0/e\}.$$

The set $P(x,t)$ is a truncated paraboloid with vertex (x,t), and if $(y,s) \in P(x,t)$ then

$$|y-x|^2 < 2(m+n)(s-t) < 2(m+n)(s-t)\log\left(\frac{c_0}{s-t}\right),$$

so that $(x,t) \in \Omega_m(y, s; c_0)$.

Because γ is a union of finitely many line segments, there is a positive number $c_1 < c_0/e$, independent of (x,t), such that if (x,t) and $(y, t+c_1)$ both belong to γ, then $(y, t+c_1) \in P(x,t)$. Choose points $(x_2, t_2), ..., (x_l, t_l)$ inductively, such that $t_j = t_1 + (j-1)c_1$ and $(x_j, t_j) \in \gamma$, for all $j \in \{2, ..., l\}$, and such that $t_l < t_0 \leq t_l + c_1$. Note that $(x_j, t_j) \in P(x_{j-1}, t_{j-1})$ for all $j \in \{2, ..., l\}$. Since $(x_1, t_1) \in \Lambda$, we have $(x_1, t_1) \in \Omega_m(y, s; c_0)$ for all $(y, s) \in P(x_1, t_1)$. Therefore, because the kernel $K_{m,c_0}(y-z, s-r)$ is positive whenever $(z,r) \in \Omega_m(y, s; c_0)$, it follows that $\mathcal{V}_m(u; y, s; c_0) = +\infty$ for all $(y, s) \in P(x_1, t_1)$. Therefore (1.28) shows that $u(y,s) = +\infty$ for all such points (y, s), and in particular that u is not integrable on any neighbourhood of (x_2, t_2).

Proceeding stepwise along γ, we deduce successively that u is not integrable on any neighbourhood of $(x_2, t_2), ..., (x_l, t_l)$, and finally that $u(x_0, t_0) = +\infty$. So our hypothesis that $u(x_0, t_0) < +\infty$ implies that u is locally integrable on Λ. Now Theorem 1.29 and (1.28) show that u is a temperature on Λ. □

We deduce a general form of the *Harnack Inequality for Temperatures* from the Harnack Monotone Convergence Theorem. The standard form follows as a corollary.

THEOREM 1.32. *Let μ be a measure on E, and let S be the the support of μ. Let K be a compact subset of E, such that for each point $(x,t) \in K$ there is a point $(x_0, t_0) \in S$ with $(x,t) \in \Lambda(x_0, t_0; E)$. Then there is a constant κ, which depends only on E, μ and K, such that*

$$\max_K u \leq \kappa \int_S u\, d\mu$$

for every nonnegative temperature u on E.

PROOF. Suppose that, given E, μ and K, there is no such constant. Then for each integer $k \geq 1$, there is a nonnegative temperature u_k on E such that

$$\max_K u_k \geq k^3 \int_S u_k \, d\mu.$$

Putting $v_k = u_k (k^2 \int_S u_k \, d\mu)^{-1}$, we obtain

$$\max_K v_k \geq k \quad \text{and} \quad \int_S v_k \, d\mu \leq k^{-2}.$$

Consider the function $v = \sum_{k=1}^\infty v_k$. Since

$$\int_S \sum_{k=1}^\infty v_k \, d\mu = \sum_{k=1}^\infty \int_S v_k \, d\mu \leq \sum_{k=1}^\infty k^{-2} < +\infty,$$

the series is convergent μ-almost everywhere on S, and hence on a dense subset of S. Applying Theorem 1.31 to the sequence of partial sums of the series, we see that if the series is convergent at a point (x_0, t_0) then it is convergent on $\Lambda(x_0, t_0; E)$ to a temperature. It follows that v is a temperature on the set

$$\bigcup_{(x,t) \in S} \Lambda(x, t; E),$$

which contains K. This contradicts the fact that $\sup_K v \geq \max_K v_k \geq k$ for all k. $\square$

COROLLARY 1.33. *If $(x_0, t_0) \in E$, and K is a compact subset of $\Lambda(x_0, t_0; E)$, then there is a constant κ, which depends only on E, (x_0, t_0) and K, such that*

$$\max_K u \leq \kappa u(x_0, t_0)$$

for every nonnegative temperature u on E.

PROOF. In Theorem 1.32, take μ to be the unit mass at (x_0, t_0). $\square$

In the Harnack inequality of the above corollary, there is a time lag between the latest time in the compact subset K and the time t_0. This is generally unavoidable, but can be overcome in the following case.

COROLLARY 1.34. *Let E contain the compact set*

$$K = \{(x, t) : |x - x_0|^2 \leq \alpha(t - t_0), \, t_0 + \beta \leq t \leq t_0\},$$

for some negative numbers α and β. Then there is a constant κ, which depends only on E, (x_0, t_0) and K, such that

$$\max_K u \leq \kappa u(x_0, t_0)$$

for every nonnegative temperature u on E.

PROOF. Temperatures are translation invariant, so we take $(x_0, t_0) = (0, 0)$. Given any number γ such that $\beta < \gamma < 0$, we define a sequence $\{s_j\}$ by putting

$$s_0 = \beta, \qquad s_j = \frac{\gamma^j}{\beta^{j-1}} \quad \text{for all} \quad j \in \mathbb{N}.$$

Then, for each integer $j \geq 0$, we put $K_j = K \cap [s_j, s_{j+1}]$, so that

$$K \backslash \{(0,0)\} = \bigcup_{j=0}^{\infty} K_j.$$

Since K is a compact subset of E, we can find a positive number δ such that the open set D of points distant less than δ from K, is contained in E. Given a nonnegative temperature u on E, we define

$$u_j(x,t) = u\left(\left(\frac{\beta}{\gamma}\right)^{j/2} x, \left(\frac{\beta}{\gamma}\right)^j t\right)$$

for all $j \geq 1$. Since temperatures are invariant under parabolic dilation, each u_j is a temperature on an open superset of D. Therefore, by Corollary 1.33, there is a constant κ, which depends only on D, (x_0, t_0) and K, such that

$$\max_{K_0} u_j \leq \kappa u_j(0,0) = \kappa u(0,0)$$

for all j. Now

$$\max_{K_0} u_j = \max\left\{u\left(\left(\frac{\beta}{\gamma}\right)^{j/2} x, \left(\frac{\beta}{\gamma}\right)^j t\right) : |x|^2 \leq \alpha t, \beta \leq t \leq \gamma\right\}$$

$$= \max\left\{u(y,s) : |y|^2 \leq \alpha s, \left(\frac{\gamma}{\beta}\right)^j \beta \leq s \leq \left(\frac{\gamma}{\beta}\right)^j \gamma\right\}$$

$$= \max\{u(y,s) : |y|^2 \leq \alpha s, s_j \leq s \leq s_{j+1}\}$$

$$= \max_{K_j} u,$$

so that

$$\max_K u = \sup_{K \backslash \{0\}} u = \sup_j \left(\max_{K_j} u\right) \leq \kappa u(0,0).$$

$\square$

The next theorem is concerned with families of temperatures that may not be sequences.

DEFINITION 1.35. A family $\mathcal{F}$ of functions on E is said to be *upward-directed* if, for each pair $u, v \in \mathcal{F}$, there exists a $w \in \mathcal{F}$ such that $u \vee v \leq w$. Similarly, $\mathcal{F}$ is said to be *downward-directed* if $u, v \in \mathcal{F}$ implies that there is $w \in \mathcal{F}$ such that $u \wedge v \geq w$.

An increasing sequence of functions is clearly an example of an upward-directed family, and so the following result generalizes and strengthens Theorem 1.31.

THEOREM 1.36. *Let $\mathcal{F}$ be an upward-directed family of temperatures on E, and let $u = \sup \mathcal{F}$. Suppose that there is a point $p_0 = (x_0, t_0) \in E$ such that $u(p_0) < +\infty$, and let α and β be negative numbers such that the set*

$$P(x_0, t_0; \alpha, \beta) = \{(x,t) : |x - x_0|^2 \leq \alpha(t - t_0), t_0 + \beta \leq t \leq t_0\}$$

is contained in E. Then u is a temperature on $\Lambda(p_0; E)$, and its restriction to $P(x_0, t_0; \alpha, \beta)$ is continuous.

PROOF. Let K be any compact subset of $\Lambda(p_0; E)$. For each positive integer k, we can find a function $u_k \in \mathcal{F}$ such that
$$u(p_0) - u_k(p_0) < \frac{1}{k}.$$
Since $\mathcal{F}$ is upward-directed, given any function $v \in \mathcal{F}$ and $k \in \mathbb{N}$, we can find a temperature $w_k \in \mathcal{F}$ such that $u_k \vee v \leq w_k$ on E. By the Harnack inequalities for temperatures in Corollaries 1.33 and 1.34, there is a positive constant κ, depending only on E, p_0, K and $P(x_0, t_0; \alpha, \beta)$, such that
$$w_k(p) - u_k(p) \leq \kappa(w_k(p_0) - u_k(p_0))$$
for all $p \in K \cup P(x_0, t_0; \alpha, \beta)$ and all k. Hence
$$v(p) - u_k(p) \leq w_k(p) - u_k(p) \leq \kappa(w_k(p_0) - u_k(p_0)) \leq \kappa(u(p_0) - u_k(p_0)) < \frac{\kappa}{k}$$
for all such p and k. Therefore
$$u(p) - u_k(p) = \sup\{v(p) - u_k(p) : v \in \mathcal{F}\} \leq \frac{\kappa}{k}$$
for all such p and k, so that the sequence $\{u_k\}$ converges uniformly to u on the set $K \cup P(x_0, t_0; \alpha, \beta)$. Therefore u is a temperature on $\Lambda(p_0; E)$, by Theorem 1.30, and the restriction of u to $P(x_0, t_0; \alpha, \beta)$ is continuous. □

DEFINITION 1.37. Let $(x_0, t_0) \in E$, and let u be a function on $\Lambda(x_0, t_0; E)$. Suppose that, given any number $\alpha < 0$, we can find a number $\beta < 0$ such that the restriction of u to the set $P(x_0, t_0; \alpha, \beta)$ of Theorem 1.36 is continuous. Then we say that u is *parabolically continuous* at (x_0, t_0).

Thus the temperature in Theorem 1.36 is parabolically continuous at (x_0, t_0).

1.8. Equicontinuous Families of Temperatures

The main result of this section is closely related to the Harnack theorems. It gives a condition for a family of temperatures to contain a sequence that converges to a temperature. A key element in the proof is the concept of equicontinuity, which we now define.

DEFINITION 1.38. A family $\mathcal{F}$ of real-valued functions on a set $S \subseteq \mathbb{R}^{n+1}$ is said to be *equicontinuous* at a point $p \in S$ if, for each $\epsilon > 0$, there exists a neighbourhood U of p such that
$$|f(p) - f(q)| < \epsilon \quad \text{whenever} \quad f \in \mathcal{F} \quad \text{and} \quad q \in S \cap U.$$
The family $\mathcal{F}$ is said to be *equicontinuous on* S if it is equicontinuous at each point of S. It is said to be *uniformly equicontinuous on* S if, for each $\epsilon > 0$, there exists a $\delta > 0$ such that
$$|f(p) - f(q)| < \epsilon \quad \text{whenever} \quad f \in \mathcal{F} \quad \text{and} \quad p, q \in S \quad \text{with} \quad |p - q| < \delta.$$

To prove our theorem, we need a relatively simple version of the Arzelà-Ascoli Theorem. That theorem comes in various forms, which have different levels of complexity, so we prove the version we require as a lemma.

LEMMA 1.39. *Let $S \subseteq \mathbb{R}^{n+1}$, and let $\{f_k\}$ be a sequence of functions in $C(S)$ that is both equicontinuous and uniformly bounded. Then $\{f_k\}$ has a subsequence $\{f_{k_j}\}$ that converges locally uniformly to a function $f \in C(S)$.*

PROOF. We begin by taking a countable dense subset R of S, and arranging it as a sequence $\{p_i\}$. Since the sequence of real numbers $\{f_k(p_1) : k \in \mathbb{N}\}$ is bounded, it has a convergent subsequence $\{f_k(p_1) : k \in \mathbb{J}_1\}$, say. Next, the sequence of real numbers $\{f_k(p_2) : k \in \mathbb{J}_1\}$ is also bounded, and so it has a convergent subsequence $\{f_k(p_2) : k \in \mathbb{J}_2\}$, say. We proceed inductively. For each $m \in \mathbb{N}$, a convergent sequence of real numbers $\{f_k(p_m) : k \in \mathbb{J}_m\}$ is chosen, and then the fact that the sequence $\{f_k(p_{m+1}) : k \in \mathbb{J}_m\}$ is bounded implies that it has a convergent subsequence $\{f_k(p_{m+1}) : k \in \mathbb{J}_{m+1}\}$, say. We now choose a sequence of positive integers $\{k_m\}$ such that $k_m < k_{m+1}$ and $k_m \in \mathbb{J}_m$ for all $m \in \mathbb{N}$. Since $\mathbb{J}_{m+1} \subseteq \mathbb{J}_m$ for all m, we have $k_j \in \mathbb{J}_m$ for all $j \geq m$. Therefore, given any point $p_i \in R$, the sequence $\{f_{k_m}(p_i) : m \geq i\}$ is a subsequence of $\{f_k(p_i) : k \in \mathbb{J}_i\}$, and hence $\{f_{k_m}(p_i)\}$ converges. Thus the sequence $\{f_{k_m}\}$ converges at every point of R.

We put $g_m = f_{k_m}$ for all m. Since the sequence $\{g_m\}$ is equicontinuous on S, given $p \in S$ and $\epsilon > 0$, we can find a neighbourhood U_p of p such that

$$|g_m(q) - g_m(p)| < \epsilon/3 \quad \text{for all} \quad q \in S \cap U_p \quad \text{and all} \quad m \in \mathbb{N}.$$

Since R is dense in S, we can find a point $q \in U_p$ such that the sequence $\{g_m(q)\}$ converges. So there is a positive integer N such that

$$|g_i(q) - g_j(q)| < \epsilon/3 \quad \text{for all} \quad i, j \geq N.$$

It follows that

$$|g_i(p) - g_j(p)| \leq |g_i(p) - g_i(q)| + |g_i(q) - g_j(q)| + |g_j(q) - g_j(p)| < \epsilon$$

for all $i, j \geq N$. Therefore $\{g_m(p)\}$ is a Cauchy sequence in $\mathbb{R}$, and so there is a number $f(p)$ such that $g_m(p) \to f(p)$ as $k \to \infty$. Thus there is a real-valued function f on S to which $\{g_m\}$ converges pointwise.

We still need to show that $f \in C(S)$ and that $g_m \to f$ locally uniformly on S. Let K be any compact subset of S, and let $\epsilon > 0$. For any $p \in K$, we choose a neighbourhood V_p of p such that

$$|g_m(q) - g_m(p)| < \epsilon/9 \quad \text{for all} \quad q \in S \cap V_p \quad \text{and all} \quad m \in \mathbb{N}.$$

Then, given any $q \in S \cap V_p$, we choose l so large that both $|f(q) - g_l(q)| < \epsilon/9$ and $|g_l(p) - f(p)| < \epsilon/9$. It follows that

$$|f(q) - f(p)| \leq |f(q) - g_l(q)| + |g_l(q) - g_l(p)| + |g_l(p) - f(p)| < \epsilon/3.$$

Thus $|f(q) - f(p)| < \epsilon$ for all $q \in S \cap V_p$, so that f is continuous at p. Since K is arbitrary, f is continuous on S. Since K is compact, we can find a finite set $\{p_1, ..., p_\nu\} \subseteq K$ such that the family of sets $\{V_{p_1}, ..., V_{p_\nu}\}$ covers K. Now we choose a number M so large that

$$|f(p_j) - g_m(p_j)| < \epsilon/3 \quad \text{for all} \quad m \geq M \quad \text{and all} \quad j \in \{1, ..., \nu\}.$$

Given any point $q \in K$, there is some neighbourhood V_{p_μ} which contains q, and so

$$|f(q) - g_m(q)| \leq |f(q) - f(p_\mu)| + |f(p_\mu) - g_m(p_\mu)| + |g_m(p_\mu) - g_m(q)| < \epsilon$$

whenever $m \geq M$. Hence the convergence of g_m to f is uniform on K. □

THEOREM 1.40. *Let $\mathcal{F}$ be a family of temperatures on E that is locally uniformly bounded. If K is a compact subset of E, then $\mathcal{F}$ is uniformly equicontinuous on K. Moreover, each sequence $\{u_k\}$ in $\mathcal{F}$ has a subsequence which converges uniformly on K. Furthermore, if $\{v_k\}$ is any convergent sequence in $\mathcal{F}$, then its limit is a temperature on E.*

PROOF. Let D be a bounded open set such that $K \subseteq D$ and $\overline{D} \subseteq E$. Then $\mathcal{F}$ is uniformly bounded on D, so that there is a number M such that $|u(p)| \leq M$ whenever $p \in D$ and $u \in \mathcal{F}$. We choose an integer $m \geq 5$, and a positive number c such that the closures of the modified heat balls $\overline{\Omega}_m(p;c) \subseteq D$ for all $p \in K$. Recall from Section 1.6 that the function

$$(x,t;z,r) \mapsto \lambda_{m,c}(|x-z|, t-r),$$

restricted to $K \times \mathbb{R}^{n+1}$, is continuous and has support in $K \times D$. It is therefore uniformly continuous, so that, given $\epsilon > 0$, we can find $\delta > 0$ such that

$$|\lambda_{m,c}(|x-z|,t-r) - \lambda_{m,c}(|y-z|,s-r)| < \epsilon \quad \text{for all} \quad (z,r) \in D,$$

whenever $(x,t), (y,s) \in K$ and $|(x,t) - (y,s)| < \delta$. Therefore, by Theorem 1.25,

$$|u(x,t) - u(y,s)| = |\mathcal{V}_m(u;x,t;c) - \mathcal{V}_m(u;y,s;c)|$$

$$= \left| \int\!\!\int_{\mathbb{R}^{n+1}} \lambda_{m,c}(|x-z|,t-r) u(z,r)\, dz\, dr \right.$$

$$\left. - \int\!\!\int_{\mathbb{R}^{n+1}} \lambda_{m,c}(|y-z|,s-r) u(z,r)\, dz\, dr \right|$$

$$\leq \int\!\!\int_D |\lambda_{m,c}(|x-z|,t-r) - \lambda_{m,c}(|y-z|,s-r)||u(z,r)|\, dz\, dr$$

$$< \epsilon M \int\!\!\int_D dz\, dr$$

whenever $(x,t), (y,s) \in K$ and $|(x,t) - (y,s)| < \delta$. Hence the family $\mathcal{F}$ is uniformly equicontinuous on K.

If $\{u_k\}$ is a sequence in $\mathcal{F}$, it is now seen to be both uniformly bounded and equicontinuous on K. Therefore it has a subsequence which converges uniformly on K, by Lemma 1.39.

Let $\{v_k\}$ be any convergent sequence in $\mathcal{F}$, and let v be its limit. Let V be any bounded open set such that $\overline{V} \subseteq E$. Then there is a subsequence $\{v_{k_i}\}$ which converges uniformly on $\overline{V}$, necessarily to v. By Theorem 1.30, v is a temperature on V. It follows that v is a temperature on E. □

1.9. Notes and Comments

Theorem 1.3 was proved by Smyrnélis [**62**], and Theorem 1.5 can be found in Watson [**87**].

The direct part of Theorem 1.6 was proved for the case $n = 1$ by Pini [**60**], as well as the converse part under a stronger hypothesis. The general case was proved by Watson [**69**] under less stringent hypotheses, and the treatment in the text comes from Watson [**87**]. For the case of temperatures, the direct part of Theorem 1.6 was proved for arbitrary n by Fulks [**24**], Smyrnélis [**62**], and Kuptsov [**45**], apparently independently of each other; and the converse part, with u assumed to belong to $C(E)$ but with (1.8) assumed to hold whenever $\overline{\Omega}(x_0, t_0; c) \subseteq E$, was proved by

Pini [**60**] for $n = 1$, and by Fulks [**24**] for general n.

Theorem 1.8 was proved under milder smoothness conditions, for $n = 1$ only, by Pini [**60**]. The extension to arbitrary n, under still milder conditions, was made by Watson [**69**].

Example 1.12 was proved by Garofalo & Lanconelli [**26**] (in a more general context), by Watson [**80**], and by Brzezina [**10, 11**]. The proof in the text comes from Watson [**87**], as does Example 1.13.

Theorem 1.11 was proved, under milder conditions, by Watson [**80, 81, 84**], using three different methods. The method in the text is the simplest of the three. There is a related **open question**, which we now describe. For any $r \geq 1$, the L^r-mean value over heat spheres is defined for a nonnegative function u by $\mathcal{M}_r(u; x_0, t_0; c) = \mathcal{M}(u^r; x_0, t_0; c)^{1/r}$. If $r > 1$, does the mean $\mathcal{M}_r$ have a convexity property similar to that of $\mathcal{M}$, if u is a nonnegative subtemperature?

The authors who first calculated the mean values of smooth subtemperatures over heat balls, all chose different kernels. In the context of Theorem 1.16 this is of no consequence, and so is ignored in these remarks. The theorem was first proved, with the converse hypothesis strengthened to have (1.21) hold whenever $\overline{\Omega}(x_0, t_0; c) \subseteq E$, by Pini [**60**] for $n = 1$. Then the case of temperatures alone was proved for general n, with a similar strengthening of the converse hypothesis, by Smyrnélis [**62**]. Subsequently Watson [**69**] proved the full theorem under a milder smoothness condition. A different proof that (1.20) holds for temperatures, one which does not involve the means over heat spheres, is given by Evans [**17**].

Suzuki & Watson [**65**] have considered the inverse mean value property. That is, they have given conditions under which a mean value property of the type

$$u(x_0, t_0) = \tau(c) \int\int_D \frac{|x_0 - x|^2}{4(t_0 - t)} u(x, t)\, dx\, dt,$$

for certain temperatures u, implies that $D = \Omega(x_0, t_0; c)$.

Theorem 1.18 was proved, in the case $n = 1$ and with a milder smoothness condition, by Pini [**60**]. For general n it was proved by Watson [**80**], under a still milder smoothness condition, as was Theorem 1.19.

The results in Section 1.6 are due to Watson [**87**].

If D is a bounded domain and $(x_0, t_0) \in \overline{D}$, then a function K on D is called a mean value density at (x_0, t_0) if $K > 0$ almost everywhere on D and

$$u(x_0, t_0) = \int\int_D K(x, t) u(x, t)\, dx\, dt$$

for every temperature u on a neighbourhood of $\overline{D}$. Thus a heat ball has a mean value density which is unbounded, and a modified heat ball $\Omega_m(x_0, t_0; c)$ has one which is bounded if $m \geq 3$. In [**66**], Suzuki & Watson considered the existence of mean value densities, of bounded ones, and of ones which are bounded away from zero. In particular, they proved that a heat ball has no bounded mean value density. However, they found no conditions under which D would have a density which is bounded away from zero, and whether there is such a domain remains an **open question**.

Various Harnack theorems for temperatures have been proved using various methods, by Hadamard [**31**], Pini [**59**], Moser [**52, 53**], Bauer [**5**], Glagoleva [**27**], and Kuptsov [**45**]. The general form of the Harnack inequality, given in Theorem 1.32, is due to Bauer [**5**]. The idea of deducing the Harnack inequality from the

Harnack monotone convergence theorem, is taken from Doob [**14**].

Glagoleva [**28**] gives some interesting applications of the Harnack inequality to temperatures (and solutions to other parabolic equations) on the half-space $\mathbb{R}^n \times\,]-\infty, 0[$. Some of her work is also given in Landis [**49**]. Related work can be found in Watson [**77**].

CHAPTER 2

The Poisson Integral for a Circular Cylinder

Any approach to heat potential theory requires a class of domains on which we can produce a temperature that takes given continous values on some specified part of the boundary. In this chapter, we show this can be done for the class of circular cylinders, using the method of double layer heat potentials. This leads to the Poisson integral formula for a circular cylinder, by which a temperature on any neighbourhood of the closure of a circular cylinder is represented as an integral over the specified part of the boundary. The integral is with respect to the caloric measure at a point, a concept we shall meet in a more general context in Chapter 8. The Poisson integral gives us a characterization of temperatures in the class of real-valued continuous functions. It also enables us to define another integral mean value, this time over part of the boundary of a heat cylinder, which is a circular cylinder with a homogeneity property compatible with parabolic dilation. This mean value gives us another characterization of temperatures in the class of real-valued continuous functions, one which foreshadows the definition of a general subtemperature in Chapter 3. We also give characterizations of temperatures in terms of all the integral mean we have introduced so far.

2.1. The Cauchy Problem on a Half-Space

Let f be a bounded, continuous function on the hyperplane $\mathbb{R}^n \times \{a\}$. The *Cauchy Problem* on the half-space $\mathbb{R}^n \times \,]a, \infty[$ consists of finding a temperature u on $\mathbb{R}^n \times \,]a, \infty[$ such that $u(x,t) \to f(y)$ as $(x,t) \to (y, a+)$, for every $y \in \mathbb{R}^n$. The Cauchy Problem is sometimes called the *Initial Value Problem*. In Section 2.2, we consider the Dirichlet Problem on a circular cylinder, our treatment of which depends on being able to solve the Cauchy Problem.

We begin by proving a basic fact about the fundamental temperature W.

LEMMA 2.1. *If δ is an arbitrary positive number, $y_0 \in \mathbb{R}^n$ and $a \in \mathbb{R}$, then*

$$\int_{|y_0 - y| > \delta} W(x - y, t - a)\, dy \to 0 \quad \text{as} \quad (x, t) \to (y_0, a+).$$

PROOF. Suppose that $|x - y_0| \leq \delta/2$. Then, for all y such that $|y_0 - y| > \delta$, we have

$$|x - y| \geq |y - y_0| - |y_0 - x| > \delta/2 \geq |x - y_0|,$$

so that

$$|y_0 - y| \leq |y_0 - x| + |x - y| \leq 2|x - y|.$$

Hence

$$\exp\left(-\frac{|x-y|^2}{4(t-a)}\right) \leq \exp\left(-\frac{|y_0 - y|^2}{16(t-a)}\right).$$

Therefore, writing $s = 4(t-a)$, we obtain

$$\int_{|y_0-y|>\delta} W(x-y,t-a)\,dy \leq \int_{|y_0-y|>\delta} (\pi s)^{-\frac{n}{2}} \exp\left(-\frac{|y_0-y|^2}{4s}\right)\,dy,$$

so that a change to polar coordinates, with $r = |y_0 - y|$, gives

$$\int_{|y_0-y|>\delta} W(x-y,t-a)\,dy \leq (\pi s)^{-\frac{n}{2}} \omega_n \int_\delta^\infty \exp\left(-\frac{r^2}{4s}\right) r^{n-1}\,dr.$$

Putting $\eta = r^2/s$, we deduce that

$$\int_{|y_0-y|>\delta} W(x-y,t-a)\,dy \leq \left(\frac{\omega_n}{2\pi^{\frac{n}{2}}}\right) \int_{\frac{\delta^2}{s}}^\infty e^{-\frac{\eta}{4}} \eta^{\frac{n}{2}-1}\,d\eta.$$

As $t \to a+$ we have $s \to 0+$, and therefore $\delta^2/s \to +\infty$, so that the last integral tends to zero. Hence, given $\epsilon > 0$, we can find $\gamma > 0$ such that

$$\int_{|y_0-y|>\delta} W(x-y,t-a)\,dy < \epsilon$$

whenever $0 < t - a < \gamma$ and $|x - y_0| \leq \delta/2$. □

THEOREM 2.2. *Let $y_0 \in \mathbb{R}^n$, let $a \in \mathbb{R}$, and let f be a bounded, measurable function on $\mathbb{R}^n$. Put*

$$u(x,t) = \int_{\mathbb{R}^n} W(x-y,t-a)f(y)\,dy.$$

Then u is a bounded temperature on $\mathbb{R}^n \times]a,\infty[$ and, if f is continuous at y_0, then $u(x,t) \to f(y_0)$ as $(x,t) \to (y_0, a+)$. In particular, if f is continuous on the whole of $\mathbb{R}^n$, then u is a solution of the Cauchy problem with initial values given by f.

PROOF. We use Theorem 1.29 to show that u is a temperature. For each point $(x,t) \in \mathbb{R}^n \times]a,\infty[$, Lemma 1.1 shows that

$$|u(x,t)| \leq \int_{\mathbb{R}^n} W(x-y,t-a)|f(y)|\,dy \leq \sup_{\mathbb{R}^n} |f|,$$

so that u is bounded and, by Theorem 1.25,

$$\mathcal{V}_5(|u|;x,t;c) \leq \sup_{\mathbb{R}^n} |f| \cdot \mathcal{V}_5(1;x,t;c) = \sup_{\mathbb{R}^n} |f|$$

whenever $c < t$. Therefore the order of the integrals can be changed, to give

$$\mathcal{V}_5(u;x,t;c) = \int_{\mathbb{R}^n} \mathcal{V}_5(W(\,\cdot\, - y, \,\cdot\, - a); x,t;c)f(y)\,dy$$
$$= \int_{\mathbb{R}^n} W(x-y,t-a)f(y)\,dy$$
$$= u(x,t).$$

Hence u is a temperature on $\mathbb{R}^n \times]a,\infty[$, by Theorem 1.29.

By Lemma 1.1,

$$\int_{\mathbb{R}^n} W(x-y,t-a)f(y)\,dy - f(y_0) = \int_{\mathbb{R}^n} W(x-y,t-a)(f(y)-f(y_0))\,dy,$$

so that

$$(2.1) \quad \left|\int_{\mathbb{R}^n} W(x-y,t-a)f(y)\,dy - f(y_0)\right| \leq \int_{\mathbb{R}^n} W(x-y,t-a)|f(y)-f(y_0)|\,dy.$$

Since f is continuous at y_0, given $\epsilon > 0$ we can find $\delta > 0$ such that $|f(y)-f(y_0)| < \epsilon$ whenever $|y-y_0| \leq \delta$. Furthermore, since f is bounded on $\mathbb{R}^n$, there is a number M such that $|f(y)| \leq M$ for all y, and hence $|f(y)-f(y_0)| \leq 2M$ whenever $|y-y_0| > \delta$. It therefore follows from inequality (2.1) that

$$\left| \int_{\mathbb{R}^n} W(x-y, t-a) f(y) \, dy - f(y_0) \right|$$
$$\leq \int_{|y-y_0|\leq \delta} W(x-y, t-a) |f(y) - f(y_0)| \, dy$$
$$+ \int_{|y-y_0|>\delta} W(x-y, t-a) |f(y) - f(y_0)| \, dy$$
$$< \epsilon \int_{|y-y_0|\leq \delta} W(x-y, t-a) \, dy + 2M \int_{|y-y_0|>\delta} W(x-y, t-a) \, dy$$
$$< \epsilon + 2M \int_{|y-y_0|>\delta} W(x-y, t-a) \, dy$$

by Lemma 1.1. As $(x,t) \to (y_0, a+)$, the last integral tends to zero, by Lemma 2.1. Therefore we can find $\gamma > 0$ such that

$$\left| \int_{\mathbb{R}^n} W(x-y, t-a) f(y) \, dy - f(y_0) \right| < 2\epsilon$$

whenever $|x-y_0|^2 + (t-a)^2 < \gamma^2$ and $t > a$. $\square$

2.2. The Dirichlet Problem on a Circular Cylinder

We consider an open ball B in $\mathbb{R}^n$, and a bounded time interval $]a,b[$, as in Section 1.5. We denote by D the circular cylinder $D = B \times \,]a,b[\, \subseteq \mathbb{R}^{n+1}$, by L the lateral surface $\partial B \times \,]a,b]$ of ∂D, and by I the initial surface $\overline{B} \times \{a\}$, so that the normal boundary $\partial_n D = L \cup I$. The *Dirichlet Problem* on D consists of showing that, for any continuous function f on $\partial_n D$, there is a temperature u on D which has a continuous extension by f to $\partial_n D$. The Dirichlet Problem is sometimes called the *First Boundary Value Problem*, or the *First Initial-Boundary Value Problem* if the roles of the different variables is being emphasized.

It transpires that the function u is actually a temperature on $\overline{D}\backslash\partial_n D$, which means that $u \in C^{2,1}(\overline{D}\backslash\partial_n D)$ and satisfies the mean value equalities of Theorems 1.6, 1.16 and 1.29 on $\overline{D}\backslash\partial_n D$.

We shall prove the following theorem.

THEOREM 2.3. *Let $f \in C(\partial_n D)$. Then there is a function $u \in C(\overline{D})$ such that u is a temperature on $\overline{D}\backslash\partial_n D$ and $u = f$ on $\partial_n D$.*

The proof is long and complicated, so we extract parts of it as lemmas. We first show that it is enough to prove the first part of the theorem for an arbitrarily small time interval $]a,b[$.

LEMMA 2.4. *Suppose that there is a positive number ϵ with the property that, given any function $g \in C((\partial B \times [a, a+\epsilon]) \cup I)$, there exists a temperature v_g on*

$B\times\,]a,a+\epsilon[$ such that $v_g(p) \to g(q)$ as $p \to q$ with $p \in B\times\,]a,a+\epsilon[$, for all $q \in (\partial B \times [a,a+\epsilon]) \cup I$. Then, given any function $f \in C(L \cup I)$, there is a temperature $u = u_f$ on D such that $u(p) \to f(q)$ as $p \to q$ with $p \in D$, for all $q \in L \cup I$.

PROOF. We first extend f to a bounded continuous function on $(\partial B\times[a,\infty[)\cup I$, which we also denote by f. We choose a positive integer k such that $b-a < k\epsilon$, and put $\delta = (b-a)/k$. By hypothesis, there is a temperature u_1 on $B\times\,]a,a+\epsilon[$ such that $u_1(p) \to f(q)$ as $p \to q$ with $p \in B\times\,]a,a+\epsilon[$, for all $q \in (\partial B \times [a,a+\epsilon]) \cup I$.

The proof procedes inductively. Suppose that, for some positive integer j, there is a temperature u_j on $B\times\,]a,a+(j-1)\delta+\epsilon[$ such that $u_j(p) \to f(q)$ as $p \to q$ with $p \in B\times\,]a,a+(j-1)\delta+\epsilon[$, for all $q \in (\partial B \times [a,a+(j-1)\delta+\epsilon]) \cup I$. Put

$$f_j(y,s) = \begin{cases} u_j(y, s+j\delta) & \text{if } (y,s) \in B \times \{a\}, \\ f(y, s+j\delta) & \text{if } (y,s) \in \partial B \times [a, a+\epsilon]. \end{cases}$$

Then f_j is a continuous function on $(\partial B \times [a,a+\epsilon]) \cup I$. Therefore, by hypothesis, there is a temperature v_j on $B\times\,]a,a+\epsilon[$ such that $v_j(p) \to f_j(q)$ as $p \to q$ with $p \in B\times\,]a,a+\epsilon[$, for all $q \in (\partial B \times [a,a+\epsilon]) \cup I$. Now if $(y,s) \in B \times \{a\}$ then, as $(x,t) \to (y,s)$ with $(x,t) \in B\times\,]a,a+\epsilon-\delta[$, we have

$$u_j(x,t+j\delta) - v_j(x,t) \to u_j(y,s+j\delta) - u_j(y,s+j\delta) = 0;$$

and if $(y,s) \in \partial B \times [a, a+\epsilon-\delta]$, then

$$u_j(x,t+j\delta) - v_j(x,t) \to f(y,s+j\delta) - f(y,s+j\delta) = 0.$$

So $u_j(x,t+j\delta) = v_j(x,t)$ for all $(x,t) \in B\times\,]a,a+\epsilon-\delta[$, by the boundary uniqueness theorem (Theorem 1.24). So if we define

$$u_{j+1}(x,t) = \begin{cases} u_j(x,t) & \text{if } (x,t) \in B\times\,]a,a+(j-1)\delta+\epsilon[, \\ v_j(x,t+j\delta) & \text{if } (x,t) \in B\times\,]a+j\delta, a+j\delta+\epsilon[, \end{cases}$$

then u_{j+1} is well-defined and is a temperature on $B\times\,]a,a+j\delta+\epsilon[$ such that $u_{j+1}(p) \to f(q)$ as $p \to q$ with $p \in B\times\,]a,a+j\delta+\epsilon[$, for all $q \in (\partial B\times[a,a+j\delta+\epsilon])\cup I$.

Since a temperature u_1 exists, it follows inductively that u_j exists for all j. Since $b - a = k\delta$ and $\delta < \epsilon$, we have $a + (k-1)\delta + \epsilon = a + k\delta + (\epsilon - \delta) > b$, so that the temperature u_k is defined on (a superset of) D, and satisfies $u_k(p) \to f(q)$ as $p \to q$ with $p \in D$, for all $q \in L \cup I$. □

We next show that it is enough to prove the first part of the theorem with the additional hypothesis that $f = 0$ on I.

LEMMA 2.5. Suppose that, given any function $g \in C(L \cup I)$ such that $g = 0$ on I, there is a temperature u_g on D such that

$$\lim_{p \to q,\, p \in D} u_g(p) = g(q) \quad \text{for all} \quad q \in L \cup I.$$

Then, given an arbitrary function $f \in C(L \cup I)$, there is a temperature u_f on D such that

$$\lim_{p \to q,\, p \in D} u_f(p) = f(q) \quad \text{for all} \quad q \in L \cup I.$$

PROOF. Given an arbitrary function $f \in C(L \cup I)$, restrict it to I, then extend that restriction to the whole of $\mathbb{R}^n \times \{a\}$ in such a way that the extension $\bar{f}$ remains bounded and continuous. Let v denote the solution to the Cauchy problem on $\mathbb{R}^n \times [a, \infty[$ with boundary function $\bar{f}$ on $\mathbb{R}^n \times \{a\}$. Define $g = f - v$ on $L \cup I$. Then g is continuous, and $g(\cdot, a) = 0$. So, by hypothesis, there is a temperature u_g on D such that
$$\lim_{p \to q, \, p \in D} u_g(p) = g(q) \quad \text{for all} \quad q \in L \cup I.$$
Now the function $u_f = v + u_g$ is a temperature on D which satisfies
$$\lim_{p \to q, \, p \in D} u_f(p) = f(q) \quad \text{for all} \quad q \in L \cup I.$$
□

2.3. Double Layer Heat Potentials

Given a function f with first order partial derivates defined on a subset of $\mathbb{R}^n$, and a unit vector ν in $\mathbb{R}^n$, we let
$$\nabla_y f = \left(\frac{\partial f}{\partial y_1}, \dots, \frac{\partial f}{\partial y_n} \right)$$
denote the *gradient* of f, and
$$\langle \nabla_y f, \nu \rangle = \sum_{i=1}^{n} \frac{\partial f}{\partial y_i} \nu_i$$
denote the *directional derivative* of f in the direction of ν.

Given a continuous function g on $L \cup I$ such that $g = 0$ on I, we obtain a temperature u_g on D such that
$$\lim_{p \to q, \, p \in D} u_g(p) = g(q) \quad \text{for all} \quad q \in L \cup I$$
in the form of a *double layer heat potential*

(2.2) $$u(x,t) = \int_a^t \int_{\partial B} \langle \nabla_y W(x-y, t-s), \nu_y \rangle \rho(y,s) \, d\sigma(y) ds,$$

where (x, t) belongs to the strip $S(a, b) = \mathbb{R}^n \times]a, b]$, ν_y denotes the outward unit normal to the surface ∂B at y, σ denotes the surface measure on ∂B, and ρ is called the *density* of the potential. Then we can invoke Lemma 2.5 to show that the Dirichlet problem for an arbitrary continuous function f on $L \cup I$ has a solution.

To show that $u(x,t)$ exists for all $(x,t) \in S(a,b)$, and investigate its limits as $(x,t) \to (y_0, s_0)$ for $(y_0, s_0) \in L$, we need to rearrange formula (2.2) and introduce an auxiliary function u_0. First, if $t - s > 0$, then
$$\langle \nabla_y W(x-y, t-s), \nu_y \rangle = \frac{-|x-y|}{2^{n+1} \pi^{\frac{n}{2}} (t-s)^{\frac{n}{2}+1}} \exp\left(-\frac{|x-y|^2}{4(t-s)} \right) \left\langle \frac{y-x}{|y-x|}, \nu_y \right\rangle.$$

Next, if $d\omega(y)$ denotes the solid angle subtended by the element $d\sigma(y)$ of the surface ∂B, as seen from the point x, then
$$d\omega(y) = \frac{d\sigma(y)}{|x-y|^{n-1}} \left\langle \frac{y-x}{|y-x|}, \nu_y \right\rangle.$$

Therefore we can write formula (2.2) as

$$(2.3) \quad u(x,t) = \frac{-1}{2^{n+1}\pi^{\frac{n}{2}}} \int_{\partial B} \left(\int_a^t \frac{|x-y|^n}{(t-s)^{\frac{n}{2}+1}} \exp\left(-\frac{|x-y|^2}{4(t-s)}\right) \rho(y,s)\, ds \right) d\omega(y).$$

Using the formula (2.3), we can show that $u(x,t)$ exists for all $(x,t) \in \mathbb{R}^{n+1}\setminus L$. The function $r \mapsto r^{\frac{n}{2}+1}e^{-r}$ is bounded on $]0,\infty[$. Taking $r = |x-y|^2/[4(t-s)]$, we see that there is a constant K such that

$$\frac{|x-y|^n}{(t-s)^{\frac{n}{2}+1}} \exp\left(-\frac{|x-y|^2}{4(t-s)}\right) \leq \frac{K}{|x-y|^2}.$$

Therefore the kernel in (2.3) is bounded if $|x-y|$ is bounded away from zero. Hence $u(x,t)$ exists and is finite for all $(x,t) \in S(a,b)$ outside an arbitrary neighbourhood of L, and therefore for all $(x,t) \in S(a,b)\setminus L$. We shall soon show its existence for every $(x,t) \in S(a,b)$.

The double layer heat potential (2.2) satisfies the heat equation on $S(a,b)\setminus L$. This is because, for each point $(y,s) \in L$, the function $(x,t) \mapsto W(x-y,t-s)$ is infinitely differentiable on $\mathbb{R}^{n+1}\setminus\{(y,s)\}$ and is a temperature there. So, for all $(x,t) \in S(a,b)\setminus L$, we have

$$\Theta_{x,t} u(x,t) = \int_a^t \int_{\partial B} \langle \nabla_y(\Theta_{x,t} W(x-y,t-s)), \nu_y \rangle \rho(y,s)\, d\sigma(y)ds = 0.$$

The next lemma introduces the auxiliary function u_0.

LEMMA 2.6. *If*

$$u_0(x,t) = \frac{-1}{2^{n+1}\pi^{\frac{n}{2}}} \int_{\partial B} \left(\int_{-\infty}^t \frac{|x-y|^n}{(t-s)^{\frac{n}{2}+1}} \exp\left(-\frac{|x-y|^2}{4(t-s)}\right) ds \right) d\omega(y)$$

is the double layer heat potential with density 1 and a replaced by $-\infty$, then there is a positive number κ such that

$$u_0(x,t) = \begin{cases} -2\kappa & \text{if } x \in B, \\ -\kappa & \text{if } x \in \partial B, \\ 0 & \text{if } x \in (\mathbb{R}^n \setminus B). \end{cases}$$

PROOF. In the inner integral, we make the change of variable

$$\eta = \frac{|x-y|}{2\sqrt{t-s}}, \qquad d\eta = \frac{|x-y|}{4(t-s)^{3/2}} ds,$$

to obtain

$$\int_{-\infty}^t \frac{|x-y|^n}{(t-s)^{\frac{n}{2}+1}} \exp\left(-\frac{|x-y|^2}{4(t-s)}\right) ds$$

$$= 2^{n+1} \int_{-\infty}^t \left(\frac{|x-y|}{2\sqrt{t-s}}\right)^{n-1} \exp\left(-\left(\frac{|x-y|}{2\sqrt{t-s}}\right)^2\right) \frac{|x-y|}{4(t-s)^{3/2}} ds$$

$$= 2^{n+1} \int_0^\infty \eta^{n-1} e^{-\eta^2}\, d\eta$$

$$(2.4) \qquad = 2^{n+1}\alpha,$$

say. It follows that
$$u_0(x,t) = -\frac{\alpha}{\pi^{\frac{n}{2}}} \int_{\partial B} d\omega(y).$$

Let ω_n denote the $(n-1)$-dimensional surface area of the unit sphere in $\mathbb{R}^n$, and put
$$\kappa = \frac{\alpha \omega_n}{2\pi^{\frac{n}{2}}}.$$

Then we have the result of the lemma. $\square$

We are now in a position to discuss the behaviour of the double layer heat potential at points of the lateral boundary L.

LEMMA 2.7. *If ρ is a continuous function on $\overline{L}$, then the double layer heat potential (2.2) is defined and finite everywhere on $S(a,b)$. Its restriction to L is continuous, and has a continuous extension by 0 to $\overline{L}$. Furthermore, for each point $q_0 = (y_0, s_0) \in L$, the limit as $p = (x,t) \to q_0$ with $p \in D$ exists and*

$$\lim_{p \to q_0, p \in D} u(x,t) = u(y_0, s_0) - \kappa \rho(y_0, s_0)$$

(2.5)
$$= \int_a^{s_0} \int_{\partial B} \langle \nabla_y W(y_0 - y, s_0 - s), \nu_y \rangle \rho(y,s)\, d\sigma(y)\, ds - \kappa \rho(y_0, s_0),$$

where κ is the same as in Lemma 2.6.

PROOF. We begin by rearranging the formula (2.3) for the double layer heat potential, using the auxiliary function u_0 in Lemma 2.6.

Let ρ be continuous on $\overline{L}$, and let $(y_0, s_0) \in L$. Then

$$u(x,t) - \rho(y_0, s_0) u_0(x,t)$$
$$= \frac{-1}{2^{n+1}\pi^{\frac{n}{2}}} \int_{\partial B} \left(\int_a^t \frac{|x-y|^n}{(t-s)^{\frac{n}{2}+1}} \exp\left(-\frac{|x-y|^2}{4(t-s)}\right) (\rho(y,s) - \rho(y_0, s_0))\, ds \right) d\omega(y)$$

(2.6)
$$+ \frac{\rho(y_0, s_0)}{2^{n+1}\pi^{\frac{n}{2}}} \int_{\partial B} \left(\int_{-\infty}^a \frac{|x-y|^n}{(t-s)^{\frac{n}{2}+1}} \exp\left(-\frac{|x-y|^2}{4(t-s)}\right) ds \right) d\omega(y).$$

We show that the function $u - \rho(y_0, s_0) u_0$ is defined and finite on $S(a,b)$, and is continuous at (y_0, s_0). Because u_0 is defined and finite on $S(a,b)$, this will show that u is too. Moreover, the restriction of u to L is $v + \rho(y_0, s_0)u_0 = v - \kappa\rho(y_0, s_0)$, by Lemma 2.6, and this is continuous at (y_0, s_0). So the restriction of u to L is continuous. Furthermore, due to the absolute continuity of the integral (2.3) as a set function, $u(x,t) \to 0$ as $t \to a+$. So the restriction of u to L has a continuous extension by 0 to $\overline{L}$. Finally, the fact that

$$u(x,t) - \rho(y_0, s_0) u_0(x,t) \to u(y_0, s_0) - \rho(y_0, s_0) u_0(y_0, s_0),$$

together with Lemma 2.6, will imply that, as $(x,t) \to (y_0, s_0)$ from inside D,

$$u(x,t) + 2\kappa\rho(y_0, s_0) \to u(y_0, s_0) + \kappa\rho(y_0, s_0),$$

That is, $u(x,t) \to u(y_0, s_0) - \kappa\rho(y_0, s_0)$. This will prove the lemma.

We begin with the second iterated integral in (2.6). The function $r \mapsto r^{\frac{n}{2}} e^{-r}$

is bounded on $]0, \infty[$. Taking $r = |x-y|^2/[4(t-s)]$, we see that there is a constant K such that
$$\frac{|x-y|^n}{(t-s)^{\frac{n}{2}+1}} \exp\left(-\frac{|x-y|^2}{4(t-s)}\right) \leq \frac{K}{t-s}.$$
Therefore the kernel is bounded if $t-s$ is bounded away from zero. If the point $(x_0, t_0) \in S(a,b)$, then whenever $|t_0 - t| < \frac{1}{2}(t_0 - a)$ we have $t - s \geq t - a \geq \frac{1}{2}(t_0 - a)$, so that the kernel is bounded on a neighbourhood of (x_0, t_0). Therefore u is finite on that neighbourhood, and Lebesgue's Dominated Convergence Theorem shows that u is continuous at (x_0, t_0).

We now consider the first iterated integral in (2.6). Since ρ is continuous on L, given $\epsilon > 0$ we can find a relative neighbourhood N of (y_0, s_0) in L such that
$$|\rho(y,s) - \rho(y_0, s_0)| < \frac{\epsilon}{2^{n+1} \alpha \omega_n}$$
whenever $(y,s) \in N$; here α is the same as in (2.4). Now
$$\left| \int\int_N \frac{|x-y|^n}{(t-s)^{\frac{n}{2}+1}} \exp\left(-\frac{|x-y|^2}{4(t-s)}\right) (\rho(y,s) - \rho(y_0, s_0))\, ds\, d\omega(y) \right|$$
$$\leq \frac{\epsilon}{2^{n+1} \alpha \omega_n} \int\int_N \frac{|x-y|^n}{(t-s)^{\frac{n}{2}+1}} \exp\left(-\frac{|x-y|^2}{4(t-s)}\right)\, ds\, d|\omega|(y)$$
$$\leq \frac{\epsilon}{2^{n+1} \alpha \omega_n} \int_{\partial B} \left(\int_{-\infty}^{t} \frac{|x-y|^n}{(t-s)^{\frac{n}{2}+1}} \exp\left(-\frac{|x-y|^2}{4(t-s)}\right)\, ds \right) d|\omega|(y)$$
$$< \frac{\epsilon}{\omega_n} \int_{\partial B} d|\omega|(y) \qquad \text{(by (2.4))}$$
(2.7) $\qquad \leq \epsilon.$

For brevity, we now denote the first iterated integral in (2.6) by
$$\int_L \phi(p, q)\, d\sigma(q),$$
where $p = (x, t)$, $q = (y, s)$, $\phi(p, q) = 0$ if $s \geq t$,
$$\phi(p, q) = \frac{|x-y|^n}{(t-s)^{\frac{n}{2}+1}} \exp\left(-\frac{|x-y|^2}{4(t-s)}\right) (\rho(y,s) - \rho(y_0, s_0)) \quad \text{if} \quad s < t,$$
and $d\sigma(q) = ds\, d\omega(y)$. Let U be any bounded, relative neighbourhood of the point $q_0 = (y_0, s_0)$ in $S(a,b)$ such that $\overline{U} \cap L \subseteq N$. Then ϕ is continuous on the compact set $\overline{U} \times \overline{L \setminus N}$, and therefore both bounded and uniformly continuous there. It follows that the integral
$$\int_{L \setminus N} \phi(p, q)\, d\sigma(q)$$
is defined and finite for all $p \in U$. Therefore, in view of (2.7), the integral
$$\Phi(p) = \int_L \phi(p, q)\, d\sigma(q)$$
is defined and finite for all $p \in U$. Since q_0 is any point of L, and U is any bounded, relative neighbourhood of q such that $\overline{U} \cap L \subseteq N$, it follows that $\Phi(p)$ is defined and finite for all $p \in S(a, b)$. Furthermore, given $\epsilon > 0$, we can find $\gamma > 0$ such that $B(q_0, \gamma) \cap L \subseteq N$ and
(2.8) $\qquad |\phi(p, q) - \phi(q_0, q)| < \dfrac{\epsilon}{\int_{L \setminus N} d|\sigma|(y)}$

whenever $p \in \overline{U}$, $|p - q_0| < \gamma$, and $q \in \overline{L \backslash N}$. It follows that

$$|\Phi(p) - \Phi(q_0)| = \left| \int_L \phi(p, q) \, d\sigma(q) - \int_L \phi(q_0, q) \, d\sigma(q) \right|$$

$$\leq \left| \int_N \phi(p, q) \, d\sigma(q) \right| + \left| \int_N \phi(q_0, q) \, d\sigma(q) \right| + \left| \int_{L \backslash N} \left(\phi(p, q) - \phi(q_0, q) \right) d\sigma(q) \right|$$

$$< 2\epsilon + \int_{L \backslash N} |\phi(p, q) - \phi(q_0, q)| \, d|\sigma|(q) \qquad \text{(by (3.6))}$$

$$< 3\epsilon$$

(by (2.8)), whenever $p \in \overline{U}$ and $|p - q_0| < \gamma$. Thus Φ is continuous at q_0; that is, the first iterated integral in (2.6), as a function of (x, t), is continuous at (y_0, s_0). $\square$

We are now in a position to prove Theorem 2.3. We first recall the statement of the theorem.

Given any function $f \in C(L \cup I)$, there is a function $u \in C(\overline{D})$ such that u is a temperature on $\overline{D} \backslash (L \cup I)$ and $u = f$ on $L \cup I$.

PROOF. It suffices to prove that there is a temperature u on D such that

$$\lim_{p \to q, \, p \in D} u(p) = f(q) \quad \text{for all} \quad q \in L \cup I.$$

For if the weaker statement has beed proved, we can extend $D = B \times {]}a, b{[}$ to $D^* = B \times {]}a, b^*{[}$ with $b^* > b$, and f to $f^* \in C((\partial B \times [a, b^*]) \cup I)$, and apply that result to D^* and f^*.

Let g be a continuous function on $L \cup I$ such that $g = 0$ on I. We seek a temperature $u = u_g$ on D such that

$$(2.9) \qquad \lim_{p \to q, \, p \in D} u_g(p) = g(q) \quad \text{for all} \quad q \in L \cup I$$

in the form of a double layer heat potential u given by (2.3). We already know that such a function is a temperature on D.

In order to satisfy (2.9) at a point $(y_0, s_0) \in L$, Lemma 2.7 shows that we need

$$g(y_0, s_0) = u(y_0, s_0) - \kappa \rho(y_0, s_0),$$

or

$$(2.10) \quad \rho(y_0, s_0) = \frac{1}{\kappa} \int_a^{s_0} \int_{\partial B} \langle \nabla_y W(y_0 - y, s_0 - s), \nu_y \rangle \rho(y, s) \, d\sigma(y) \, ds - \frac{1}{\kappa} g(y_0, s_0).$$

We show that this integral equation for ρ has a solution using the contraction mapping principle.

Given any number $\epsilon \in {]}0, b-a{[}$, we put $L_\epsilon = \partial B \times [a, a+\epsilon]$. For any continuous function φ on L_ϵ, we put

$$(A\varphi)(x, t) = \frac{1}{\kappa} \int_a^t \int_{\partial B} \langle \nabla_y W(x - y, t - s), \nu_y \rangle \varphi(y, s) \, d\sigma(y) \, ds - \frac{1}{\kappa} g(x, t)$$

for all $(x,t) \in L_\epsilon$. In view of Lemma 2.7, the function $A\varphi$ is continuous on L_ϵ. If ψ is also continuous on L_ϵ, then

$$\max_{L_\epsilon} |A\varphi - A\psi|$$

(2.11)
$$\leq \max_{L_\epsilon} |\varphi - \psi| \left(\frac{1}{\kappa} \int_a^{a+\epsilon} \int_{\partial B} |\langle \nabla_y W(x-y, t-s), \nu_y \rangle| \, d|\sigma|(y) \, ds \right).$$

In view of (2.4), we have

$$\int_{-\infty}^t \int_{\partial B} |\langle \nabla_y W(x-y, t-s), \nu_y \rangle| \, d|\sigma|(y) \, ds$$

$$= \frac{1}{2^{n+1} \pi^{\frac{n}{2}}} \int_{-\infty}^t \int_{\partial B} \frac{|x-y|^n}{(t-s)^{\frac{n}{2}+1}} \exp\left(-\frac{|x-y|^2}{4(t-s)}\right) d|\omega|(y) \, ds$$

$$= \frac{\alpha}{\pi^{\frac{n}{2}}} \int_{\partial B} d|\omega|(y)$$

$$\leq \frac{\alpha \omega_n}{\pi^{\frac{n}{2}}}.$$

Therefore, since the integral as a set function is absolutely continuous, we obtain

$$\beta(\epsilon) \equiv \frac{1}{\kappa} \int_a^{a+\epsilon} \int_{\partial B} |\langle \nabla_y W(x-y, t-s), \nu_y \rangle| \, d|\sigma|(y) \, ds \to 0$$

as $\epsilon \to 0$. Therefore we can choose $\epsilon < b - a$ such that $\beta(\epsilon) < 1$. With this choice of ϵ, we obtain from (2.11) the inequality

$$\max_{L_\epsilon} |A\varphi - A\psi| \leq C \max_{L_\epsilon} |\varphi - \psi|$$

with $C = \beta(\epsilon) < 1$, and with C independent of g. Therefore A is a contraction mapping on the complete metric space of continuous functions on L_ϵ with the supremum metric. Hence, by the contraction mapping principle, there is a unique continuous function ρ on L_ϵ such that $A\rho = \rho$. Thus the integral equation (2.10) has a unique solution ρ if $s_0 < a + \epsilon$. For this function ρ, the double layer heat potential

$$u_g(x,t) = \int_a^t \int_{\partial B} \langle \nabla_y W(x-y, t-s), \nu_y \rangle \rho(y,s) \, d\sigma(y) \, ds$$

is a temperature on $D_\epsilon \equiv B \times]a, a+\epsilon[$ which satisfies

$$\lim_{p \to q, \, p \in D} u_g(p) = g(q) \quad \text{for all} \quad q \in L_\epsilon \cup I,$$

and so solves the Dirichlet problem on D_ϵ for the function g.

By Lemma 2.5, given an arbitrary continuous function f on $L_\epsilon \cup I$, there is a temperature u_f on D_ϵ such that $\lim_{p \to q, \, p \in D_\epsilon} u_f(p) = f(q)$ for all $q \in L_\epsilon \cup I$. Lemma 2.4 now shows that the Dirichlet problem on D itself is solvable. □

2.4. The Poisson Integral and the Caloric Measure

Theorem 2.3 does not show how the temperature u is derived from the function f on $\partial_n D$. In this section we show that, for each point $p \in \overline{D} \setminus \partial_n D$, the value of $u(p)$ can be written as an integral of f with respect to a measure on $\partial_n D$. This is important because it will allow us to use the Lebesgue convergence theorems.

2.4. THE POISSON INTEGRAL AND THE CALORIC MEASURE

THEOREM 2.8. *Let $f \in C(\partial_n D)$, and let u_f be the temperature on $\overline{D}\backslash\partial_n D$ associated with f by Theorem 2.3. Then, given any point $p \in \overline{D}\backslash\partial_n D$, there is a unique positive Borel measure μ_p on $\partial_n D$ such that*

$$(2.12) \qquad u_f(p) = \int_{\partial_n D} f \, d\mu_p.$$

PROOF. We show that the mapping $f \mapsto u_f(p)$ is a positive linear functional on the Banach space $C(\partial_n D)$ with the supremum norm.

If $f \geq 0$, then the boundary minimum principle (Theorem 1.21) shows that $u_f \geq 0$.

If $\alpha \in \mathbb{R}$, then αf is continuous on $\partial_n D$, and so there is a temperature $u_{\alpha f}$ associated with αf by Theorem 2.3. Furthermore,

$$\lim_{p \to q,\, p \in D} \bigl(u_{\alpha f}(p) - \alpha u_f(p)\bigr) = \lim_{p \to q,\, p \in D} u_{\alpha f}(p) - \alpha \lim_{p \to q,\, p \in D} u_f(p) = 0$$

for all $q \in \partial_n D$, so that the boundary uniqueness principle (Theorem 1.24) shows that $u_{\alpha f} = \alpha u_f$ on D.

If g is another continuous function on $\partial_n D$, then so is $f + g$, and hence there is a temperature u_{f+g} associated with $f + g$ by Theorem 2.3. Also,

$$\lim_{p \to q,\, p \in D} \bigl(u_{f+g}(p) - u_f(p) - u_g(p)\bigr)$$
$$= \lim_{p \to q,\, p \in D} u_{f+g}(p) - \lim_{p \to q,\, p \in D} u_f(p) - \lim_{p \to q,\, p \in D} u_g(p)$$
$$= 0$$

for all $q \in \partial_n D$, so that $u_{f+g} - u_f - u_g = 0$ on D, by the boundary uniqueness principle.

Thus, given any $p \in \overline{D}\backslash\partial_n D$, the mapping $f \mapsto u_f(p)$ is a positive linear functional on $C(\partial_n D)$. It now follows from the Riesz representation theorem that there is a unique positive Borel measure μ_p on $\partial_n D$ such that

$$u_f(p) = \int_{\partial_n D} f \, d\mu_p.$$

□

DEFINITION 2.9. *The measure μ_p in (2.12) is called the Caloric Measure at p for D, and the integral is called the Poisson Integral of f.*

Since temperatures are invariant under translation and parabolic dilation, the caloric measure has similar properties. To see this, let $f \in C(\partial_n D)$, and let u_f be the Poisson integral of f. Take a translation of the cylinder D to another cylinder $D_0 = D + \{p_0\}$, and define a function f_0 on $\partial_n D_0$ by putting $f_0(q) = f(q - p_0)$. If u_{f_0} is the Poisson integral of f_0, then for $p \in \overline{D}\backslash\partial_n D$ we have

$$u_{f_0}(p + p_0) = \int_{\partial_n D_0} f_0(q)\, d\mu_{p+p_0}(q) = \int_{\partial_n D_0} f(q - p_0)\, d\mu_{p+p_0}(q) = \int_{\partial_n D} f(q)\, d\nu_p(q),$$

where ν_p is the translation of μ_{p+p_0} from $\partial_n D_0$ to $\partial_n D$. Putting $v_{f_0}(p) = u_{f_0}(p + p_0)$, we get a temperature v_{f_0} on D with continuous boundary values f on $\partial_n D$. So $v_{f_0}(p) = u_f(p)$ by the boundary uniqueness principle, and hence

$$u_f(p) = \int_{\partial_n D} f(q)\, d\nu_p(q),$$

for any $f \in C(\partial_n D)$. So, by the uniqueness of the measure in Theorem 2.8, $\mu_p = \nu_p$.

For the parabolic dilation, we can now take
$$D = \{(y, s) : |y| < \sqrt{c}, -b < s < 0\}$$
without loss of generality, and dilate it to another cylinder
$$D_1 = \{(y, s) : |y| < \sqrt{ac}, -ab < s < 0\}.$$
Let u_f be as before, and define a function f_1 on $\partial_n D_1$ by putting
$$f_1(y, s) = f\left(\frac{y}{\sqrt{a}}, \frac{s}{a}\right).$$
If u_{f_1} is the Poisson integral of f_1, then for $(x, t) \in \overline{D} \backslash \partial_n D$ we have
$$u_{f_1}(x\sqrt{a}, ta) = \int_{\partial_n D_1} f_1(y, s) \, d\mu_{(x\sqrt{a}, ta)}(y, s)$$
$$= \int_{\partial_n D_1} f\left(\frac{y}{\sqrt{a}}, \frac{s}{a}\right) d\mu_{(x\sqrt{a}, ta)}(y, s)$$
$$= \int_{\partial_n D} f(y, s) \, d\chi_{(x,t)}(y, s),$$
where $\chi_{(x,t)}$ is the parabolic dilation of $\mu_{(x\sqrt{a}, ta)}$ from $\partial_n D_1$ to $\partial_n D$. Putting $v_{f_1}(x, t) = u_{f_1}(x\sqrt{a}, ta)$, we get a temperature v_{f_1} on D with continuous boundary values f on $\partial_n D$. So $v_{f_1} = u_f$, and hence
$$u_f(x, t) = \int_{\partial_n D} f(y, s) \, d\chi_{(x,t)}(y, s),$$
for any $f \in C(\partial_n D)$. So, by the uniqueness of the measure in Theorem 2.8, $\mu_{(x,t)} = \chi_{(x,t)}$.

We need some information about the sets of caloric measure zero.

LEMMA 2.10. *Let $p_0 = (x_0, t_0) \in \overline{D} \backslash \partial_n D$, and let μ_{p_0} be the caloric measure at p_0 for D. Then*

(a) $\mu_{p_0}(\{(y, s) \in \partial_n D : s \geq t_0\}) = 0$,

and

(b) for any relatively open subset V of $\{(y, s) \in \partial_n D : s < t_0\}$, we have $\mu_{p_0}(V) > 0$.

PROOF. (a) Let $D = B \times {]}a, b{[}$, where B is an open ball in $\mathbb{R}^n$. We choose a number $b^* > b$, and put $D^* = B \times {]}a, b^*{[}$. We also choose a decreasing sequence $\{f_k\}$ of functions in $C(\partial_n D^*)$ such that $f_k(y, s) = 1$ if $s \geq t_0$, $f_k(y, s) = 0$ if $s \leq t_0 - \frac{1}{k}(t_0 - a)$, and $f_k(y, s) \to 0$ as $k \to \infty$ whenever $t_0 - \frac{1}{k}(t_0 - a) < s < t_0$. Let u_k be the function in $C(\overline{D})$ associated with the restriction of f_k to $\partial_n D$ by Theorem 2.3, and u_k^* be that in $C(\overline{D}^*)$ associated with f_k itself. Then $u_k = f_k = u_k^*$ on $\partial_n D$, so that $u_k = u_k^*$ on $\overline{D}$ by the boundary uniqueness principle. Since $\{f_k\}$ is a decreasing sequence, so are $\{u_k\}$ and $\{u_k^*\}$. Consider the restriction of u_k^* to the set $\{(x, t) \in \overline{D}^* : t \leq t_0 - \frac{1}{k}(t_0 - a)\}$. Since $u_k^* = 0$ on the normal boundary, the boundary uniqueness principle shows that $u_k^* = 0$ throughout that cylinder. Put $u^* = \lim_{k \to \infty} u_k^*$ on D^*, and let $T = \{(y, s) \in \partial_n D : s \geq t_0\}$. The Harnack monotone convergence theorem can be applied to the increasing sequence $\{u_1^* - u_k^*\}$ of nonnegative temperatures, to show that $u_1^* - u^*$ is a temperature on D^*. Hence

u^* is a temperature also. Since $u_k^*(x,t) = 0$ whenever $t \leq t_0 - \frac{1}{k}(t_0 - a)$, we have $u^*(x,t) = 0$ whenever $t < t_0$ and so, by continuity, whenever $t \leq t_0$. Since $u^* = \lim_{k\to\infty} u_k$ on $\overline{D}\backslash\partial_n D$, it now follows from Lebesgue's monotone convergence theorem that

$$0 = \lim_{k\to\infty} u_k(x_0, t_0) = \int_{\partial_n D} \lim_{k\to\infty} f_k \, d\mu_{p_0} = \int_T d\mu_{p_0} = \mu_{p_0}(T).$$

(b) We choose a function $f \in C(\partial_n D)$ such that $f \geq 0$ on $\partial_n D$, $f = 0$ except on V, and $f(q_0) = 1$ for some point $q_0 \in V$. Let u be the function in $C(\overline{D})$ associated with f by Theorem 2.3, so that $u = f$ on $\partial_n D$. Then $u \geq 0$ by the minimum principle, and $u(p) \to 1$ as $p \to q_0$. If we had $\mu_{p_0}(V) = 0$, it would follow from Theorem 2.8 that

$$u(p_0) = \int_{\partial_n D} f \, d\mu_{p_0} = \int_V f \, d\mu_{p_0} = 0,$$

which implies that $u(x,t) = 0$ whenever $t < t_0$ (Theorem 1.21), contrary to the fact that $u(p) \to 1$ as $p \to q_0$. $\square$

2.5. Characterizations of Temperatures

We first characterize temperatures in terms of the Poisson Integral.

THEOREM 2.11. *If u is a temperature on E, and D is a circular cylinder with $\overline{D} \subseteq E$, then u has the representation*

$$u(p) = \int_{\partial_n D} u \, d\mu_p$$

for all $p \in \overline{D}\backslash\partial_n D$, where μ_p is the caloric measure at p for D.

Conversely, suppose that $u \in C(E)$ and that, for each point $p_0 \in E$, there is a circular cylinder D containing p_0 such that $\overline{D} \subseteq E$ and u has the representation

$$u(p) = \int_{\partial_n D} u \, d\mu_p$$

for all $p \in D$, where μ_p is the caloric measure at p for D. Then u is a temperature on E.

PROOF. Suppose that u is a temperature on E, and that D is a circular cylinder with $\overline{D} \subseteq E$. Let f be the restriction of u to $\partial_n D$. By Theorem 2.3, there is a function $u_f \in C(\overline{D})$ such that u_f is a temperature on $\overline{D}\backslash\partial_n D$ and $u_f = f$ on $\partial_n D$. By Theorem 2.8, u_f has the representation

$$u_f(p) = \int_{\partial_n D} f \, d\mu_p$$

for all $p \in \overline{D}\backslash\partial_n D$. The functions u and u_f belong to $C(\overline{D}\backslash\partial_n D)$, are temperatures on D, and are equal on $\partial_n D$. Therefore, by Theorem 1.24, $u = u_f$ on D and hence, by continuity, on $\overline{D}$. So u has the required representation.

To prove the converse, we take any point $p_0 \in E$, and note that there is some circular cylinder D containing p_0, such that $\overline{D} \subseteq E$ and

$$u(p) = \int_{\partial_n D} u \, d\mu_p$$

for all $p \in D$. Let f be the restriction of u to $\partial_n D$. Then, by Theorems 2.3 and 2.8, there is a function $u_f \in C(\overline{D})$ such that u_f is a temperature on D, and

$$u_f(p) = \int_{\partial_n D} f \, d\mu_p$$

for all $p \in D$. So $u = u_f$ on D. Hence u is a temperature on a neighbourhood of the arbitrary point p_0, and therefore on E. $\square$

The Poisson integral representation gives another mean value characterization of temperatures, as follows.

DEFINITION 2.12. For each $(x,t) \in \mathbb{R}^{n+1}$ and $c > 0$, we put

$$\Delta(x,t;c) = B(x,\sqrt{c}) \times \,]t-c,t[,$$

where $B(x,\sqrt{c})$ denotes the open ball in $\mathbb{R}^n$ with centre x and radius $\sqrt{c}$. We call $\Delta(x,t;c)$ the *Heat Cylinder* with *centre* (x,t) and *radius* c.

The *mean value over normal boundary of the heat cylinder* is defined, for any function u such that the integral exists, by

$$\mathcal{L}(u;x,t;c) = \int_{\partial_n \Delta(x,t;c)} u \, d\mu_{(x,t)},$$

where $\mu_{(x,t)}$ is the caloric measure at (x,t) for $\Delta(x,t;c)$.

Since the caloric measure is invariant under translation and parabolic dilation (see Section 2.4), the mean $\mathcal{L}(u;x,t;c)$ depends only on u, (x,t) and c.

Note that, by taking $u = 1$ in Theorem 2.11, we obtain $\mathcal{L}(1;x,t;c) = 1$ for all (x,t) and c.

Although the caloric measure is not given explicitly, the mean $\mathcal{L}(u;x,t;c)$ has the advantage that it forms part of the solution of the Dirichlet problem for $\Delta(x,t;c)$.

We can give characterizations of temperatures in terms of the means $\mathcal{L}(u;x,t;c)$, or $\mathcal{M}(u;x,t;c)$, or $\mathcal{V}(u;x,t;c)$, or $\mathcal{V}_m(u;x,t;c)$ for any integer $m \geq 1$. These are stronger than those given in Chapter 1, because they require only continuity and not smoothness. The proof depends on showing that a continuous function which possesses a weak mean value property also satisfies the maximum principle.

THEOREM 2.13. *Let $D = B \times \,]a,b[$ be an arbitrary circular cylinder in $\mathbb{R}^{n+1}$, and let $u \in C(D \cup \partial_n D)$. If, given any point $(x,t) \in D$ and $\epsilon > 0$, we can find a positive number $c < \epsilon$ such that either*

   (a) $u(x,t) \leq \mathcal{L}(u;x,t;c)$,

or

   (b) $u(x,t) \leq \mathcal{M}(u;x,t;c)$,

or

   (c) $u(x,t) \leq \mathcal{V}(u;x,t;c)$,

or, for some integer $m \geq 1$,

   (d) $u(x,t) \leq \mathcal{V}_m(u;x,t;c)$,

holds, then u satisfies the maximum principle of Theorem 1.21. That is, if there is *a point $(x_0, t_0) \in D$ such that $u(x_0, t_0) \geq u(x, t)$ whenever $(x, t) \in D$ and $t < t_0$, then $u(x_0, t_0) = u(x, t)$ for all such points (x, t); consequently*

$$\sup_{D \cup \partial_n D} u = \max_{\partial_n D} u.$$

PROOF. Suppose that there is a point $(x_0, t_0) \in D$ such that $u(x_0, t_0) \geq u(x, t)$ whenever $(x, t) \in D$ and $t < t_0$. Put $M = u(x_0, t_0)$, and let (x_1, t_1) be any point of D such that $t_1 < t_0$. Join (x_0, t_0) and (x_1, t_1) with a closed line segment γ, and put

$$S = \{s : \text{there is a point } (y, s) \in \gamma \text{ with } u(y, s) = M\}.$$

Then $S \neq \emptyset$ because $t_0 \in S$, and S is lower bounded by t_1. Put $s^* = \inf S$. If $\overline{\Delta}(x_0, t_0; c) \subseteq E$ then, because Lemma 2.10 shows that the caloric measure at (x_0, t_0) of $\partial_n \Delta(x_0, t_0; c) \setminus \Lambda(x_0, t_0; E)$ is zero, we have $u \leq M$ almost everywhere on $\partial_n \Delta(x_0, t_0; c)$ with respect to that measure. Therefore, if condition (a) holds, we can find a number $c < t_0 - t_1$ such that

$$M = u(x_0, t_0) \leq \mathcal{L}(u; x_0, t_0; c) \leq \mathcal{L}(M; x_0, t_0; c) = M.$$

By Lemma 2.10, this implies that $u = M$ on a dense subset of $\partial_n \Delta(x_0, t_0; c)$, and the continuity of u then shows that $u \equiv M$ on $\partial_n \Delta(x_0, t_0; c)$. Since $c < t_0 - t_1$, the set $\gamma \cap (\partial_n \Delta(x_0, t_0; c)) \neq \emptyset$, so that there is a point $s_1 \in S$ such that $s_1 < t_0$. Similar arguments are valid if conditions (b), (c), or (d) are satisfied. Hence $s^* < t_0$.

Suppose that $t_1 < s^* < t_0$. Then there is a sequence of points $\{(z_k, r_k)\}$ on γ such that $u(z_k, r_k) = M$ for all k and $r_k \to s^*$ as $k \to \infty$. The continuity of u now implies that there is a point (y^*, s^*) on γ such that $u(y^*, s^*) = M$. If condition (a) holds, we can find $c < s^* - t_1$ such that $u \equiv M$ on $\partial_n \Delta(y^*, s^*; c)$, and therefore a point $s_2 \in S$ such that $s_2 < s^*$. Similar arguments are valid if conditions (b), (c), or (d) are satisfied, so we have a contradiction. Hence $s^* = t_1$, and $u(x_1, t_1) = M$.

Thus $u(x, t) = M$ for all $(x, t) \in D$ such that $t < t_0$. Now it follows, as in the last paragraph of the proof of Theorem 1.21, that

$$\sup_{D \cup \partial_n D} u = \max_{\partial_n D} u.$$

$\square$

We now characterize temperatures in terms of the various means.

THEOREM 2.14. *If u is a temperature on E and $(x, t) \in E$, then*
(a) $u(x, t) = \mathcal{L}(u; x, t; c)$ whenever $\overline{\Delta}(x, t; c) \subseteq E$,
and
(b) $u(x, t) = \mathcal{M}(u; x, t; c)$ whenever $\overline{\Omega}(x, t; c) \subseteq E$,
and
(c) $u(x, t) = \mathcal{V}(u; x, t; c)$ whenever $\overline{\Omega}(x, t; c) \subseteq E$,
and, for any integer $m \geq 1$,
(d) $u(x, t) = \mathcal{V}_m(u; x, t; c)$ whenever $\overline{\Omega}_m(x, t; c) \subseteq E$.
Conversely, if $u \in C(E)$ and, given any point $(x, t) \in E$ and $\epsilon > 0$, we can find a positive number $c < \epsilon$ such that either
(a) $u(x, t) = \mathcal{L}(u; x, t; c)$,
or
(b) $u(x, t) = \mathcal{M}(u; x, t; c)$,
or

(c) $u(x,t) = \mathcal{V}(u; x, t; c)$,
or, for some integer $m \geq 1$,
(d) $u(x,t) = \mathcal{V}_m(u; x, t; c)$,
holds, then u is a temperature on E.

PROOF. If u is a temperature on E, then (a) follows from Theorem 2.11, (b) from Theorem 1.6, (c) from Theorem 1.16, and (d) from Theorem 1.25.

Conversely, suppose that $u \in C(E)$, and let D be an arbitrary circular cylinder such that $\overline{D} \subseteq E$. Let f denote the restriction of u to $\partial_n D$. By Theorem 2.3, there is a function $u_f \in C(\overline{D})$ which is a temperature on D and satisfies $u_f = f$ on $\partial_n D$. By Theorem 2.11, whenever $\Delta(x, t; c)$ is a heat cylinder such that $\overline{\Delta}(x, t; c) \subseteq D$, the equality

$$u_f(x,t) = \mathcal{L}(u_f; x, t; c)$$

holds. By Theorems 1.6 and 1.16, whenever $\Omega(x,t;c)$ is a heat ball such that $\overline{\Omega}(x,t;c) \subseteq D$, the equalities

$$u_f(x,t) = \mathcal{M}(u_f; x, t; c) = \mathcal{V}(u_f; x, t; c)$$

hold. By Theorem 1.25, whenever $m \geq 1$ and $\Omega_m(x,t;c)$ is a modified heat ball such that $\overline{\Omega}_m(x,t;c) \subseteq D$, the equality

$$u_f(x,t) = \mathcal{V}_m(u_f; x, t; c)$$

holds. Therefore, if $v = u - u_f$ on $\overline{D}$, then v satisfies the same condition (a), (b), (c), or (d) that u satisfies on E. Hence, by applying Theorem 2.13 to both v and $-v$, we obtain

$$0 = \min_{\partial_n D} v = \inf_{D \cup \partial_n D} v \leq \sup_{D \cup \partial_n D} v = \max_{\partial_n D} v = 0,$$

so that $u = u_f$ on D. Thus u is a temperature on any circular cylinder D such that $\overline{D} \subseteq E$, and hence on E. □

As a consequence of Theorem 2.14, we can show that a temperature on $B \times \,]a,b]$ satisfies the various mean value formulas even at points of $B \times \{b\}$.

COROLLARY 2.15. *Let $D = B \times \,]a,b[$ be a circular cylinder, let $x \in B$, and let u be a temperature on $\overline{D} \backslash \partial_n D$. Then*
(a) $u(x,b) = \mathcal{L}(u; x, b; c)$ whenever $\overline{\Delta}(x,b;c) \subseteq \overline{D} \backslash \partial_n D$,
and
(b) $u(x,b) = \mathcal{M}(u; x, b; c)$ whenever $\overline{\Omega}(x,b;c) \subseteq \overline{D} \backslash \partial_n D$,
and
(c) $u(x,b) = \mathcal{V}(u; x, b; c)$ whenever $\overline{\Omega}(x,b;c) \subseteq \overline{D} \backslash \partial_n D$,
and, for any integer $m \geq 1$,
(d) $u(x,b) = \mathcal{V}_m(u; x, b; c)$ whenever $\overline{\Omega}_m(x,b;c) \subseteq \overline{D} \backslash \partial_n D$.

PROOF. We give details only for part (a), as the proofs of the other parts are similar.

Given a closed heat cylinder $\overline{\Delta}(x,b;c) \subseteq \overline{D} \backslash \partial_n D$, we choose an open ball B^* in $\mathbb{R}^n$ such that $\overline{\Delta}(x,b;c) \subseteq B^* \times \,]a,b]$ and $\overline{B^*} \subseteq B$. We now choose a number $a^* > a$ such that $\overline{\Delta}(x,b;c) \subseteq B^* \times \,]a^*,b]$, and any number $b^* > b$. The restriction of u to $\partial_n(B^* \times \,]a^*,b[)$ is continuous and real-valued, and so we can extend it to a function $f \in C(\partial_n(B^* \times \,]a^*,b^*[))$. We now put $D^* = B^* \times \,]a^*,b^*[$, and denote by u^*

the element of $C(\overline{D^*})$ that is a temperature on $\overline{D^*}\backslash\partial_n D^*$ and equal to f on $\partial_n D^*$ (whose existence is guaranteed by Theorem 2.3). By the boundary uniqueness theorem for temperatures (Theorem 1.24), $u = u^*$ on $B^* \times\,]a^*, b[$ and hence, by continuity, on $B^* \times\,]a^*, b]$. We now apply Theorem 2.14 on D^*, and deduce that

$$u(x, b) = u^*(x, b) = \mathcal{L}(u^*; x, b; c) = \mathcal{L}(u; x, b; c),$$

as required. □

2.6. Extensions of some Harnack Theorems

Theorem 2.3 combines with the Harnack theorems of Section 1.7 to give results about temperatures on $\overline{D}\backslash\partial_n D$, for any circular cylinder D.

LEMMA 2.16. *Let $D = B \times\,]c, b[$ be a circular cylinder, let K be a compact subset of D, let $p_0 = (x_0, t_0) \in B \times\,]c, b]$, and let α and β be negative numbers such that the compact set*

$$P = P(p_0; \alpha, \beta) = \{(x, t) : |x - x_0|^2 \leq \alpha(t - t_0),\ t_0 + \beta \leq t \leq t_0\}$$

is contained in $B \times\,]c, b]$. Then there is a constant κ, which depends only on D, K and P, such that

$$\max_{K \cup P} u \leq \kappa u(p_0)$$

for every nonnegative temperature u on $B \times\,]c, b]$.

PROOF. Let B^* be an open ball in $\mathbb{R}^n$ such that $K \cup P \subseteq B^* \times\,]c, b]$ and $\overline{B^*} \subseteq B$. Choose a number c^* such that $c < c^* < b$ and $K \cup P \subseteq B^* \times\,]c^*, b]$. Given any nonnegative temperature u on $B \times\,]c, b]$, the restriction of u to $\partial_n(B^* \times\,]c^*, b[)$ is continuous and real-valued. Therefore, given any number $b^* > b$, we can extend that restriction to a nonnegative function $f \in C(\partial_n(B^* \times\,]c^*, b^*[))$. We now put $D^* = B^* \times\,]c^*, b^*[$, and denote by u^* the element of $C(\overline{D^*})$ that is a temperature on $\overline{D^*}\backslash\partial_n D^*$ and equal to f on $\partial_n D^*$ (whose existence is guaranteed by Theorem 2.3). By the boundary uniqueness theorem for temperatures (Theorem 1.24), $u = u^*$ on $B^* \times\,]c^*, b[$ and hence, by continuity, on $B^* \times\,]c^*, b]$. The temperature u^* is nonnegative on D^*, because if there was a point $q \in D^*$ such that $u^*(q) < 0$ we would have

$$\sup_{D^* \cup \partial_n D^*} (-u^*) > 0 \geq \max_{\partial_n D^*}(-f) = \max_{\partial_n D^*}(-u^*),$$

contrary to the maximum principle of Theorem 1.21. We now apply the Harnack inequalities of Corollaries 1.33 and 1.34 on D^*. Thus, there is a constant κ, which depends only on D, K and P, such that

$$\max_{K \cup P} v \leq \kappa v(p_0)$$

for every nonnegative temperature v on D^*. It follows that

$$\max_{K \cup P} u = \max_{K \cup P} u^* \leq \kappa u^*(p_0) = u(p_0),$$

as required. □

LEMMA 2.17. *Let $D = B \times\,]c, b[$ be a circular cylinder, let $\mathcal{F}$ be an upward-directed family of temperatures on $B \times\,]c, b]$, and let $u = \sup \mathcal{F}$. If there is a point $p_0 = (x_0, t_0) \in B \times\,]c, b]$ such that $u(p_0) < +\infty$, then u is a temperature on $B \times\,]c, t_0[$ and parabolically continuous at p_0.*

PROOF. Let K be a compact subset of $B \times \,]c, t_0[$, and for any given $\alpha < 0$ let the number $\beta < 0$ be chosen such that the set
$$P = P(p_0; \alpha, \beta) = \{(x,t) : |x - x_0|^2 \leq \alpha(t - t_0),\ t_0 + \beta \leq t \leq t_0\}$$
is contained in $B \times \,]c, b]$. For each positive integer k, we can find a function $u_k \in \mathcal{F}$ such that
$$u(p_0) - u_k(p_0) < \frac{1}{k}.$$
Since $\mathcal{F}$ is upward-directed, given any function $v \in \mathcal{F}$ and $k \in \mathbb{N}$, we can find a temperature $w_k \in \mathcal{F}$ such that $u_k \vee v \leq w_k$ on E. By Lemma 2.16, there is a positive constant κ, depending only on E, K and P, such that
$$w_k(p) - u_k(p) \leq \kappa(w_k(p_0) - u_k(p_0))$$
for all $p \in K \cup P$ and all k. Hence
$$v(p) - u_k(p) \leq w_k(p) - u_k(p) \leq \kappa(w_k(p_0) - u_k(p_0)) \leq \kappa(u(p_0) - u_k(p_0)) < \frac{\kappa}{k}$$
for all such p and k. Therefore
$$u(p) - u_k(p) = \sup\{v(p) - u_k(p) : v \in \mathcal{F}\} \leq \frac{\kappa}{k}$$
for all such p and k, so that the sequence $\{u_k\}$ converges uniformly to u on $K \cup P$. Therefore u is a temperature on $B \times \,]c, t_0[$, by Theorem 1.30, and the restriction of u to P is continuous. Since α was chosen arbitrarily, u is parabolically continuous at p_0. □

2.7. Notes and Comments

The results in the first three sections are classical. Their treatment here is taken from Landis [49], where more general cylindrical domains are considered.

The treatment in sections 2.4 and 2.5 is based on Watson [89]. The part of Theorem 2.14 that involves the means $\mathcal{M}$ was first proved by Pini [60] in the case $n = 1$, then by Fulks [24] in the general case; and the part involving the volume means $\mathcal{V}$ was first proved by Watson [69].

CHAPTER 3

Subtemperatures and the Dirichlet Problem on Convex Domains of Revolution

In this chapter we introduce general subtemperatures, discuss their basic properties, and prove several characterizations of them. The class of real continuous subtemperatures is too narrow for our approach to the Dirichlet problem on an arbitrary open set E. We need as broad a class of functions as possible that retain the main properties we need, namely that they satisfy the maximum principle and are majorized by any temperature whose boundary values are greater. Continuity is not essential for the maximum principle, upper semicontinuity is sufficient. We begin by explaining this term.

3.1. Semicontinuous Functions

Semicontinuous functions feature in the definition of a subtemperature, so in this short section we collect the facts that we shall need about them.

DEFINITION 3.1. Let S be a subset of $\mathbb{R}^{n+1}$. An extended real-valued function f on S is called *upper semicontinuous* on S if the set $\{p \in S : f(p) < a\}$ is a relatively open subset of S for each $a \in \mathbb{R}$. A function g is called *lower semicontinuous* on S if $-g$ is upper semicontinuous.

Thus an extended real-valued function h on S is continuous if and only if it is both upper semicontinuous and lower semicontinuous.

If f is upper semicontinuous on S, and $T \subseteq S$, then the restriction of f to T is upper semicontinuous on T.

If f and g are both upper semicontinuous on S, then
$$\{p \in S : (f \vee g)(p) < a\} = \{p \in S : f(p) < a\} \cap \{p \in S : g(p) < a\}$$
is a relatively open subset of S for each $a \in \mathbb{R}$. Hence $f \vee g$ is upper semicontinuous on S.

LEMMA 3.2. *An extended real-valued function f is upper semicontinuous on S if and only if*

(3.1)
$$\limsup_{q \to p} f(q) \leq f(p)$$

whenever p is a limit point of S in S.

PROOF. Suppose that $p \in S$ and is a limit point of S such that (3.1) holds. If $a \in \mathbb{R}$ is such that $a > f(p)$, then (3.1) shows that we can find $\delta > 0$ such that

$f(q) < a$ whenever $|p - q| < \delta$ and $q \in S$. So p is a relative interior point of the set $\{q \in S : f(q) < a\}$. Since any isolated point q of S is a relative interior point, the set $\{q \in S : f(q) < a\}$ is relatively open.

Conversely, suppose that $\{q \in S : f(q) < a\}$ is relatively open for all $a \in \mathbb{R}$. Let p be a limit point of S in S. If $f(p) = +\infty$, then (3.1) holds. Otherwise, for any real number $a > f(p)$, the set $\{q \in S : f(q) < a\}$ is relatively open and contains p. So there is a $\delta > 0$ such that $f(q) < a$ whenever $|p - q| < \delta$ and $q \in S$. Hence $\limsup_{q \to p} f(q) \leq a$ for every $a > f(p)$, and so (3.1) holds. □

DEFINITION 3.3. When the inequality (3.1) holds at p, we say that f is *upper semicontinuous* at p.

Note that, if f and g are both upper semicontinuous at the same point p, then so is $f + g$.

The notions of upper and lower semicontinuity split the notion of continuity into two pieces. It is familiar that a real-valued continuous function on a compact set is bounded, and attains its supremum and infimum. For a function which is merely upper semicontinuous, half of that remains true.

LEMMA 3.4. *If f is an upper semicontinuous function on a compact set K, and $f(p) < +\infty$ for all $p \in K$, then f is upper bounded and attains its supremum.*

PROOF. Put $M = \sup_K f$. Choose a sequence $\{q_k\}$ of points in K such that $f(q_k) \to M$. Since K is compact, the sequence $\{q_k\}$ has a subsequence $\{q_{j_k}\}$ which converges to a point $p_0 \in K$. So $M = \lim f(q_{j_k}) \leq f(p_0)$, and this implies that $f(p_0) = M$, which in turn implies that $M < +\infty$. □

The limit of a decreasing sequence of continuous functions may not itself be continuous, but as the next lemma shows it must be upper semicontinuous. Indeed, this holds if the functions in the sequence are merely upper semicontinuous.

LEMMA 3.5. *If $\{f_k\}$ is a decreasing sequence of upper semicontinuous functions on a set $S \subseteq \mathbb{R}^{n+1}$, then its pointwise limit f is also upper semicontinuous.*

PROOF. Given any $a \in \mathbb{R}$, the set $S_k = \{p \in S : f_k(p) < a\}$ is relatively open for all k. Since the sequence $\{f_k\}$ is decreasing,

$$\{p \in S : f(p) < a\} = \bigcup_{k=1}^{\infty} S_k.$$

That union is also relatively open, and so f is upper semicontinuous. □

Conversely, functions that are both upper bounded and upper semicontinuous can always be expressed as pointwise limits of decreasing sequences of real-valued continuous functions.

LEMMA 3.6. *Let $S \subseteq \mathbb{R}^{n+1}$, and let f be an extended real-valued function on S. If f is upper semicontinuous and upper bounded on S, then there is a decreasing sequence $\{f_k\}$ of uniformly continuous, real-valued functions on $\mathbb{R}^{n+1}$ that converges pointwise to f on S.*

PROOF. We first extend f to the whole of $\mathbb{R}^{n+1}$ by putting

$$\bar{f}(p) = \begin{cases} f(p) & \text{if } p \in S, \\ \limsup_{q \to p,\, q \in S} f(q) & \text{if } p \in \overline{S} \backslash S, \\ -\infty & \text{if } p \in \mathbb{R}^{n+1} \backslash \overline{S}. \end{cases}$$

Then $\bar{f}$ is upper semicontinuous and upper bounded on $\mathbb{R}^{n+1}$. We prove the lemma with $\bar{f}$ in place of f.

If $\bar{f} = -\infty$, then we define $f_k = -k$. Otherwise we put

$$f_k(p) = \sup\{\bar{f}(q) - k|p - q| : q \in \mathbb{R}^{n+1}\}$$

for every $p \in \mathbb{R}^{n+1}$. Then each f_k is real-valued, $f_k(p) \geq \bar{f}(p) - k|p - p| = \bar{f}(p)$ for all p, and the sequence $\{f_k\}$ is decreasing. Given any points $p, q, r \in \mathbb{R}^{n+1}$, we have

$$f_k(r) \geq \bar{f}(q) - k|r - q| \geq \bar{f}(q) - k|r - p| - k|p - q|,$$

so that

$$f_k(r) + k|r - p| \geq \bar{f}(q) - k|p - q|,$$

and hence

$$f_k(r) + k|r - p| \geq f_k(p).$$

Interchanging p and r, we deduce that

$$|f_k(p) - f_k(r)| \leq k|r - p|,$$

so that f_k is uniformly continuous on $\mathbb{R}^{n+1}$.

We now show that $\{f_k\}$ converges pointwise to $\bar{f}$ on $\mathbb{R}^{n+1}$. Let $p \in \mathbb{R}^{n+1}$ and $\epsilon > 0$. Since $\bar{f}$ is upper semicontinuous and upper bounded, we can find $\delta > 0$ such that $\bar{f}(q) < \bar{f}(p) + \epsilon$ whenever $|p - q| < \delta$. Therefore

$$\sup\{\bar{f}(q) - k|p - q| : |p - q| < \delta\} \leq \bar{f}(p) + \epsilon,$$

and

$$\sup\{\bar{f}(q) - k|p - q| : |p - q| \geq \delta\} \leq \sup\{\bar{f}(q) : q \in \mathbb{R}^{n+1}\} - k\delta,$$

so that

$$\bar{f}(p) \leq f_k(p) \leq \max\{\bar{f}(p) + \epsilon, \sup \bar{f} - k\delta\} = \bar{f}(p) + \epsilon$$

for all sufficiently large k. Hence $f_k \to \bar{f}$ pointwise on $\mathbb{R}^{n+1}$. □

3.2. Subtemperatures

We can define subtemperatures in terms of any of the means $\mathcal{L}$, $\mathcal{M}$, $\mathcal{V}$ or $\mathcal{V}_m$ for any $m \in \mathbb{N}$. We choose the means $\mathcal{L}$ because, although the measure in $\mathcal{L}$ is not given explicitly, $\mathcal{L}$ has the greater advantage of being part of a Poisson integral.

DEFINITION 3.7. Let w be an extended real-valued function on an open set E. We call w a *subtemperature* on E if it satisfies the following four conditions.

(δ_1) $-\infty \leq w(p) < +\infty$ for all $p \in E$.
(δ_2) w is upper semicontinuous on E.
(δ_3) w is finite on a dense subset of E.
(δ_4) Given any point $p \in E$ and positive number ϵ, there is a positive number $c < \epsilon$ such that the closed heat cylinder $\overline{\Delta}(p; c)$ is a subset of E and the inequality $w(p) \leq \mathcal{L}(w; p; c)$ holds.

This definition of subtemperature, when restricted to continuous real-valued functions, appears to conflict with that of a real continuous subtemperature given in Section 1.5. The discrepancy is only apparent, and is resolved in Theorem 3.51 below.

If w is a subtemperature on E, and V is an open subset of E, then w is a subtemperature on V.

We call a function v a *supertemperature* on E if $-v$ is a subtemperature on E. Therefore all results about subtemperatures easily imply dual results about supertemperatures. Note that v is a temperature on E if and only if it is both a subtemperature and a supertemperature on E, in view of Theorem 2.14.

Let W denote the fundamental temperature, defined in Section 1.1. Given any two points $p = (x, t)$ and $q = (y, s)$, we put $G(p; q) = W(x - y, t - s)$. For each fixed q, the function $p \mapsto G(p; q)$ is a temperature on $\mathbb{R}^{n+1} \backslash \{q\}$ and a supertemperature on $\mathbb{R}^{n+1}$. (It is obviously real valued and lower semicontinuous, and because it is a temperature on $\mathbb{R}^{n+1} \backslash \{q\}$, Theorem 2.14 shows that it satisfies the equality $G(p; q) = \mathcal{L}(G(.; q); p; c)$ whenever $\overline{\Delta}(p; c) \subseteq \mathbb{R}^{n+1} \backslash \{q\}$. The equality is trivially satisfied if $p = q$). The function $G(.; q)$ is called the *fundamental supertemperature with pole at* q.

If f is a real-valued, decreasing, left continuous function on an open subset U of $\mathbb{R}$, and $w(x, t) = f(t)$ for all $(x, t) \in \mathbb{R}^n \times U$, then w is a subtemperature on $\mathbb{R}^n \times U$.

We refer to any function w that satisfies condition (δ_1) as *upper finite*. If w is both upper finite and upper semicontinuous, then Lemma 3.4 shows that w is locally upper bounded, so that the means $\mathcal{L}(w; p; c)$ (for $\overline{\Delta}(p; c) \subseteq E$) always exist and are never $+\infty$.

If w is a subtemperature on E, and a is a positive number, then aw is a subtemperature on E.

Note that, if v and w are both subtemperatures on E, we cannot immediately conclude that $v + w$ is also a subtemperature. The conditions (δ_3) and (δ_4) are both too weak for that. If either v or w is a temperature, then $v + w$ is a subtemperature, in view of Theorem 2.14. The general case is given in Corollary 3.57 below.

If u is a temperature on E, and ϕ is a convex function defined on an interval containing $u(E)$, then $\phi \circ u$ is a subtemperature on E. This follows from the fact that $\phi \circ u \in C(E)$ (see Lemma 1.9), and Jensen's inequality. The identity $\mathcal{L}(1; p; c) = 1$ shows that Jensen's inequality is applicable, so that whenever the closed heat cylinder $\overline{\Delta}(p; c) \subseteq E$, we have

$$(\phi \circ u)(p) = \phi(\mathcal{L}(u; p; c)) \leq \mathcal{L}(\phi \circ u; p; c).$$

We now present more substantial examples of subtemperatures, along similar lines.

EXAMPLE 3.8. Suppose that u is a temperature on E, that v is a positive temperature on E, and that ϕ is a convex function on $\mathbb{R}$. Then the function w, defined on E by

$$w(p) = v(p)\phi\left(\frac{u(p)}{v(p)}\right),$$

is a subtemperature on E. In particular, if $\alpha \geq 1$, then $|u|^\alpha$, $v^{1-\alpha}$ and $|u|^\alpha v^{1-\alpha}$ are all subtemperatures.

To prove this, we first observe that the last part of Lemma 1.9 that ϕ is the supremum of the affine functions χ such that $\chi \leq \phi$. That is, for all $r \in \mathbb{R}$,
$$\phi(r) = \sup\{ar + b\},$$
where the supremum is taken over the set of all real numbers a and b such that $as + b \leq \phi(s)$ for all $s \in \mathbb{R}$. Therefore, whenever $p \in E$,
$$w(p) = v(p)\sup\left\{a\frac{u(p)}{v(p)} + b\right\} = \sup\{au(p) + bv(p)\},$$
where the suprema are taken over the same set as before. Each function $au + bv$ is a temperature on E; and because u, v and ϕ are all continuous, so is w. Thus $w \in C(E)$, and since $(au+bv)(p) = \mathcal{L}(au+bv;p;c) \leq \mathcal{L}(w;p;c)$ by Theorem 2.14, w satisfies $w(p) \leq \mathcal{L}(w;p;c)$, whenever $\overline{\Delta}(p;c) \subseteq E$.

EXAMPLE 3.9. Suppose that u is a subtemperature on E, that v is a positive temperature on E, and that ϕ is an increasing convex function on $\mathbb{R}$, extended to $-\infty$ by putting $\phi(-\infty) = \lim_{r \to -\infty} \phi(r)$. Then the function w, defined on E by
$$w(p) = v(p)\phi\left(\frac{u(p)}{v(p)}\right),$$
is a subtemperature on E. In particular, if $\alpha \geq 1$ and $u^+ = u \vee 0$, then $(u^+)^\alpha$ and $(u^+)^\alpha v^{1-\alpha}$ are subtemperatures.

The proof is broadly similar to that of Example 3.8. Note that, since ϕ is increasing, if a and b are such that $as+b \leq \phi(s)$ for all $s \in \mathbb{R}$, then $a \geq 0$. Therefore, by Lemma 1.9, $\phi(r) = \sup\{ar + b\}$ for all $r \in \mathbb{R}$, and $w(p) = \sup\{au(p) + bv(p)\}$ for all $p \in E$, where the suprema are taken over the set of all numbers $a \geq 0$ and $b \in \mathbb{R}$ such that $as + b \leq \phi(s)$ for all $s \in \mathbb{R}$. Since $a \geq 0$, the function $au + bv$ is a subtemperature on E. Furthermore, ϕ is increasing and continuous (by Lemma 1.9), v is continuous, and u/v is upper semicontinuous and upper finite, so that for each $p \in E$ we have
$$\limsup_{q \to p} w(q) = v(p)\phi\left(\limsup_{q \to p} \frac{u(q)}{v(q)}\right) \leq w(p),$$
and w is upper finite. In addition, whenever $\overline{\Delta}(p;c)$ is contained in E, we have $(au + bv)(p) \leq \mathcal{L}(au+bv;p;c) \leq \mathcal{L}(w;p;c)$ so that $w(p) \leq \mathcal{L}(w;p;c)$. Hence w is a subtemperature.

In order to prove a very flexible characterization of subtemperatures, we shall introduce a wider class of functions that lack the finiteness condition (δ_3). These functions will also be useful when we come to consider the Dirichlet problem on arbitrary open sets in Chapter 8.

DEFINITION 3.10. Let w be an extended real-valued function on an open set E. We call w a *hypotemperature* on E if it satisfies conditions (δ_1), (δ_2) and (δ_4) of the definition of a subtemperature.

Recall, from Section 1.7, that if $p_0 = (x_0, t_0)$ is any point in an open set E, we denote by $\Lambda(p_0; E) \equiv \Lambda(x_0, t_0; E)$ the set of points p that are lower than p_0 relative to E, in the sense that there is a polygonal path $\gamma \subseteq E$ joining p_0 to p, along which the temporal variable is strictly decreasing.

THEOREM 3.11 (The Strong Maximum Principle). *Let w be a hypotemperature on an open set E. If there is a point $(x_0, t_0) \in E$ such that $w(x_0, t_0) \geq w(x, t)$ for all $(x, t) \in \Lambda(x_0, t_0; E)$, then $w(x_0, t_0) = w(x, t)$ for all such points (x, t).*

PROOF. The result is trivially true if $w(x_0, t_0) = -\infty$, so we suppose otherwise. We put $M = w(x_0, t_0)$, and let (x_1, t_1) be an arbitrary point of $\Lambda(x_0, t_0; E)$. Let γ be a polygonal path in E that connects (x_0, t_0) to (x_1, t_1), along which the temporal variable is strictly decreasing. Put

$$S = \{s : \text{there is a point } (y, s) \in \gamma \text{ with } w(y, s) = M\}.$$

Then $S \neq \emptyset$ because $t_0 \in S$, and S is lower bounded by t_1. Put $s^* = \inf S$. If $\overline{\Delta}(x_0, t_0; c) \subseteq E$, then because Lemma 2.10 shows that the caloric measure at (x_0, t_0) of $\partial_n \Delta(x_0, t_0; c) \setminus \Lambda(x_0, t_0; E)$ is zero, we have $w \leq M$ almost everywhere on $\partial_n \Delta(x_0, t_0; c)$ with respect to that measure. So, since w satisfies condition (δ_4), there is a number $c < t_0 - t_1$ such that

$$M = w(x_0, t_0) \leq \mathcal{L}(w; x_0, t_0; c) \leq \mathcal{L}(M; x_0, t_0; c) = M,$$

and hence $w = M$ on a dense subset of $\partial_n \Delta(x_0, t_0; c)$, by Lemma 2.10. Therefore, for any point $(y, s) \in \partial_n \Delta(x_0, t_0; c)$ such that $s < t_0$, the upper semicontinuity of w shows that

$$M = \limsup_{(x,t) \to (y,s)} w(x, t) \leq w(y, s) \leq M.$$

Since $c < t_0 - t_1$, the set $\gamma \cap \partial_n \Delta(x_0, t_0; c) \neq \emptyset$, so that we can find a point $(y_1, s_1) \in \gamma$ such that $s_1 < t_0$ and $w(y_1, s_1) = M$. Therefore $s^* < t_0$.

Suppose that $t_1 < s^* < t_0$. There is a sequence of points $\{(z_k, r_k)\}$ on γ such that $w(z_k, r_k) = M$ for all k, and $r_k \to s^*$ as $k \to \infty$. This implies first that there is a point $(y^*, s^*) \in \gamma$, and second that, since w is upper semicontinuous,

$$M = \lim_{k \to \infty} w(z_k, r_k) \leq w(y^*, s^*) \leq M.$$

Hence $s^* \in S$. Therefore, since w satisfies condition (δ_4), there is $c < s^* - t_1$ such that

$$M = w(y^*, s^*) \leq \mathcal{L}(w; y^*, s^*; c) \leq \mathcal{L}(M; y^*, s^*; c) = M,$$

so that $w = M$ on a dense subset of $\partial_n \Delta(y^*, s^*; c)$ which, as before, implies that there is a point $(y_2, s_2) \in \gamma \cap \partial_n \Delta(y^*, s^*; c)$ such that $s_2 < s^*$ and $w(y_2, s_2) = M$. This contradicts the definition of s^*, so it is not possible to have $t_1 < s^*$, and hence $t_1 = s^*$.

Since $t_1 = \inf S$, there is a sequence $\{(\xi_k, \eta_k)\}$ of points on γ such that $w(\xi_k, \eta_k) = M$ for all k, and $\eta_k \to t_1$ as $k \to \infty$. Therefore

$$M = \lim_{k \to \infty} w(\xi_k, \eta_k) \leq w(x_1, t_1) \leq M,$$

so that $w(x_1, t_1) = M$. Thus $w = M$ throughout $\Lambda(x_0, t_0; E)$. □

COROLLARY 3.12. *Let w be a hypotemperature on E. Given any point (x_0, t_0) in E, there is a point $(x_1, t_1) \in \Lambda(x_0, t_0; E)$ such that $w(x_0, t_0) \leq w(x_1, t_1)$.*

PROOF. If $w(x_0, t_0) \geq w(x, t)$ for all $(x, t) \in \Lambda(x_0, t_0; E)$, then Theorem 3.11 shows that $w(x_0, t_0) = w(x, t)$ for all such points (x, t). The only other possibility is that there is a point $(x_1, t_1) \in \Lambda(x_0, t_0; E)$ such that $w(x_0, t_0) < w(x_1, t_1)$. □

For the case of hypotemperatures on a circular cylinder, a boundary maximum principle can be deduced following the proof of the last part of Theorem 1.21. For a general open set E, it is not so easy. We shall need to use the *Hausdorff Maximality Theorem*, an equivalent of the Axiom of Choice. For this purpose, we recall some definitions from set theory.

A set S is said to be *partially ordered* by a binary relation $\prec$ if, for all points $p, q, r \in S$,

(a) $p \prec q$ and $q \prec r$ implies $p \prec r$,
(b) $p \prec p$,
(c) $p \prec q$ and $q \prec p$ implies $p = q$.

A subset T of a partially ordered set S is said to be *totally ordered* if every pair $p, q \in T$ satisfies either $p \prec q$ or $q \prec p$. In addition, T is called *maximal* if, whenever any member of $S\setminus T$ is adjoined to T, the resultant set is not totally ordered.

The assertion of the Hausdorff Maximality Theorem is that *every nonempty partially ordered set contains a maximal totally ordered subset*.

We can now establish the *boundary maximum principle* for hypotemperatures on an arbitrary open set.

THEOREM 3.13. *Let w be a hypotemperature on an open set E, and suppose that*

$$(3.2) \qquad \limsup_{k \to \infty} w(p_k) \leq A$$

for every sequence $\{p_k\}$ in E that satisfies $p_{k+1} \in \Lambda(p_k; E)$ for all k, and which tends either to a boundary point of E or to the point at infinity. Then $w(p) \leq A$ for all $p \in E$.

PROOF. Given any number $\alpha > A$, we put $S_\alpha = \{p \in E : w(p) \geq \alpha\}$. If $S_\alpha = \emptyset$ for all α, there is nothing to prove. If $S_\alpha \neq \emptyset$ for some α, we define a partial order $\prec$ on S_α by putting $p \prec q$ if $p \in \Lambda(q; E) \cup \{q\}$. By the Hausdorff Maximality Theorem, S_α contains a maximal totally ordered subset T_α. We put $t^* = \inf\{t : \text{there is a point } (x, t) \in T_\alpha\}$. Since T_α is totally ordered, there is a sequence $\{p_i\} = \{(x_i, t_i)\}$ of points of T_α such that $p_{i+1} \in \Lambda(p_i; E) \cup \{p_i\}$ for all i, and $t_i \to t^*$ as $i \to \infty$.

If the sequence $\{p_i\}$ has a cluster point in ∂E, or is unbounded, then it contains infinitely many points. It therefore has a subsequence $\{p_{i_k}\}$ that converges to a point of ∂E, or tends to the point at infinity, such that $p_{i_{k+1}} \in \Lambda(p_{i_k}; E)$ for all k. Hence, by (3.2),

$$\alpha \leq \limsup_{k \to \infty} w(p_{i_k}) \leq A < \alpha,$$

a contradiction. Therefore $\{p_i\}$ is contained in some compact subset of E. Hence $t^* > -\infty$, and $\{p_i\}$ has a subsequence $\{p_{i_j}\}$ that converges to a point $p^* = (x^*, t^*)$ in $E \cap \overline{T}_\alpha$. Put $q_j = (y_j, s_j) = p_{i_j}$ for all j. Then $q_{j+1} \in \Lambda(q_j; E) \cup \{q_j\}$ for all j. Since $p^* \in E \cap \overline{T}_\alpha$ and $w \geq \alpha$ on T_α, the upper semicontinuity of w implies that $w(p^*) \geq \alpha$, so that $p^* \in S_\alpha$. Furthermore, p^* is the centre of some euclidean ball $B(p^*, \delta) \subseteq E$, and there exists some number N such that $q_j \in B(p^*, \delta)$ for all $j \geq N$. It follows that $p^* \prec q_j$ for all $j \geq N$. Since T_α is totally ordered and $q_j \to p^* = (x^*, t^*)$, for each point $p \in T_\alpha \setminus \{p^*\}$ there is some $j \geq N$ such that $q_j \prec p$. Hence $p^* \prec p$ for all $p \in T_\alpha$, so that $T_\alpha \cup \{p^*\}$ is totally ordered. Since

T_α is maximal, this shows that $p^* \in T_\alpha$. By Corollary 3.12, there is some point $p' \in \Lambda(p^*; E)$ such that $w(p') \geq w(p^*) \geq \alpha$. This implies first that $p' \in S_\alpha$, then that $p' \in T_\alpha$. Now we have another contradiction, because $t^* = \inf\{t : (x,t) \in T_\alpha\}$ and $p' \in \Lambda(x^*, t^*; E)$.

Thus if $S_\alpha \neq \emptyset$, we obtain a contradiction in every possible situation. We conclude that $S_\alpha = \emptyset$ for all α, so that $w(p) \leq A$ for all $p \in E$. □

For the case of a circular cylinder, Theorem 3.13 gives a predictable result, as follows.

COROLLARY 3.14. *Let w be a hypotemperature on a circular cylinder D. If*
$$\limsup_{p \to q} w(p) \leq A$$
for all $q \in \partial_n D$, then $w(p) \leq A$ for all $p \in D$.

PROOF. If $\{p_k\}$ is a sequence in D that satisfies $p_{k+1} \in \Lambda(p_k; D)$ for all k, and tends to a boundary point of D, then $\{p_k\}$ tends to a point $q \in \partial_n D$. So
$$\limsup_{k \to \infty} w(p_k) \leq \limsup_{p \to q} w(p) \leq A.$$
□

REMARK 3.15. Putting $p_k = (x_k, t_k)$ in Theorem 3.13, we remark that the condition $p_{k+1} \in \Lambda(p_k; E)$ for all k, is stronger than $t_{k+1} < t_k$ for all k. For example, if
$$E = B \times \left(\bigcup_{j=1}^{\infty} \left] \frac{1}{j+1}, \frac{1}{j} \right[\right),$$
where B is an open ball in $\mathbb{R}^n$, then $B \times \{0\} \subseteq \partial E$ and, for each $x \in B$, the sequence $\{(x, (k+\frac{1}{2})^{-1})\}$ converges to $(x, 0)$. But there is no sequence $\{p_k\}$ in E that converges to $(x, 0)$ and satisfies $p_{k+1} \in \Lambda(p_k; E)$ for all k.

If E is unbounded, we must treat the point at infinity in the same way as points of ∂E in Theorem 3.13. In some cases, the condition that $p_{k+1} \in \Lambda(p_k; E)$ implies that the point at infinity can be ignored, even when there are sequences $\{(x_k, t_k)\}$ in E such that $t_{k+1} < t_k$ for all k and $|x_k| \to \infty$ as $k \to \infty$. For example, if
$$E = \{(x,t) \in \mathbb{R}^2 : \sin x < t < 2 + \sin x\},$$
then $(2k\pi, k^{-1}) \in E$ for all k, but there is no sequence $\{p_k\}$ in E that tends to the point at infinity and satisfies $p_{k+1} \in \Lambda(p_k; E)$ for all k.

Our next theorem generalizes Theorem 1.23, characterizes hypotemperatures in terms of being majorized by temperatures on circular cylinders, and strengthens condition (δ_4). To prove it, we need a lemma that refines the condition of upper semicontinuity.

LEMMA 3.16. *Let w be a hypotemperature on an open set E, and let (y,s) be a point in E. Then*
$$w(y,s) = \limsup_{(x,t) \to (y, s-)} w(x,t).$$

PROOF. We put $q = (y, s)$ and $l = \limsup_{(x,t) \to (y,s-)} w(x, t)$. Since w is upper semicontinuous and upper finite, we have $l \leq w(q) < +\infty$. Given any number $L > l$, we can find a heat cylinder $\Delta(q; c_0)$ such that $w(p) \leq L$ for all $p \in \Delta(q; c_0)$. Now condition (δ_4) shows that there is a positive number $c < c_0$ such that
$$w(q) \leq \mathcal{L}(w; q; c) \leq \mathcal{L}(L; q; c) = L$$
since, in view of Lemma 2.10, $w \leq L$ almost everywhere on $\partial_n \Delta(q; c_0)$ with respect to the caloric measure at q. Thus $w(q) \leq L$ whenever $l < L$, so that $w(q) \leq l$. Hence $w(q) = l$. □

THEOREM 3.17. *Let w be an upper finite and upper semicontinuous function on an open set E. Consider the following property: Whenever D is a circular cylinder such that $\overline{D} \subseteq E$, and v is a function in $C(\overline{D})$ that is a temperature on D and satisfies $v \geq w$ on $\partial_n D$, then $v \geq w$ on $\overline{D}$.*

If w is a hypotemperature on E, then the stated property holds.

Conversely, if the stated property holds, then the inequality $w(p) \leq \mathcal{L}(w; p; c)$ holds whenever the closed heat cylinder $\overline{\Delta}(p; c) \subseteq E$, so that w is a hypotemperature on E.

PROOF. Let w be a hypotemperature on E, and let D be a circular cylinder such that $\overline{D} \subseteq E$. If $v \in C(\overline{D})$, is a temperature on D, and satisfies $v \geq w$ on $\partial_n D$, then $w - v$ is a hypotemperature on D, in view of Theorem 2.14. Furthermore, whenever $q \in \partial_n D$ we have
$$\limsup_{p \to q,\, p \in D} (w(p) - v(p)) \leq w(q) - v(q) \leq 0,$$
so that $w(p) \leq v(p)$ for all $p \in D$, by Corollary 3.14. Finally, if $q \in \overline{D}$ but $q \notin D \cup \partial_n D$, Lemma 3.16 shows that
$$w(q) - v(q) = \limsup_{p \to q,\, p \in D} w(p) - \lim_{p \to q,\, p \in D} v(p) = \limsup_{p \to q,\, p \in D} (w(p) - v(p)) \leq 0.$$

Conversely, suppose that w has the stated property. If $\Delta(p; c)$ is a heat cylinder such that $\overline{\Delta}(p; c) \subseteq E$, then the restriction of w to $\partial_n \Delta(p; c)$ is upper semicontinuous and upper finite, and therefore upper bounded, by Lemma 3.4. Therefore, by Lemma 3.6, we can find a sequence $\{f_k\}$ in $C(\partial_n \Delta(p; c))$ that decreases to w on $\partial_n \Delta(p; c)$. For each k, let v_k be the Poisson integral of f_k on $\overline{\Delta}(p; c) \backslash \partial_n \Delta(p; c)$, and let $v_k = f_k$ on $\partial_n \Delta(p; c)$. Then $v_k \in C(\overline{\Delta}(p; c))$, v_k is a temperature on $\overline{\Delta}(p; c) \backslash \partial_n \Delta(p; c)$, and $v_k \geq w$ on $\partial_n \Delta(p; c)$. So our hypothesis implies that $v_k \geq w$ on $\overline{\Delta}(p; c)$. In particular,
$$w(p) \leq \lim_{k \to \infty} v_k(p) = \mathcal{L}(\lim_{k \to \infty} f_k; p; c) = \mathcal{L}(w; p; c)$$
by Lebesgue's monotone convergence theorem. □

COROLLARY 3.18. *If v and w are hypotemperatures on E, then $w \vee v$ is too.*

PROOF. Conditions (δ_1) and (δ_2) obviously hold for $w \vee v$, and (δ_4) holds because
$$(w \vee v)(p) \leq \mathcal{L}(w; p; c) \vee \mathcal{L}(v; p; c) \leq \mathcal{L}(w \vee v; p; c)$$
for all values of c such that $\overline{\Delta}(p; c) \subseteq E$. □

COROLLARY 3.19. *If v and w are hypotemperatures on E, and either one is real-valued, then $v + w$ is a hypotemperature on E.*

PROOF. Conditions (δ_1) and (δ_2) obviously hold for $v+w$, and (δ_4) follows from Theorem 3.17. □

COROLLARY 3.20. *If w is a subtemperature on E and $p_0 \in E$, then*
$$w(p_0) = \lim_{c \to 0+} \mathcal{L}(w; p_0; c).$$

PROOF. Let $A > w(p_0)$. Since w is upper semicontinuous, we can find a neighbourhood N of p_0 such that $w(p) < A$ for all $p \in N$. Hence, whenever $\overline{\Delta}(p_0; c) \subseteq N$, Theorem 3.17 implies that
$$w(p_0) \leq \mathcal{L}(w; p_0; c) \leq \mathcal{L}(A; p_0; c) = A.$$
Therefore
$$w(p_0) \leq \liminf_{c \to 0+} \mathcal{L}(w; p_0; c) \leq \limsup_{c \to 0+} \mathcal{L}(w; p_0; c) \leq A,$$
which implies the result. □

THEOREM 3.21. *Let w be a subtemperature on an open set E, and let D be a circular cylinder such that $\overline{D} \subseteq E$. Then the Poisson integral u of the restriction of w to $\partial_n D$ exists, and the function $\pi_D w$, defined on E by putting*
$$\pi_D w = \begin{cases} u & \text{on } \overline{D} \backslash \partial_n D, \\ w & \text{on } E \backslash (\overline{D} \backslash \partial_n D), \end{cases}$$
has the following properties:
  (a) $\pi_D w$ is a subtemperature on E,
  (b) $\pi_D w \geq w$ on E,
  (c) $\pi_D w$ is a temperature on $\overline{D} \backslash \partial_n D$,
  (d) $\pi_D w = w$ on $\partial_n D \cup (E \backslash \overline{D})$,
  (e) if $v \in C(\overline{D})$, $v \geq w$ on $\overline{D}$, and v is a temperature on D, then $v \geq \pi_D w$ on $\overline{D}$.

PROOF. Let $D = B \times\,]a, b[$, where B is an open ball in $\mathbb{R}^n$ and $]a, b[$ is a bounded interval in $\mathbb{R}$. Choose a number $b^* > b$ such that the cylinder $D^* = B \times\,]a, b^*[$ also has its closure contained in E. Since w is upper semicontinuous and upper finite on the compact set $\partial_n D^*$, Lemma 3.4 shows that it is upper bounded on $\partial_n D^*$. Therefore Lemma 3.6 shows that we can find a decreasing sequence $\{f_k\}$ of functions in $C(\partial_n D^*)$ such that $f_k \to w$ on $\partial_n D^*$ as $k \to \infty$. For each k, we put u_k equal to the Poisson integral of f_k on $\overline{D^*} \backslash \partial_n D^*$, and u_k equal to f_k on $\partial_n D^*$. Then $u_k \in C(\overline{D^*})$ and u_k is a temperature on D^*. Since $\{f_k\}$ is a decreasing sequence, so is $\{u_k\}$. We put $u = \lim_{k \to \infty} u_k$. By Theorem 3.17, $w \leq u_k$ on $\overline{D^*}$ for all k, and hence $w \leq u$. Since u is the limit of a decreasing sequence of continuous functions, it is upper semicontinuous on $\overline{D^*}$ by Lemma 3.5; and since $f_k \to w$ on $\partial_n D^*$, $u = w$ there. Lebesgue's monotone convergence theorem now shows that u is the Poisson integral of the restriction of w to $\partial_n D^*$. Furthermore, $u_1 - u$ is the limit of the increasing sequence $\{u_1 - u_k\}$ of nonnegative temperatures on D^*, and $u_1(p) - u(p) \leq u_1(p) - w(p) < +\infty$ on a dense subset of D^*, so that the Harnack monotone convergence theorem shows that $u_1 - u$ is a temperature on D^*. Hence, in particular, the restriction of u to $\overline{D}$ is a temperature on $\overline{D} \backslash \partial_n D$, is the Poisson integral of the restriction of w to $\partial_n D$ on $\overline{D} \backslash \partial_n D$, in view of Lemma 2.10, and $u(p) = \mathcal{L}(u; p; c)$ whenever $\overline{\Delta}(p; c) \subseteq \overline{D} \backslash \partial_n D$, by Theorem 2.14.

We now define the function $\pi_D w$ as in the statement of the theorem, and show that $\pi_D w$ is a subtemperature on E. Since $w < +\infty$ on E, and u is the limit of a decreasing sequence of functions in $C(\overline{D})$, $\pi_D w$ is upper finite on E. Since $u \geq w$ on $\overline{D}$, and w satisfies condition (δ_3), $\pi_D w$ also satisfies that condition, and $\pi_D w \geq w$ on E. Furthermore, $\pi_D w$ is certainly upper semicontinuous at points outside $B \times \{b\}$; and if $q \in B \times \{b\}$, then

$$\lim_{p \to q,\, p \in D} u(p) = u(q) \geq w(q) \geq \limsup_{p \to q,\, p \notin D} w(p),$$

which implies that $\pi_D w$ is upper semicontinuous at q. It remains to prove that $\pi_D w$ satisfies condition (δ_4). If $p \in E$ but $p \notin \overline{D} \backslash \partial_n D$, then whenever the closed heat cylinder $\overline{\Delta}(p; c) \subseteq E$, we have

$$\pi_D w(p) = w(p) \leq \mathcal{L}(w; p; c) \leq \mathcal{L}(\pi_D w; p; c).$$

On the other hand, we have already shown that $u(p) = \mathcal{L}(u; p; c)$ whenever $\overline{\Delta}(p; c)$ is contained in $\overline{D} \backslash \partial_n D$, so that $\pi_D w(p) = \mathcal{L}(\pi_D w; p; c)$ for such values of p and c. Hence $\pi_D w$ is a subtemperature on E.

It only remains to prove part (e). Suppose that $v \in C(\overline{D})$, $v \geq w$ on $\overline{D}$, and v is a temperature on D. Given any $\epsilon > 0$, the sequence $\{f_k\}$ decreases to the limit $w < v + \epsilon$ on $\partial_n D$. Therefore the sequence of sets $\{S_k\}$, defined by

$$S_k = \{q \in \partial_n D : f_k(q) < v(q) + \epsilon\}$$

is expanding to the union $\partial_n D$. Both f_k and v are continuous on $\partial_n D$, so that each set S_k is relatively open. Therefore, since $\partial_n D$ is compact, there is a number κ such that $S_k = \partial_n D$ whenever $k > \kappa$. Thus $f_k(q) < v(q) + \epsilon$ for all $q \in \partial_n D$ if $k > \kappa$. This implies, using the maximum principle, that $u_k(q) < v(q) + \epsilon$ for all $q \in \overline{D}$ if $k > \kappa$. Therefore $u < v + \epsilon$ on $\overline{D}$ for any $\epsilon > 0$, and so $u \leq v$. $\square$

COROLLARY 3.22. *Let w be a subtemperature on an open set E, and let $a \in \mathbb{R}$ be such that $(\mathbb{R}^n \times \{a\}) \cap E \neq \emptyset$. Then $\{(x, a) : w(x, a) > -\infty\}$ is a dense subset of $(\mathbb{R}^n \times \{a\}) \cap E$.*

PROOF. Suppose, on the contrary, that there is an open ball B in $\mathbb{R}^n$ such that $w(x, a) = -\infty$ for all $x \in B$. We choose a circular cylinder $D = B \times]a, b[$ such that $\overline{D} \subseteq E$, and a point $p_0 \in D$. By Theorem 3.21, the Poisson integral u of the restriction of w to $\partial_n D$ exists and is a temperature on D, so that $u(p_0) > -\infty$. However, if μ_{p_0} denotes the caloric measure at p_0 for D, then Lemma 2.10 shows that $\mu_{p_0}(B \times \{a\}) > 0$, so that $u(p_0) = -\infty$ and we have a contradiction. $\square$

COROLLARY 3.23. *If w is a subtemperature on E and $p_0 \in E$, then the function $c \mapsto \mathcal{L}(w; p_0; c)$ is increasing on the set of $c > 0$ such that $\overline{\Delta}(p_0; c) \subseteq E$.*

PROOF. We suppose that $\overline{\Delta}(p_0; c)$ is a subset of E, and that $0 < b < c$. We write $\Delta(b) = \Delta(p_0; b)$. By Theorem 3.21, the function $\pi_{\Delta(b)} w$ is a subtemperature on E and is equall to w on a superset of $\partial_n \Delta(c)$. Therefore

$$w(p_0) \leq \mathcal{L}(w; p_0; b) = \pi_{\Delta(b)} w(p_0) \leq \mathcal{L}(\pi_{\Delta(b)} w; p_0; c) = \mathcal{L}(w; p_0; c).$$

$\square$

REMARK 3.24. The function $\pi_D w$ in Theorem 3.21 is not unique in having the properties (a)-(e). For example, let w be the characteristic function of the closed half-space $\mathbb{R}^n \times]-\infty, 0]$, and let $D = B(0, r) \times]0, b[$. Then w itself has all these properties, but its restriction to $\overline{D}\backslash\partial_n D$ is identically zero, whereas the Poisson integral of its restriction to $\partial_n D$ is

$$u(p) = \int_{B(0,r)\times\{0\}} d\mu_p = \mu_p(B(0,r)\times\{0\}) > 0$$

for all $p \in \overline{D}\backslash\partial_n D$, by Lemma 2.10.

3.3. The Dirichlet Problem on Convex Domains of Revolution

Let $x_0 \in \mathbb{R}$ and $a, b \in \mathbb{R}^n$. A *Convex Domain of Revolution* is any open set that has the form

$$R = R(x_0; \rho; a, b) = \{(x,t) \in \mathbb{R}^{n+1} : |x - x_0| < \rho(t),\ a < t < b\}$$

for some continuous concave function $\rho : [a,b] \to [0, +\infty[$.

The convex domains of revolution include the heat balls, the modified heat balls, and the circular cylinders.

We have seen, in Chapter 2, that the Dirichlet problem on a circular cylinder $B(x_0, r) \times]a, b[$ does not require boundary values to be taken on the upper boundary $B(x_0, r) \times \{b\}$, and so no values are specified there. Similarly, the Dirichlet problem on a convex domain of revolution does not involve boundary values on the set $T = \{(x,b) : |x - x_0| < \rho(b)\}$, if $T \neq \emptyset$. The case of the boundary points of the form (x, b) that lie outside T is very delicate. Sometimes a boundary value can be taken at such a point, sometimes not, depending on the geometry of $R(x_0; \rho; a, b)$ in a neighbourhood of the point. Because of this, and because we need the the set of boundary points relevant to the Dirichlet problem to be closed, a boundary value is specified at that point. Corresponding to the normal boundary of a circular cylinder, we define the *normal boundary* of a convex domain of revolution R to be

$$\partial_n R = \partial R \backslash T.$$

Note that $\partial_n R$ is compact.

The *Dirichlet Problem* on a convex domain of revolution R consists of showing that, for an arbitrary function $f \in C(\partial_n R)$, there is a function $u_f \in C(R \cup \partial_n R)$ that is a temperature on R and coincides with f on $\partial_n R$.

We show that this problem has a solution, except when the left hand derivative $\rho'_-(b) = -\infty$. That condition means that the hyperplane $\mathbb{R}^n \times \{b\}$ is tangental to $\partial_n R$ at points of the form (x, b) that lie outside T, which can mean that continuous boundary values are not taken at such points.

We use the *Perron-Wiener-Brelot method*, or PWB method. This is the same method that we shall later use to solve the generalized Dirichlet problem, which involves arbitrary open sets and general boundary functions. It is easier and clearer without those added complications, and we need the solutions for convex domains of revolution to prove several theorems about subtemperatures that we require for the general case.

The following concept and theorem form the basis of our approach.

DEFINITION 3.25. A nonempty family $\mathcal{F}$ of supertemperatures on an open set E, is called a *saturated family* if it satisfies the two conditions:

(a) if $v, w \in \mathcal{F}$, then $v \wedge w \in \mathcal{F}$;

(b) if $w \in \mathcal{F}$, D is a circular cylinder such that $\overline{D} \subseteq E$, and $\pi_D w$ is the function defined in Theorem 3.21, then $\pi_D w \in \mathcal{F}$.

3.3. THE DIRICHLET PROBLEM ON CONVEX DOMAINS OF REVOLUTION

THEOREM 3.26. *If $\mathcal{F}$ is a saturated family of supertemperatures on an open set E, and the function $u = \inf \mathcal{F}$ satisfies $u(p_0) > -\infty$ at some point $p_0 \in E$, then u is a temperature on $\Lambda(p_0; E)$.*

PROOF. Let q_0 be any point of E such that $u(q_0) > -\infty$. Let D be any circular cylinder such that $q_0 \in D$ and $\overline{D} \subseteq E$. For each supertemperature $w \in \mathcal{F}$, we let $\pi_D w$ be the function defined in Theorem 3.21, so that $\pi_D w$ is a supertemperature on E, and $\pi_D w \leq w$ on E. Since $\mathcal{F}$ is saturated we have $\pi_D w \in \mathcal{F}$, and therefore $u = \inf\{\pi_D w : w \in \mathcal{F}\}$. If $v, w \in \mathcal{F}$, then $v \wedge w \in \mathcal{F}$ because $\mathcal{F}$ is saturated, and so the family $\mathcal{F}$ is downward-directed. Furthermore, an application of the minimum principle on D shows that $\pi_D(v \wedge w) \leq \pi_D v \wedge \pi_D w$, and therefore the family $\{\pi_D w : w \in \mathcal{F}\}$ is also downward-directed. Since $\pi_D w$ is a temperature on D for all $w \in \mathcal{F}$, it follows from Theorem 1.36 that u is a temperature on $\Lambda(q_0; D)$.

Now let p_* be any point of $\Lambda(p_0; E)$, and let γ be a polygonal path in E that connects p_0 to p_*, along which the temporal variable is strictly decreasing. For each point $p = (x, t) \in \gamma$ and positive number c, we put

$$D(p; c) = B(x, c) \times]t - c, t + c[\quad \text{and} \quad \Lambda(p; c) = B(x, c) \times]t - c, t[= \Lambda(p; D(p; c)).$$

Since γ is a compact subset of the open set E, we can find $c_0 > 0$ such that $\overline{D}(p; c_0) \subseteq E$ for all $p \in \gamma$. We now let m be the integer such that the length of γ lies in the interval $]mc_0/2, (m+1)c_0/2]$. Since $u(p_0) > -\infty$, we know that u is a temperature on $\Lambda(p_0; c_0)$. The length of that portion of γ which is contained in $\Lambda(p_0; c_0)$ is at least c_0, and so there is a point $p_1 \in \gamma \cap \Lambda(p_0; c_0)$ such that the length of that portion of γ which lies between p_0 and p_1 is $c_0/2$. Since $u(p_1) > -\infty$, u is a temperature on $\Lambda(p_1; c_0)$. The length of γ contained in $\Lambda(p_1; c_0)$ is at least c_0, and so there is a point $p_2 \in \gamma \cap \Lambda(p_1; c_0)$ such that the length of γ between p_1 and p_2 is $c_0/2$. Repeating this argument m times, we find that there is a point $p_m \in \gamma$ such that u is a temperature on $\Lambda(p_m; c_0)$ and $p_* \in \Lambda(p_m; c_0)$. Thus u is a temperature on a neighbourhood of p_*, and hence on $\Lambda(p_0; E)$. □

The idea of the PWB method is as follows. Given $f \in C(\partial_n R)$, it is easy to find supertemperatures on R that have limiting values greater than or equal to f at all points of $\partial_n R$. If we can show that they form a saturated family, then their infimum will be a temperature, and with luck its boundary values will also be greater than or equal to f. Doing the same with subtemperatures, but with boundary values less than or equal to f, should give a temperature with similar boundary values. If the two temperatures are equal, the problem may be solved.

We note that the boundary maximum principle for hypotemperatures on a convex domain of revolution, takes the same form as it does on a circular cylinder (Corollary 3.14), with the same proof.

DEFINITION 3.27. Let R be a convex domain of revolution, and let $f \in C(\partial_n R)$. The *Upper Class* $\mathfrak{U}_f$, determined by f, consists of all supertemperatures v on R that are upper bounded and satisfy

$$\liminf_{p \to q} v(p) \geq f(q)$$

for all $q \in \partial_n R$.

Note that, by the boundary minimum principle, $v \geq \min f$ on R. Note also that, because $v \wedge (\max f)$ is also a supertemperature, by Corollary 3.18, the upper boundedness condition on v is no real restriction.

DEFINITION 3.28. The *Lower Class* $\mathfrak{L}_f$, determined by f, consists of all lower bounded subtemperatures u on R that satisfy
$$\limsup_{p \to q} u(p) \leq f(q)$$
for all $q \in \partial_n R$.

Note that neither class is empty, because $\mathfrak{U}_f$ contains the constant function $\max f$, and $\mathfrak{L}_f$ contains $\min f$.

DEFINITION 3.29. The *Upper PWB Solution* for f on R is the function U_f given by
$$U_f(p) = \inf\{v(p) : v \in \mathfrak{U}_f\},$$
and the *Lower PWB Solution* is given by
$$L_f(p) = \sup\{u(p) : u \in \mathfrak{L}_f\}.$$
Both functions are bounded.

DEFINITION 3.30. If $U_f = L_f$, and is a temperature on R, then we put $S_f = U_f$ and call it the *PWB Solution* for f on R.

We shall show that every function $f \in C(\partial_n R)$ has a PWB solution on R, and then investigate the boundary values of S_f to see if it actually solves the Dirichlet problem. First we show that, if the Dirichlet problem for f has a solution, then it is given by S_f.

LEMMA 3.31. *Let R be a convex domain of revolution, and let $f \in C(\partial_n R)$. If $u \in \mathfrak{L}_f$ and $v \in \mathfrak{U}_f$, then $u \leq v$ on R. Consequently $L_f \leq U_f$ on R.*

PROOF. The function u is a bounded subtemperature on R, and v is a bounded supertemperature, so that the difference $u - v$ is a subtemperature, by Corollary 3.19. Furthermore, whenever $q \in \partial_n R$, we have
$$\limsup_{p \to q}(u - v)(p) \leq \limsup_{p \to q} u(p) - \liminf_{p \to q} v(p) \leq 0,$$
and so it follows from the boundary maximum principle that $u \leq v$ on R. Thus any function $u \in \mathfrak{L}_f$ satisfies $u \leq U_f$, and therefore $L_f \leq U_f$. □

THEOREM 3.32. *Let R be a convex domain of revolution, and let $f \in C(\partial_n R)$. If there is a temperature u_f on R such that*
$$\lim_{p \to q} u_f(p) = f(q)$$
for all $q \in \partial_n R$, then f has a PWB-solution and it is u_f.

PROOF. By the boundary maximum principle, we have $\min f \leq u_f \leq \max f$ on R. Therefore $u_f \in \mathfrak{L}_f \cap \mathfrak{U}_f$, and so $U_f \leq u_f \leq L_f$. Since $L_f \leq U_f$ by Lemma 3.31, we deduce that $U_f = u_f = L_f$. Since u_f is a temperature on R, $S_f = u_f$. □

LEMMA 3.33. *Let R be a convex domain of revolution, and let $f \in C(\partial_n R)$. Then both L_f and U_f are temperatures on R.*

PROOF. Let $v, w \in \mathfrak{U}_f$. Then $v \wedge w$ is an upper bounded supertemperature on R, by Corollary 3.18, and
$$\liminf_{p \to q}(v \wedge w)(p) = (\liminf_{p \to q} v(p)) \wedge (\liminf_{p \to q} w(p)) \geq f(q)$$
for all $q \in \partial_n R$. Therefore $v \wedge w \in \mathfrak{U}_f$. Next, if $v \in \mathfrak{U}_f$ and D is a circular cylinder such that $\overline{D} \subseteq R$, then the function $\pi_D v$ of Theorem 3.21, is a supertemperature on R, is upper bounded on R, and satisfies
$$\liminf_{p \to q} \pi_D v(p) = \liminf_{p \to q} v(p) \geq f(q)$$
for all $q \in \partial_n R$. Therefore $\pi_D v$ belongs to $\mathfrak{U}_f$. Thus $\mathfrak{U}_f$ is a saturated family of supertemperatures on R. Furthermore, since $v \geq \min f$ for every $v \in \mathfrak{U}_f$, it follows from Theorem 3.26 that U_f is a temperature on R. Dually, L_f is also a temperature. □

DEFINITION 3.34. Let R be a convex domain of revolution, and let $f \in C(\partial_n R)$. If f has a PWB solution on R, we say that f is *resolutive*.

LEMMA 3.35. *Let R be a convex domain of revolution, let $f, g \in C(\partial_n R)$, and let $\alpha \in \mathbb{R}$.*
(a) The constant function α is resolutive, and $S_\alpha = \alpha$ on R.
(b) $U_{f+\alpha} = U_f + \alpha$ and $L_{f+\alpha} = L_f + \alpha$. If f is resolutive, then $f + \alpha$ is resolutive and $S_{f+\alpha} = S_f + \alpha$.
(c) If $\alpha > 0$, then $U_{\alpha f} = \alpha U_f$ and $L_{\alpha f} = \alpha L_f$. If f is resolutive, then αf is resolutive and $S_{\alpha f} = \alpha S_f$.
(d) If $f \leq g$, then $U_f \leq U_g$ and $L_f \leq L_g$.
(e) $U_{-f} = -L_f$. If f is resolutive, then $-f$ is resolutive and $S_{-f} = -S_f$.
(f) $U_{f+g} \leq U_f + U_g$ and $L_{f+g} \geq L_f + L_g$. If f and g are resolutive, then $f + g$ is resolutive and $S_{f+g} = S_f + S_g$.

PROOF. (a) This is a special case of Theorem 3.32.
(b) If $v \in \mathfrak{U}_f$ then $v + \alpha \in \mathfrak{U}_{f+\alpha}$, and conversely. So $U_{f+\alpha} = U_f + \alpha$. Similarly, $L_{f+\alpha} = L_f + \alpha$. If f is resolutive, then $L_f = U_f$ and is a temperature, so that $U_{f+\alpha} = U_f + \alpha = L_f + \alpha = L_{f+\alpha}$ and is also a temperature.
(c) If $v \in \mathfrak{U}_f$ then $\alpha v \in \mathfrak{U}_{\alpha f}$, and conversely. So $U_{\alpha f} = \alpha U_f$. Similarly, $L_{\alpha f} = \alpha L_f$. If f is resolutive, then $L_f = U_f$ and is a temperature, so that $U_{\alpha f} = \alpha U_f = \alpha L_f = L_{\alpha f}$ and is also a temperature.
(d) If $v \in \mathfrak{U}_g$, then $v \in \mathfrak{U}_f$, so that $\mathfrak{U}_f$ is the infimum over a more inclusive class of functions, and so $U_f \leq U_g$. Similarly, if $u \in \mathfrak{L}_f$ then $u \in \mathfrak{L}_g$, so that $L_f \leq L_g$.
(e) If $v \in \mathfrak{U}_f$ then $-v \in \mathfrak{L}_{-f}$, and conversely. So $U_f = -L_{-f}$. Similarly, $L_f = -U_{-f}$. If f is resolutive, then $L_f = U_f$ and is a temperature, so that $-U_{-f} = -L_{-f}$ and is also a temperature.
(f) If $v \in \mathfrak{U}_f$ and $w \in \mathfrak{U}_g$, then Corollary 3.19 implies that $v + w \in \mathfrak{U}_{f+g}$. So for each function $w \in \mathfrak{U}_g$ we have $U_f + w \geq U_{f+g}$. Therefore $U_f + U_g \geq U_{f+g}$. Now the inequality $L_{f+g} \geq L_f + L_g$ follows from part (e). If f and g are resolutive, then (using Lemma 3.31)
$$S_f + S_g = L_f + L_g \leq L_{f+g} \leq U_{f+g} \leq U_f + U_g = S_f + S_g,$$
which shows that $L_{f+g} = U_{f+g} = S_f + S_g$. □

In order to show that every function in $C(\partial_n R)$ is resolutive, we first obtain a class of resolutive functions that admits every real continuous function as a limit of a uniformly convergent sequence in the class. Then we show that the limit of a uniformly convergent sequence of resolutive functions is itself resolutive.

To accomplish the first part of our plan, we use the Stone-Weierstrass theorem. In this result, there is a class of functions $\mathcal{F}$ contained in a space $C(K)$ which the term *separates the points of K* is applied to. This means that, whenever $p, q \in K$ and $p \neq q$, there is some function $f \in \mathcal{F}$ such that $f(p) \neq f(q)$. Specialized to the case of $\mathbb{R}^{n+1}$, the Stone-Weierstrass theorem reads:

Let K be a compact subset of $\mathbb{R}^{n+1}$, and let $\mathcal{F}$ be a linear subspace of $C(K)$ which contains the constant function 1, separates the points of K, and contains $f \vee g$ whenever it contains f and g. Then $\mathcal{F}$ is dense in $C(K)$ with respect to the supremum norm.

In our case $K = \partial_n R$, and $\mathcal{F}$ has to be a class of resolutive functions. We already know that the constant functions are resolutive, but they do not separate points. The last condition, that $\mathcal{F}$ contains $f \vee g$ whenever it contains f and g, suggests that boundary values of subtemperatures may be useful. This is, indeed, the case.

LEMMA 3.36. *If R is a convex domain of revolution, and w is a function in $C(R \cup \partial_n R)$ that is a subtemperature on R, then the restriction of w to $\partial_n R$ is resolutive.*

PROOF. Let f denote the restriction of w to $\partial_n R$. By Lemma 3.33, the lower PWB solution L_f is a temperature on R. Furthermore $u \in \mathfrak{L}_f$, so that $u \leq L_f$ on R. Therefore
$$\liminf_{p \to q} L_f(p) \geq \lim_{p \to q} u(p) = f(q)$$
for all $q \in \partial_n R$, so that $L_f \in \mathfrak{U}_f$, and hence $L_f \geq U_f$. But we always have $L_f \leq U_f$, by Lemma 3.31, so f is resolutive. $\square$

LEMMA 3.37. *Let R be a convex domain of revolution, and let $\{f_j\}$ be a sequence of resolutive functions in $C(\partial_n R)$. If $\{f_j\}$ converges uniformly on $\partial_n R$ to a function f, then f is resolutive and the sequence $\{S_{f_j}\}$ converges uniformly on R to S_f.*

PROOF. Note that $f \in C(\partial_n R)$, so that U_f and L_f are temperatures on R, by Lemma 3.33. Given $\epsilon > 0$, we can find a number N such that $f_j - \epsilon < f < f_j + \epsilon$ on $\partial_n R$ whenever $j > N$. Therefore, by Lemma 3.35,
$$U_{f_j} - \epsilon = U_{f_j - \epsilon} \leq U_f \leq U_{f_j + \epsilon} = U_{f_j} + \epsilon$$
on R whenever $j > N$. Hence the sequence $\{S_{f_j}\} = \{U_{f_j}\}$ converges uniformly on R to U_f. A similar argument with the lower solutions shows that the sequence $\{S_{f_j}\} = \{L_{f_j}\}$ converges uniformly on R to L_f. Hence $U_f = L_f$, and the lemma is established. $\square$

THEOREM 3.38. *If R is a convex domain of revolution, then every function in $C(\partial_n R)$ is resolutive.*

PROOF. Let $\mathcal{G}$ denote the class of those functions in $C(R \cup \partial_n R)$ that are also subtemperatures on R, let $\mathcal{D}$ denote the class of differences $u - v$ of functions in $\mathcal{G}$, and let $\mathcal{F}$ denote the class of restrictions to $\partial_n R$ of the functions in $\mathcal{D}$. Then $\mathcal{F}$

is a linear subspace of $C(\partial_n R)$ that contains the constant functions. By Lemmas 3.36 and 3.35, the restrictions to $\partial_n R$ of the functions in $\mathcal{G}$ are resolutive, and the functions in $\mathcal{F}$ are all resolutive. Furthermore, for any point (x_0, t_0) such that $\overline{R} \subseteq \mathbb{R}^n \times]t_0, +\infty[$, the class $\mathcal{G}$ contains the function $(x,t) \mapsto -W(x - x_0, t - t_0)$, and so $\mathcal{F}$ separates points of $\partial_n R$. Finally, if $u, v \in \mathcal{G}$ then Corollaries 3.18 and 3.19 imply that $u \wedge v, u + v \in \mathcal{G}$, so that if $u_1, u_2, v_1, v_2 \in \mathcal{G}$ the function

$$(u_1 - v_1) \vee (u_2 - v_2) = u_1 + u_2 - (u_2 + v_1) \wedge (u_1 + v_2) \in \mathcal{D}.$$

Thus $f \vee g \in \mathcal{F}$ whenever $f, g \in \mathcal{F}$. It now follows from the Stone-Weierstrass theorem that $\mathcal{F}$ is dense in $C(\partial_n R)$ with respect to the supremum norm. So every function in $C(\partial_n R)$ can be expressed as the uniform limit of a sequence in $\mathcal{F}$. Since every function in $\mathcal{F}$ is resolutive, it follows from Lemma 3.37 that every function in $C(\partial_n R)$ is resolutive. $\square$

3.4. Boundary Behaviour of the PWB Solution

We now show that, if R is a convex domain of revolution satisfying a certain auxiliary condition, then for any function $f \in C(\partial_n R)$, the PWB solution S_f solves the Dirichlet problem for f on R. The extra condition cannot be omitted, but we shall see later that it can be weakened.

THEOREM 3.39. *Let $R = \{(x,t) \in \mathbb{R}^{n+1} : |x - x_0| < \rho(t), a < t < b\}$ be a convex domain of revolution such that $\rho'_-(b) > -\infty$, and let $f \in C(\partial_n R)$. Then the PWB solution S_f for f on R satisfies*

$$\lim_{p \to q} S_f(p) = f(q)$$

for all $q \in \partial_n R$.

PROOF. Because ρ is concave, it follows from Lemma 1.9 that, given any point $(y_0, s_0) \in \partial_n R$, we can find a hyperplane H such that $(y_0, s_0) \in H$ and $R \cap H = \emptyset$. On the opposite side of H to R, we position a reflected heat ball

$$\Omega^*(\eta_0, \sigma_0; c_0) = \{(x,t) : W(x - \eta_0, t - \sigma_0) > \tau(c_0)\},$$

with $\sigma_0 < s_0$, so that it is tangential to H at (y_0, s_0). This is possible unless $s_0 = b$ and $H = \mathbb{R}^n \times \{b\}$. Our condition that $\rho'_-(b) > -\infty$ implies that, if $(y_0, s_0) \in \partial_n R$ and $s_0 = b$, we can find an H that is not equal to $\mathbb{R}^n \times \{b\}$. The function w, defined on R by

$$w(x,t) = \tau(c_0) - W(x - \eta_0, t - \sigma_0),$$

is a positive temperature on R such that

$$\lim_{(x,t) \to (y_0, s_0)} w(x,t) = 0,$$

and for any neighbourhood N of (y_0, s_0),

$$\inf_{R \setminus N} w > 0.$$

Given $\epsilon > 0$, we put $A = f(y_0, s_0) + \epsilon$. Since f is continuous at (y_0, s_0), we can find a neighbourhood N of (y_0, s_0) such that $f < A$ on $N \cap \partial_n R$. Since $\inf_{R \setminus N} w > 0$, we can choose a positive number α such that $\alpha \inf_{R \setminus N} w > \max f - A$. We put

$u = A + \alpha w$ on R, and note that u is a lower bounded temperature on R. Whenever $(y, s) \in (\partial_n R) \backslash N$ we have

$$\liminf_{(x,t) \to (y,s)} u(x, t) \geq A + \alpha \inf_{R \backslash N} w > \max f \geq f(y, s);$$

and whenever $(y, s) \in (\partial_n R) \cap N$ we have

$$\liminf_{(x,t) \to (y,s)} u(x, t) \geq A > f(y, s).$$

Therefore the function $v = u \wedge (\max f)$, which is a supertemperature on R by Corollary 3.18, belongs to the upper class $\mathfrak{U}_f$. Hence the upper PWB solution $U_f \leq v$ on R, which implies that

$$\limsup_{(x,t) \to (y_0,s_0)} U_f(x,t) \leq \limsup_{(x,t) \to (y_0,s_0)} u(x,t) = A + \alpha \lim_{(x,t) \to (y_0,s_0)} w(x,t) = A.$$

Hence, since f is resolutive by Theorem 3.38,

$$\limsup_{(x,t) \to (y_0,s_0)} S_f(x,t) \leq f(y_0, s_0).$$

A similar inequality holds with f replaced by $-f$, and so it follows from Lemma 3.35 that

$$\liminf_{(x,t) \to (y_0,s_0)} S_f(x,t) = -\limsup_{(x,t) \to (y_0,s_0)} S_{-f}(x,t) \geq f(y_0, s_0).$$

Hence $S_f(x,t) \to f(y_0, s_0)$ as $(x, t) \to (y_0, s_0)$. □

REMARK 3.40. Theorem 3.39 shows that, if $\kappa \in]0, +\infty[$ and R is the cone with vertex (x_0, b) given by $\{(x, t) : |x - x_0| < \kappa(b - t), a < t < b\}$, then the Dirichlet problem is solvable on R for any function $f \in C(\partial_n R)$, even though $\partial_n R = \partial R$.

COROLLARY 3.41. *Let $R = \{(x,t) \in \mathbb{R}^{n+1} : |x - x_0| < \rho(t), a < t < b\}$ be any convex domain of revolution, let $a < c < b$, and let $C = R \cap (\mathbb{R}^n \times]a, c[)$. If $f \in C(\partial_n R)$, then the PWB solution S_f for f on R satisfies $\lim_{p \to q} S_f(p) = f(q)$ for all $q \in \partial_n C$. Furthermore, the restriction to C of S_f is the PWB solution on C for the restriction of f to $\partial_n C$.*

PROOF. We choose d such that $c < d < b$, and let D denote the convex domain of revolution $\{(x,t) : |x - x_0| < \rho(t), a < t < d\}$. Since ρ is a concave function on a neighbourhood of d, we have $\rho'_-(d) > -\infty$ by Lemma 1.9. Therefore, if S_f^D denotes the PWB solution on D for the restriction of f to $\partial_n D$, Theorem 3.39 shows that

$$\lim_{p \to q, \, p \in D} S_f^D(p) = f(q)$$

for all $q \in \partial_n D$. Furthermore, $\min f \leq S_f^D \leq \max f$ on D. Now we define functions u and v on R by putting $u(p) = v(p) = S_f^D(p)$ for all $p \in R \cap \overline{C}$, and $u(p) = \min f$, $v(p) = \max f$ for all $p \in R \backslash \overline{C}$. Then u is a bounded subtemperature on R that satisfies $\limsup_{p \to q} u(p) \leq f(q)$ for all $q \in \partial_n R$, so that $u \in \mathfrak{L}_f$. Similarly $v \in \mathfrak{U}_f$. Therefore $u \leq S_f \leq v$ on R, which implies that

$$\lim_{p \to q, \, p \in \overline{C}} S_f(p) = f(q)$$

for all $q \in \partial_n C$. Since c is arbitrary, it follows that $\lim_{p \to q} S_f(p) = f(q)$ for all $q \in \partial_n C$. So the restriction to C of S_f solves the Dirichlet problem on C, and hence is the PWB solution on C for the restriction of f to $\partial_n C$, by Theorem 3.32. □

We now show that the conclusion of Theorem 3.39 may fail if the condition $\rho'_-(b) > -\infty$ is omitted. In particular, the result does not hold for a heat ball or a modified heat ball.

THEOREM 3.42. *Let $R = \{(x,t) : |x - x_0| < \rho(t), a < t < t_0\}$ be a convex domain of revolution such that $\rho(t_0) = 0$, and which contains a heat ball $\Omega(x_0, t_0; c)$ for some c. Let $f \in C(\partial R)$ and satisfy $0 = f(x_0, t_0) < f(y, s)$ for all $(y, s) \in \partial R$ such that $s < t_0$. Then the Dirichlet problem on R is not solvable for f.*

PROOF. By Theorem 3.38, f is resolutive. By Theorem 3.32, the PWB solution S_f solves the Dirichlet problem for f on R, if a solution exists. Suppose that S_f can be continuously extended by f to $\overline{R}$, and choose d such that $0 < d < c$. Since S_f is a temperature on R, and $\overline{\Omega}(x_0, t_0; d)\setminus\{(x_0, t_0)\} \subseteq \Omega(x_0, t_0; c) \subseteq R$, we can find $\alpha > 0$ such that
$$S_f(x_0, t_0 - l) = \mathcal{M}(S_f; x_0, t_0 - l; d)$$
whenever $0 < l < \alpha$. Letting $l \to 0+$, we deduce that
$$0 = \lim_{l\to 0} S_f(x_0, t_0 - l) = \lim_{l\to 0} \mathcal{M}(S_f; x_0, t_0 - l; d) = \mathcal{M}(S_f; x_0, t_0; d),$$
because S_f is continuous on $\overline{\Omega}(x_0, t_0; c)$. Since $0 = \min f \leq S_f$ on R, it follows that $S_f(x, t) = 0$ for all $(x, t) \in \partial\Omega(x_0, t_0; d)$, and hence for all $(x, t) \in R$ by the strong minimum principle. But, by Corollary 3.41,
$$\lim_{(x,t)\to (y,s)} S_f(x,t) = f(y,s) > 0$$
for all $(y, s) \in \partial R$ such that $s < t_0$. This contradiction shows that the Dirichlet problem on R is not solvable for f. □

3.5. Characterizations of Hypotemperatures and Subtemperatures

Our characterizations involve the mean values $\mathcal{M}$ and $\mathcal{V}$. For their proofs, we require more information about the means $\mathcal{M}$. This is given in the following lemma.

LEMMA 3.43. *Let u be a function defined on the heat sphere $\partial\Omega(x_0, t_0; d)$ for which the mean $\mathcal{M}(u; x_0, t_0; d)$ exists, and let $c > 0$. If $\alpha = d/c$, and the function v is defined on the heat sphere $\partial\Omega(x_0, t_0; c)$ by*
$$v(\xi, \eta) = u((1 - \sqrt{\alpha})x_0 + \sqrt{\alpha}\xi, (1 - \alpha)t_0 + \alpha\eta),$$
then
$$(3.3) \qquad \mathcal{M}(v; x_0, t_0; c) = \mathcal{M}(u; x_0, t_0; d).$$

PROOF. To prove (3.3), we need to investigate the surface measure σ_d on the heat sphere $\partial\Omega(x_0, t_0; d)$. We put $y = x_0 - x$, $s = t_0 - t$, and $r = |y|$. Then, in view of (1.17), σ_d has the form
$$d\sigma_d = \sqrt{1 + \left(\frac{\partial r}{\partial s}\right)^2}\, |\Psi'_d|\, d\theta_1 ... d\theta_{n-1} ds,$$

where $\theta_1,...,\theta_{n-1}, r$ are hyperspherical coordinates in $\mathbb{R}^n$, $r = \sqrt{2ns\log(d/s)}$, and $|\Psi_d'| = \pm r^{n-1}\cos\theta_2 \cos^2\theta_3 ... \cos^{n-2}\theta_{n-1}$. A routine calculation shows that

$$\sqrt{1+\left(\frac{\partial r}{\partial s}\right)^2} = \frac{1}{2rs}\sqrt{4r^2s^2 + (r^2-2ns)^2}.$$

Therefore, if
$$Q(y,s) = \frac{|y|^2}{\sqrt{4|y|^2s^2+(|y|^2-2ns)^2}}$$

is the kernel for the fundamental means over heat spheres $\mathcal{M}$, given in Section 1.2, we have
$$d\sigma_d = \frac{|y|}{2sQ(y,s)}|\Psi_d'|\,d\theta_1...d\theta_{n-1}ds.$$

Given $\alpha = d/c$, we now make the change of variables
$$x_0 - x = \sqrt{\alpha}(x_0-\xi), \qquad t_0 - t = \alpha(t_0-\eta),$$
which transforms $\partial\Omega(x_0,t_0;d)$ to $\partial\Omega(x_0,t_0;c)$ and gives
$$|\Psi_d'| = \pm\left(2n(t_0-t)\log\frac{d}{t_0-t}\right)^{\frac{n-1}{2}}\cos\theta_2\cos^2\theta_3...\cos^{n-2}\theta_{n-1} = \alpha^{\frac{n-1}{2}}|\Psi_c'|.$$

Hence

$\mathcal{M}(u;x_0,t_0;d)$
$$= \tau(d)\int_{\partial\Omega(d)} Q(x_0-x,t_0-t)u(x,t)\,d\sigma_d$$
$$= \tau(d)\int_{\partial\Omega(d)} \frac{|x_0-x|}{2(t_0-t)}u(x,t)|\Psi_d'|\,d\theta_1...d\theta_{n-1}dt$$
$$= \tau(c)\int_{\partial\Omega(c)} \frac{|x_0-\xi|}{2(t_0-\eta)}u(x,t)|\Psi_c'|\,d\theta_1...d\theta_{n-1}d\eta$$
$$= \tau(c)\int_{\partial\Omega(c)} Q(x_0-\xi,t_0-\eta)u((1-\sqrt{\alpha})x_0+\sqrt{\alpha}\xi,(1-\alpha)t_0+\alpha\eta)\,d\sigma_c$$
$$= \mathcal{M}(v;x_0,t_0;c).$$

$\square$

Lemma 3.43 is required only for the proof of the following crucial result.

LEMMA 3.44. *Let u belong to the class $C(\overline{\Omega}(x_0,t_0;c))$ and be a temperature on $\Omega(x_0,t_0;c)$. Then*
$$(3.4) \qquad u(x_0,t_0) = \mathcal{M}(u;x_0,t_0;d)$$
whenever $0 < d \leq c$.

PROOF. Suppose first that $d < c$. Since u is a temperature on $\Omega(x_0,t_0;c)$, and $\overline{\Omega}(x_0,t_0;d)\backslash\{(x_0,t_0)\} \subseteq \Omega(x_0,t_0;c)$, we can find $\beta > 0$ such that
$$u(x_0,t_0-l) = \mathcal{M}(u;x_0,t_0-l;d)$$
whenever $0 < l < \beta$. Letting $l \to 0+$, we deduce that
$$u(x_0,t_0) = \lim_{l\to 0+}\mathcal{M}(u;x_0,t_0-l;d) = \mathcal{M}(u;x_0,t_0;d)$$

because $u \in C(\overline{\Omega}(x_0, t_0; c))$. So (3.4) holds if $0 < d < c$.

To prove that (3.4) holds when $d = c$, we use Lemma 3.43. We take $d < c$, and put $\alpha = d/c$ and $v(\xi, \eta) = u((1 - \sqrt{\alpha})x_0 + \sqrt{\alpha}\xi, (1 - \alpha)t_0 + \alpha\eta)$. Then v is defined on $\partial\Omega(x_0, t_0; c)$, and $\mathcal{M}(v; x_0, t_0; c) = \mathcal{M}(u; x_0, t_0; d)$. As $d \to c-$ we have $\alpha \to 1-$ and $v(\xi, \eta) \to u(\xi, \eta)$. Therefore

$$u(x_0, t_0) = \lim_{d \to c-} \mathcal{M}(u; x_0, t_0; d) = \lim_{\alpha \to 1-} \mathcal{M}(v; x_0, t_0; c) = \mathcal{M}(u; x_0, t_0; c).$$

□

Our characterizations are based on the following variant of Theorem 3.17, in which circular cylinders are replaced by convex domains of revolution.

THEOREM 3.45. *Let w be an upper finite and upper semicontinuous function on an open set E. Consider the following property: Whenever R is a convex domain of revolution such that $\overline{R} \subseteq E$, and v is a function in $C(\overline{R})$ that is a temperature on R and satisfies $v \geq w$ on $\partial_n R$, then $v \geq w$ on $\overline{R}$.*

The property holds if and only if w is a hypotemperature on E.

PROOF. If w is a hypotemperature on E, then the proof that the property holds is similar to the first part of the proof of Theorem 3.17. The converse follows from Theorem 3.17. □

Another crucial part of our approach is to show that functions which satisfy the definition of a hypotemperature with $\mathcal{L}$ replaced by $\mathcal{M}$ or $\mathcal{V}$, also satisfy the maximum principle. This generalizes Theorem 1.21.

THEOREM 3.46. *Let $R = \{(x, t) : |x - x_0| < \rho(t), a < t < b\}$ be a convex domain of revolution, and let w be an upper finite and upper semicontinuous function on R. If, given any point $(x, t) \in R$ and $\epsilon > 0$, we can find a positive number $c < \epsilon$ such that either*

(a) $w(x, t) \leq \mathcal{M}(w; x, t; c)$,

or

(b) $w(x, t) \leq \mathcal{V}(w; x, t; c)$,

or, for some integer $m \geq 1$,

(c) $w(x, t) \leq \mathcal{V}_m(w; x, t; c)$,

holds, then u satisfies the maximum principle on R. That is, if there is a point $(x_0, t_0) \in R$ such that $w(x_0, t_0) \geq w(x, t)$ whenever $(x, t) \in R$ and $t < t_0$, then $w(x_0, t_0) = w(x, t)$ for all such points (x, t). Consequently, if

$$\limsup_{(x,t) \to (y,s)} w(x, t) \leq A$$

for all $(y, s) \in \partial_n R$, then $w(x, t) \leq A$ for all $(x, t) \in R$.

PROOF. Suppose that there is a point $(x_0, t_0) \in R$ such that $w(x_0, t_0) \geq w(x, t)$ whenever $(x, t) \in R$ and $t < t_0$. If $w(x_0, t_0) = -\infty$ the result is trivally true, so we suppose otherwise. We put $M = w(x_0, t_0)$, and let (x_1, t_1) be any point of R such that $t_1 < t_0$. Join (x_0, t_0) and (x_1, t_1) with a closed line segment γ, and put

$$S = \{s : \text{there is a point } (y, s) \in \gamma \text{ with } w(y, s) = M\}.$$

Then $S \neq \emptyset$ because $t_0 \in S$, and S is lower bounded by t_1. Put $s^* = \inf S$. If condition (a) holds, we can find a number $c < t_0 - t_1$ such that
$$M = w(x_0, t_0) \leq \mathcal{M}(w; x_0, t_0; c) \leq \mathcal{M}(M; x_0, t_0; c) = M.$$
This implies that $w = M$ almost everywhere on $\partial\Omega(x_0, t_0; c)$, and so the upper semicontinuity of w shows that $w \equiv M$ on $\partial\Omega(x_0, t_0; c)$. Since $c < t_0 - t_1$, the set $\gamma \cap (\partial\Omega(x_0, t_0; c)) \neq \emptyset$, so that there is a point $s_1 \in S$ such that $s_1 < t_0$. Similar arguments are valid if conditions (b) or (c) are satisfied. Hence $s^* < t_0$.

Suppose that $t_1 < s^* < t_0$. There is a sequence of points $\{(z_k, r_k)\}$ on γ such that $w(z_k, r_k) = M$ for all k, and $r_k \to s^*$ as $k \to \infty$. The upper semicontinuity of w now implies that there is a point (y^*, s^*) on γ such that $u(y^*, s^*) = M$. If condition (a) holds, we can find $c < s^* - t_1$ such that $w \equiv M$ on $\partial\Omega(y^*, s^*; c)$, and therefore a point $s_2 \in S$ such that $s_2 < s^*$. Similar arguments are valid if conditions (b) or (c) are satisfied, so we have a contradiction. Hence $s^* = t_1$, and $w(x_1, t_1) = M$ by similar reasoning to that at the beginning of this paragraph. This proves the first part of the theorem.

For the second part, we extend w to $R \cup \partial_n R$ by putting
$$w(y, s) = \limsup_{(x,t) \to (y,s)} w(x, t) \leq A$$
for all $(y, s) \in \partial_n R$. Given any α such that $a < \alpha < b$, we let R_α denote the set $\{(x, t) : |x - x_0| < \rho(t), a < t < \alpha\}$. Then w is upper semicontinuous and upper finite on $\overline{R}_\alpha$, and so has a maximum value M_α, in view of Lemma 3.4. We choose a point $(x', t') \in \overline{R}_\alpha$ such that $w(x', t') = M_\alpha$. If $(x', t') \in R$, then the first part of the theorem shows that $w(x, t) = M_\alpha$ for all $(x, t) \in \overline{R}$ such that $t \leq t'$. So there is no loss of generality in assuming that $(x', t') \in \partial_n R$, which implies that $M_\alpha \leq A$. Since this holds for all α, we have $w \leq A$ on R, as required. $\square$

We now work towards our characterization of hypotemperatures which uses the fundamental means $\mathcal{M}$. A similar characterization using the volume means $\mathcal{V}$ follows. We extract part of the proof as a lemma, because it is also needed for the proofs of subsequent theorems.

LEMMA 3.47. *Let w be an upper finite and upper semicontinuous function on an open set E. Let $\mathfrak{R}$ denote the class of convex domains of revolution R, for which both $\partial_n R = \partial R$ and the Dirichlet problem on R has a solution for every $f \in C(\partial R)$. Consider the following property: Whenever $R \in \mathfrak{R}$ is such that $\overline{R} \subseteq E$, and v is a function in $C(\overline{R})$ that is a temperature on R and satisfies $v \geq w$ on ∂R, then $v \geq w$ on $\overline{R}$.*

If the stated property holds, then the inequalities
$$w(p) \leq \mathcal{M}(w; p; d) \leq \mathcal{M}(w; p; c)$$
hold whenever $0 < d \leq c$ and $\overline{\Omega}(p; c) \subseteq E$.

PROOF. Let $\Omega(x_0, t_0; c)$ be any heat ball whose closure is contained in E. Recall that
$$\Omega(x_0, t_0; c) = \{(x, t) : |x - x_0| < \phi(t), \, t_0 - c < t < t_0\}$$
is a convex domain of revolution with
$$\phi(t) = \sqrt{2n(t_0 - t) \log \frac{c}{t_0 - t}},$$

3.5. CHARACTERIZATIONS OF HYPOTEMPERATURES AND SUBTEMPERATURES

and note that
$$\max\{\phi(t) : t_0 - c < t < t_0\} = \phi\left(t_0 - \frac{c}{e}\right).$$

Let k be a positive integer such that $1/k < c/e$, let $r = \lambda_k(t)$ be the equation of the tangent line to the curve $r = \phi(t)$ at the point $t = t_0 - \frac{1}{k}$, and let b_k denote the zero of λ_k. We put

$$\rho_k(t) = \begin{cases} \phi(t) & \text{if } t_0 - c \leq t \leq t_0 - \frac{1}{k}, \\ \lambda_k(t) & \text{if } t_0 - \frac{1}{k} \leq t \leq b_k, \end{cases}$$

and let
$$R_k = \{(x,t) : |x - x_0| < \rho_k(t),\ t_0 - c < t < b_k\}.$$

Lemma 1.9 shows that any concave curve lies below its tangent, and so each domain R_k contains $\Omega(x_0, t_0; c)$. Furthermore, for each k we have $\rho'_k(b_k) = \phi'(t_0 - \frac{1}{k}) > -\infty$, so that $R_k \in \mathfrak{R}$ in view of Theorem 3.39. Note that, if $t_0 - c \leq t \leq t_0 - \frac{1}{k}$, then the point (x,t) belongs to $\partial\Omega(x_0, t_0; c)$ if and only if it belongs to ∂R_k.

The closures $\overline{R}_k$ form a contracting sequence of sets whose intersection is the closed heat ball $\overline{\Omega}(x_0, t_0; c)$, and so there is a number k_0 such that $\overline{R}_k \subseteq E$ for all $k > k_0$. For each $k > k_0$, the function w is upper semicontinuous and upper bounded on ∂R_k, and hence we can find a decreasing sequence $\{\psi_j^{(k)}\}$ in $C(\partial R_k)$ that tends pointwise to w on ∂R_k. For each j, we put $u_j^{(k)}$ equal to the PWB solution for $\psi_j^{(k)}$ on R_k, and $u_j^{(k)} = \psi_j^{(k)}$ on ∂R_k. Then each function $u_j^{(k)} \in C(\overline{R}_k)$, by Theorem 3.39, and is a temperature on R_k. In particular, each function $u_j^{(k)} \in C(\overline{\Omega}(x_0, t_0; c))$ and is a temperature on $\Omega(x_0, t_0; c)$. Therefore, by Lemma 3.44,

$$u_j^{(k)}(x_0, t_0) = \mathcal{M}(u_j^{(k)}; x_0, t_0; d)$$

whenever $0 < d \leq c$. Furthermore, by the stated property, $w \leq u_j^{(k)}$ on $\overline{R}_k$ for all j and k. Since the sequence $\{\psi_j^{(k)}\}$ is decreasing on ∂R_k, the maximum principle shows that the sequence $\{u_j^{(k)}\}$ is also decreasing. Put $v_k = \lim_{j\to\infty} u_j^{(k)} \geq w$ on $\overline{R}_k$, for each k. Then, whenever $0 < d \leq c$, we have

$$v_k(x_0, t_0) = \lim_{j\to\infty} \mathcal{M}(u_j^{(k)}; x_0, t_0; d) = \mathcal{M}(v_k; x_0, t_0; d),$$

by Lebesgue's monotone convergence theorem.

We need to show that the sequence $\{v_k\}$ is decreasing on $\overline{\Omega}(x_0, t_0; c)$, in order to apply the monotone convergence theorem again. Let $k > k_0$. Each function $u_j^{(k)}$ belongs to $C(\overline{R}_k) \subseteq C(\overline{R}_{k+1})$, satisfies $u_j^{(k)} \geq w$ on $\overline{R}_k \supseteq \overline{R}_{k+1}$, and is a temperature on $R_k \supseteq R_{k+1}$. Given any $\epsilon > 0$ and positive integer J, the sequence $\{\psi_j^{(k+1)}\}$ decreases to the limit $w < u_J^{(k)} + \epsilon$ on ∂R_{k+1}. Therefore the sequence of sets $\{S_j\}$, defined by

$$S_j = \{q \in \partial R_{k+1} : \psi_j^{(k+1)}(q) < u_J^{(k)}(q) + \epsilon\}$$

is expanding to the union ∂R_{k+1}. Both $\psi_j^{(k+1)}$ and $u_J^{(k)}$ are continuous on ∂R_{k+1}, so that each set S_j is relatively open. Therefore, since ∂R_{k+1} is compact, there is a number j_0 such that $S_j = \partial R_{k+1}$ whenever $j > j_0$. Thus $\psi_j^{(k+1)}(q) < u_J^{(k)}(q) + \epsilon$ for all $q \in \partial R_{k+1}$ if $j > j_0$. This implies, using the maximum principle, that $u_j^{(k+1)}(q) < u_J^{(k)}(q) + \epsilon$ for all $q \in \overline{R}_{k+1}$ if $j > j_0$. Therefore $v_{k+1} < u_J^{(k)} + \epsilon$ on $\overline{R}_{k+1}$

for any $\epsilon > 0$ and positive integer J, and so $v_{k+1} \leq v_k$. Hence the sequence $\{v_k\}$ is decreasing on $\overline{\Omega}(x_0,t_0;c)$. Put $v = \lim_{k\to\infty} v_k \geq w$ on $\overline{\Omega}(x_0,t_0;c)$. Whenever $(x,t) \in \partial\Omega(x_0,t_0;c)$ and $t \leq t_0 - \frac{1}{k}$, we have

$$v_k(x,t) = \lim_{j\to\infty} \psi_j^{(k)}(x,t) = w(x,t),$$

so that $v(x,t) = w(x,t)$ for all $(x,t) \in \partial\Omega(x_0,t_0;c)\backslash\{(x_0,t_0)\}$. Hence Lebesgue's monotone convergence theorem shows that

$$w(x_0,t_0) \leq v(x_0,t_0) = \lim_{k\to\infty} \mathcal{M}(v_k;x_0,t_0;d) = \mathcal{M}(v;x_0,t_0;d)$$

whenever $0 < d \leq c$. It follows that

$$\mathcal{M}(w;x_0,t_0;d) \leq \mathcal{M}(v;x_0,t_0;d) = v(x_0,t_0) = \mathcal{M}(v;x_0,t_0;c) = \mathcal{M}(w;x_0,t_0;c)$$

whenever $0 < d \leq c$. This proves the lemma. □

The next theorem extends Theorem 1.6 to general hypotemperatures.

THEOREM 3.48. *Let w be an upper finite and upper semicontinuous function on an open set E. Suppose that, given any point $p \in E$ and $\epsilon > 0$, we can find a positive number $c < \epsilon$ such that the inequality $w(p) \leq \mathcal{M}(w;p;c)$ holds. Then w is a hypotemperature on E.*

Conversely, if w is a hypotemperature on E and $p \in E$, then the inequality $w(p) \leq \mathcal{M}(w;p;c)$ holds for all $c > 0$ such that $\overline{\Omega}(p;c) \subseteq E$.

PROOF. Suppose that, given any point $p \in E$ and $\epsilon > 0$, we can find a positive number $c < \epsilon$ such that $w(p) \leq \mathcal{M}(w;p;c)$. Let R be a convex domain of revolution such that $\overline{R} \subseteq E$. Then w satisfies the same conditions on R as it does on E. We use Theorem 3.45. Let $v \in C(\overline{R})$, be a temperature on R, and satisfy $v \geq w$ on $\partial_n R$. Then $w - v$ satisfies the same conditions on R as does w, in view of Theorem 1.6. Therefore $w - v$ satisfies the maximum principle of Theorem 3.46. Furthermore, whenever $q \in \partial_n R$ we have

$$\limsup_{p\to q,\, p\in R} (w(p) - v(p)) \leq w(q) - v(q) \leq 0,$$

so that $w(p) \leq v(p)$ for all $p \in R$. Moreover, if $q \in \overline{R}$ but $q \notin R \cup \partial_n R$, we claim that $w(q) = \limsup_{p\to q, p\in R} w(p)$. To prove this, we put $l = \limsup_{p\to q, p\in R} w(p)$, and note that $l \leq w(q) < +\infty$ since w is upper semicontinuous and upper finite. Given any number $L > l$, we can find a heat ball $\Omega(q;c_0)$ such that $w(p) \leq L$ for all $p \in \Omega(q;c_0)$. Our hypothesis shows that there is a positive number $c < c_0$ such that

$$w(q) \leq \mathcal{M}(w;q;c) \leq \mathcal{M}(L;q;c) = L.$$

Thus $w(q) \leq l$, and hence $w(q) = l$. It now follows that

$$w(q) - v(q) = \limsup_{p\to q,\, p\in R} w(p) - \lim_{p\to q,\, p\in R} v(p) = \limsup_{p\to q,\, p\in R} (w(p) - v(p)) \leq 0.$$

Hence $w(p) \leq v(p)$ for all $p \in \overline{R}$, and therefore Theorem 3.45 shows that w is a hypotemperature on E.

Now suppose, conversely, that w is a hypotemperature on E. Then Theorem 3.45 shows that w satisfies the hypotheses of Lemma 3.47, and the result follows. □

3.5. CHARACTERIZATIONS OF HYPOTEMPERATURES AND SUBTEMPERATURES

COROLLARY 3.49. *Let $w \in C^{2,1}(E)$. Then w is a subtemperature on E if and only if $\Theta w \geq 0$ on E.*

PROOF. Suppose that w is a subtemperature on E and that $p \in E$. Then the inequality $w(p) \leq \mathcal{M}(w;p;c)$ holds for all $c > 0$ such that $\overline{\Omega}(p;c) \subseteq E$, by Theorem 3.48. So $\Theta w \geq 0$ on E, by Theorem 1.6.

Conversely, if $\Theta w \geq 0$ then $w(p) \leq \mathcal{M}(w;p;c)$ holds whenever $\overline{\Omega}(p;c) \subseteq E$, by Theorem 1.6. Therefore w is a subtemperature on E, by Theorem 3.48. $\square$

Corollary 3.49 shows that the smooth subtemperatures defined in Section 1.3 are precisely the subtemperatures that belong to $C^{2,1}(E)$.

Our next theorem shows that hypotemperatures can be characterized in terms of the class $\mathfrak{R}$ of Lemma 3.47.

THEOREM 3.50. *Let w be an upper finite and upper semicontinuous function on an open set E. Let $\mathfrak{R}$ denote the class of convex domains of revolution R, for which both $\partial_n R = \partial R$ and the Dirichlet problem on R has a solution for every $f \in C(\partial R)$. Consider the following property: Whenever $R \in \mathfrak{R}$ is such that $\overline{R} \subseteq E$, and v is a function in $C(\overline{R})$ that is a temperature on R and satisfies $v \geq w$ on ∂R, then $v \geq w$ on $\overline{R}$.*

The stated property holds if and only if w is a hypotemperature on E.

PROOF. If w is a hypotemperature on E, then the stated property follows from Theorem 3.45.

Conversely, if the stated property holds then, by Lemma 3.47, the inequality $w(p) \leq \mathcal{M}(w;p;c)$ holds whenever $\overline{\Omega}(p;c) \subseteq E$. So w is a hypotemperature on E, by Theorem 3.48. $\square$

We now give our characterization of hypotemperatures in terms of the volume means $\mathcal{V}$. It is a generalization of Theorem 1.16 to arbitrary hypotemperatures.

THEOREM 3.51. *Let w be an upper finite and upper semicontinuous function on an open set E. Suppose that, given any point $p \in E$ and $\epsilon > 0$, we can find a positive number $c < \epsilon$ such that the inequality $w(p) \leq \mathcal{V}(w;p;c)$ holds. Then w is a hypotemperature on E.*

Conversely, if w is a hypotemperature on E and $p \in E$, then the inequality $w(p) \leq \mathcal{V}(w;p;c)$ holds for all $c > 0$ such that $\overline{\Omega}(p;c) \subseteq E$.

PROOF. The proof of the first part is similar to that of the first part of Theorem 3.48 (using Theorem 1.16 in place of Theorem 1.6).

Conversely, if w is a hypotemperature on E and $p \in E$ then, by Theorem 3.48, the inequality $w(p) \leq \mathcal{M}(w;p;l)$ holds for all $l > 0$ such that $\overline{\Omega}(p;l) \subseteq E$. It therefore follows from (1.18) that

$$\mathcal{V}(w;p;c) = \frac{n}{2} c^{-\frac{n}{2}} \int_0^c l^{\frac{n}{2}-1} \mathcal{M}(w;p;l)\, dl \geq \frac{n}{2} c^{-\frac{n}{2}} \int_0^c l^{\frac{n}{2}-1} w(p)\, dl = w(p)$$

whenever $\overline{\Omega}(p;c) \subseteq E$. $\square$

REMARK 3.52. Theorem 3.51 shows that the real continuous subtemperatures, defined in Section 1.5, are precisely the subtemperatures that are real-valued and continuous.

Theorem 3.51 implies the following refinement of Lemma 3.16.

COROLLARY 3.53. *If w is a hypotemperature on E and $p \in E$, then*

$$w(p) = \lim_{c \to 0+} \sup_{q \in \Omega(p;c)} w(q).$$

PROOF. We denote by l the right hand side of the equation. Since w is upper semicontinuous and upper finite, we have $l \leq w(p) < +\infty$. Given any number $L > l$, we can find a heat ball $\Omega(p; c_0)$ such that $w(q) < L$ for all $q \in \Omega(p; c_0)$. By Theorem 3.51, for all $c < c_0$ we have

$$w(p) \leq \mathcal{V}(w; p; c) \leq \mathcal{V}(L; p; c) = L.$$

Thus $w(p) \leq L$ whenever $L > l$, so that $w(p) \leq l$. Hence $w(p) = l$. □

If we use either the fundamental means $\mathcal{M}$, or the volume means $\mathcal{V}$ or $\mathcal{V}_m$ for $m \geq 1$, instead of the means $\mathcal{L}$, we can weaken the finiteness condition (δ_3) in the definition of a subtemperature. For $\mathcal{M}$ and $\mathcal{V}$, the next two theorems show this. For $\mathcal{V}_m$, the proof must be postponed until Chapter 6. Much of the proof is contained in the following lemma, which we shall use again in the next section (and Chapter 6). It is convenient to put $\mathcal{V}_0 = \mathcal{V}$ and $\Omega_0 = \Omega$ in the lemma.

LEMMA 3.54. *Let m be an integer with $m \geq 0$, let w be a locally upper bounded, extended real-valued function on an open set E, and let $(x_0, t_0) \in E$. If $w(x_0, t_0)$ is finite, and the inequality*

$$w(y, s) \leq \mathcal{V}_m(w; y, s; c)$$

holds whenever $\overline{\Omega}_m(y, s; c) \subseteq \Lambda(x_0, t_0; E) \cup \{(x_0, t_0)\}$, then w is locally integrable on $\Lambda(x_0, t_0; E)$.

PROOF. We prove the contrapositive. Suppose that w is not locally integrable on $\Lambda(x_0, t_0; E)$. Then we can find a point $(x_1, t_1) \in \Lambda(x_0, t_0; E)$ such that w is not integrable on any neighbourhood of (x_1, t_1). Join (x_0, t_0) to (x_1, t_1) by a polygonal path γ in $\Lambda(x_0, t_0; E) \cup \{(x_0, t_0)\}$ along which the temporal variable is strictly decreasing. Since γ is compact, its distance from $\mathbb{R}^{n+1} \backslash E$ is positive, and so we can find $c_0 > 0$ such that $\overline{\Omega}_m(x, t; c_0) \subseteq E$ for all $(x, t) \in \gamma$. Given $(x, t) \in \gamma$, we put

$$P(x, t) = \{(y, s) : |y - x|^2 < 2(m + n)(s - t), \ s - t < c_0/e\}.$$

The set $P(x, t)$ is a truncated paraboloid with vertex (x, t), and if $(y, s) \in P(x, t)$ then

$$|y - x|^2 < 2(m + n)(s - t) < 2(m + n)(s - t) \log\left(\frac{c_0}{s - t}\right),$$

so that $(x, t) \in \Omega_m(y, s; c_0)$.

Observe that, because γ is a union of finitely many line segments, there is a positive number $c_1 < c_0/e$, independent of (x, t), such that if $(x, t), (y, t + c_1) \in \gamma$ then $(y, t + c_1) \in P(x, t)$. Choose points $(x_2, t_2), ..., (x_l, t_l)$ inductively, such that $t_j = t_1 + (j - 1)c_1$ and $(x_j, t_j) \in \gamma$, for all $j \in \{2, ..., l\}$, and such that $t_l < t_0 \leq t_l + c_1$. Note that $(x_j, t_j) \in P(x_{j-1}, t_{j-1})$ for all $j \in \{2, ..., l\}$. Since $(x_1, t_1) \in \Lambda(x_0, t_0; E)$,

we have $(x_1, t_1) \in \Omega_m(y, s; c_0)$ for all points $(y, s) \in P(x_1, t_1)$. Therefore, if $m = 0$ we have
$$w(y, s) \leq \tau(c_0) \int\int_{\Omega(y,s;c_0)} \frac{|y - z|^2}{4(s - r)^2} w(z, r) \, dz \, dr = -\infty$$
for all points $(y, s) \in P(x_1, t_1)$ such that $y \neq x_1$. In particular, w is not integrable on any neighbourhood of (x_2, t_2). A similar argument works for every $m \geq 1$. Proceeding stepwise along γ, we deduce successively that w is not integrable on any neighbourhood of each of the points $(x_2, t_2), ..., (x_l, t_l)$. Since $t_l < t_0 \leq t_l + c_1$, we have $(x_l, t_l) \in \overline{\Omega}_m(x_0, t_0; c_0)$. Now $w(y, s) = -\infty$ for all $(y, s) \in P(x_l, t_l)$ such that $y \neq x_l$, so that $w(x_0, t_0) \leq \mathcal{V}_m(w; x_0, t_0; c_0) = -\infty$. □

COROLLARY 3.55. *If w is a hypotemperature on an open set E, and there is a point $p_0 \in E$ such that $w(p_0)$ is finite, then w is a subtemperature on $\Lambda(p_0; E)$.*

PROOF. By Theorem 3.51, if $p \in E$ the inequality $w(p) \leq \mathcal{V}(w; p; c)$ holds whenever $\overline{\Omega}(p; c) \subseteq E$. Since w is locally upper bounded on E, and $w(p_0)$ is finite, Lemma 3.54 (with $m = 0$) now shows that w is locally integrable on $\Lambda(p_0; E)$. Hence w is finite on a dense subset of $\Lambda(p_0; E)$, and is therefore a subtemperature there. □

THEOREM 3.56. *Let w be an extended real-valued function on an open set E. Then w is a subtemperature on E if and only if the following four conditions are satisfied:*
(a) $-\infty \leq w(p) < +\infty$ for all $p \in E$;
(b) w is upper semicontinuous on E;
(c) given any point $p \in E$, we can find a point $q \in E$ such that $p \in \Lambda(q; E)$ and $w(q) > -\infty$;
(d) given any point $p \in E$ and $\epsilon > 0$, we can find a positive number $c < \epsilon$ such that the inequality $w(p) \leq \mathcal{V}(w; p; c)$ holds.
Furthermore, every subtemperature on E is locally integrable on E.

PROOF. Theorem 3.51 shows that any subtemperature on E satisfies condition (d).

For the converse, conditions (a) and (b) imply that w is locally upper bounded on E, and Theorem 3.51 shows that the inequality $w(p) \leq \mathcal{V}(w; p; c)$ holds for all $c > 0$ such that $\overline{\Omega}(p; c) \subseteq E$. Therefore, by Lemma 3.54, w is locally integrable on $\Lambda(q; E)$ whenever $w(q) > -\infty$. Now condition (c) implies that w is locally integrable on E, and hence finite on a dense subset of E. Hence w is a subtemperature on E, by Theorem 3.51. □

COROLLARY 3.57. *If v and w are subtemperatures on the open set E, then $v + w$ is also a subtemperature on E.*

PROOF. Conditions (δ_1) and (δ_2) are obviously satisfied by $v + w$, and (δ_4) follows from Theorem 3.17. For (δ_3), Theorem 3.56 shows that each of v and w is finite outside a set of full measure, so that $v + w$ is too, and hence $v + w$ is finite on a dense subset of E. □

THEOREM 3.58. *Let w be an extended real-valued function on an open set E. Then w is a subtemperature on E if and only if the following four conditions are satisfied:*
(a) $-\infty \leq w(p) < +\infty$ for all $p \in E$;

(b) w is upper semicontinuous on E;

(c) given any point $p \in E$, we can find a point $q \in E$ such that $p \in \Lambda(q; E)$ and $w(q) > -\infty$;

(d) given any point $p \in E$ and $\epsilon > 0$, we can find a positive number $c < \epsilon$ such that the inequality $w(p) \leq \mathcal{M}(w; p; c)$ holds.

PROOF. Theorem 3.48 shows that any subtemperature on E satisfies condition (d).

For the converse, conditions (a), (b), (d) and Theorem 3.48 show that the inequality $w(p) \leq \mathcal{M}(w; p; c)$ holds for all $c > 0$ such that $\overline{\Omega}(p; c) \subseteq E$. Therefore formula (1.18) shows that the inequality $w(p) \leq \mathcal{V}(w; p; c)$ holds for all $c > 0$ such that $\overline{\Omega}(p; c) \subseteq E$, and the result follows from Theorem 3.56. □

3.6. Properties of Hypotemperatures

We now use Lemma 3.47 to extend Theorem 1.8 to general hypotemperatures.

THEOREM 3.59. *Let w be a hypotemperature on an open set E, and let $p_0 \in E$. Then the function $c \mapsto \mathcal{M}(w; p_0; c)$ is increasing on the set of $c > 0$ such that $\overline{\Omega}(p_0; c) \subseteq E$, and for all such c the inequality $\mathcal{V}(w; p_0; c) \leq \mathcal{M}(w; p_0; c)$ holds. Furthermore,*

$$\lim_{c \to 0+} \mathcal{M}(w; p_0; c) = \lim_{c \to 0+} \mathcal{V}(w; p_0; c) = w(p_0),$$

and if v is a hypotemperature such that $v \leq w$ almost everywhere on E, then $v \leq w$ everywhere on E.

PROOF. In view of the characterization of hypotemperatures given in Theorem 3.50, we can apply Lemma 3.47 to w. Thus $c \mapsto \mathcal{M}(w; p_0; c)$ is increasing.

The inequality now follows, because

$$\mathcal{V}(w; p_0; c) = \frac{n}{2} c^{-\frac{n}{2}} \int_0^c l^{\frac{n}{2}-1} \mathcal{M}(u; p_0; l) \, dl$$
$$\leq \frac{n}{2} c^{-\frac{n}{2}} \int_0^c l^{\frac{n}{2}-1} \mathcal{M}(u; p_0; c) \, dl$$
$$= \mathcal{M}(w; p_0; c).$$

We now consider the limits. If $A > w(p_0)$, then since w is upper semicontinuous, we can find a neighbourhood N of p_0 such that $w(p) < A$ for all $p \in N$. So, whenever $\overline{\Omega}(p_0; c) \subseteq N$, we have

$$w(p_0) \leq \mathcal{M}(w; p_0; c) \leq \mathcal{M}(A; p_0; c) = A,$$

using Theorem 3.48. Therefore

$$w(p_0) \leq \mathcal{V}(w; p_0; c) \leq \mathcal{M}(w; p_0; c) \to w(p_0)$$

as $c \to 0+$, in view of the characterization of hypotemperatures given in Theorem 3.51.

For the last part, if v is a hypotemperature and $v \leq w$ a.e. on E, then whenever $\overline{\Omega}(p; c) \subseteq E$ we have $\mathcal{V}(v; p; c) \leq \mathcal{V}(w; p; c)$. Making $c \to 0+$, we obtain $v(p) \leq w(p)$ for all $p \in E$. □

Our next result is a hypotemperature version of Theorem 1.31, the Harnack monotone convergence theorem for temperatures.

THEOREM 3.60. *Let $\{w_k\}$ be a decreasing sequence of hypotemperatures on an open set E, and let $w = \lim_{k\to\infty} w_k$. Then w is a hypotemperature on E, and if $w(p_0) > -\infty$ for some point $p_0 \in E$, then w is a subtemperature on $\Lambda(p_0; E)$.*

PROOF. We use the characterization of hypotemperatures given in Theorem 3.51. It is obvious that w is upper finite on E, and Lemma 3.5 shows that w is upper semicontinuous. For each function w_k, the inequality $w_k(p) \leq \mathcal{V}(w_k; p; c)$ holds whenever $\overline{\Omega}(p; c) \subseteq E$, by Theorem 3.51, and so the Lebesgue monotone convergence theorem shows that w has the same property. Hence w is a hypotemperature on E, and the last part follows from Corollary 3.55. □

REMARK 3.61. In Theorem 3.60, $\Lambda(p_0; E)$ may be the largest open set on which w is a subtemperature. For example, let $E = \mathbb{R}^{n+1}$ and let w_k by defined by

$$w_k(x,t) = \begin{cases} -k & \text{if } t > 0, \\ 0 & \text{if } t \leq 0. \end{cases}$$

Then each w_k is a subtemperature on $\mathbb{R}^{n+1}$, and $w(x, 0) > -\infty$ for all $x \in \mathbb{R}^n$, but the sequence $\{w_k\}$ does not converge at any point (x, t) where $t > 0$.

Using the following lemma, we can extend Theorem 3.60 to downward-directed families that are not necessarily sequences, and thus obtain a version of Theorem 1.36 for hypotemperatures.

LEMMA 3.62. *Let $\{u_\alpha : \alpha \in I\}$ be a family of upper semicontinuous and upper finite functions on an open set E. Then there is a countable subset J of I such that $\inf\{u_\alpha : \alpha \in J\} = \inf\{u_\alpha : \alpha \in I\}$ on E.*

PROOF. We put $u = \inf\{u_\alpha : \alpha \in I\}$, and note that u is upper finite on E. Since each function u_α is upper semicontinuous on E, for any $a \in \mathbb{R}$ each set $S_\alpha^a = \{p \in E : u_\alpha(p) < a\}$ is open, so that the set

$$S^a = \{p \in E : u(p) < a\} = \bigcup_{\alpha \in I} S_\alpha^a$$

is also open, which means that u is also upper semicontinuous on E. Furthermore, by Lindelöf's theorem, there is a countable subset J^a of I such that

$$\bigcup_{\alpha \in J^a} S_\alpha^a = S^a.$$

We put $J = \bigcup_{r \in \mathbb{Q}} J^r$ and $v = \inf\{u_\alpha : \alpha \in J\}$. Then $v \geq u$ because $J \subseteq I$, and J is countable because each J^r is countable. Moreover,

(3.5) $\qquad \{p \in E : u(p) < r\} = S^r = \bigcup_{\alpha \in J} S_\alpha^r = \{p \in E : v(p) < r\}$

for all $r \in \mathbb{Q}$. If there was a point $p_0 \in E$ such that $u(p_0) < v(p_0)$, then we could choose $r \in \mathbb{Q}$ such that $u(p_0) < r \leq v(p_0)$, and contradict (3.5). It follows that $v = u$ on E, as required. □

We recall, from Section 1.7, that a family $\mathcal{F}$ of functions on E is said to be downward-directed if $u, v \in \mathcal{F}$ implies that there is $w \in \mathcal{F}$ such that $u \wedge v \geq w$.

THEOREM 3.63. *Let $\mathcal{F}$ be a downward-directed family of hypotemperatures on an open set E, and let $u = \inf \mathcal{F}$. Then u is a hypotemperature on E, and if there is a point $p_0 \in E$ such that $u(p_0) > -\infty$, then u is a subtemperature on $\Lambda(p_0; E)$.*

PROOF. By Lemma 3.62, there is a sequence $\{w_k\}$ of functions in $\mathcal{F}$ such that $\inf\{w_k : k \in \mathbb{N}\} = \inf \mathcal{F}$ on E. We put $u_1 = w_1$, and for each $k \geq 1$ we choose a function $u_{k+1} \in \mathcal{F}$ such that $u_{k+1} \leq u_k \wedge w_{k+1}$, which is possible because $\mathcal{F}$ is downward-directed. Then $\{u_k\}$ is a decreasing sequence of hypotemperatures whose pointwise limit is u. The result now follows from Theorem 3.60. □

3.7. Thermic Majorants

If f and g are extended real-valued functions on a set S, and $f(p) \leq g(p)$ for all $p \in S$, then f is called a *minorant* of g, and g is called a *majorant* of f.

It is sometimes useful to know when a subtemperature has a *thermic majorant*, that is, when it has a majorant that is also a temperature. We shall employ the following concept.

THEOREM 3.64. *Let w be a subtemperature on an open set E which is majorized by a supertemperature on E, and put*

$$\mathcal{F} = \{v : v \text{ is a supertemperature on } E, v \geq w \text{ on } E\}.$$

Then $\inf \mathcal{F}$ is a temperature on E.

PROOF. If $v_1, v_2 \in \mathcal{F}$, then $v_1 \wedge v_2 \in \mathcal{F}$. Also, if $v \in \mathcal{F}$ and D is a circular cylinder such that $\overline{D} \subseteq E$, then the function $\pi_D v$ satisfies

$$\liminf_{k \to \infty}(\pi_D v - w)(p_k) = \liminf_{k \to \infty}(v - w)(p_k) \geq 0$$

whenever $\{p_k\}$ is a sequence of points of E such that $p_{k+1} \in \Lambda(p_k; E)$ for all k, and $\{p_k\}$ converges either to a point of ∂E or to the point at infinity. Since $\pi_D v - w$ is a supertemperature on E, it follows from the minimum principle that $\pi_D v \geq w$ on E. Hence $\pi_D v \in \mathcal{F}$, and so $\mathcal{F}$ is a saturated family of supertemperatures on E. Since w is finite on a dense subset of E, it follows from Theorem 3.26 that $\inf \mathcal{F}$ is a temperature on E. □

DEFINITION 3.65. The function $\inf \mathcal{F}$ of Theorem 3.64 is called the *least thermic majorant* of w on E.

Dually, if w is a *super*temperature on E that is *minorized* by a *sub*temperature on E, then w has what is called a *greatest thermic minorant E*.

Least thermic majorants of subtemperatures are additive, in the following sense.

THEOREM 3.66. *For $i \in \{1, 2\}$, let w_i be a subtemperature that has a least thermic majorant u_i on E. Then $w_1 + w_2$ has a least thermic majorant on E, namely $u_1 + u_2$.*

PROOF. Certainly $u_1 + u_2$ is a thermic majorant of $w_1 + w_2$ on E, so that $w_1 + w_2$ has a least thermic majorant u on E, by Theorem 3.64. Since $u - w_1 \geq w_2$ on E, the supertemperature $u - w_1$ majorizes w_2 on E, and so $u - w_1 \geq u_2$ on E, by Theorem 3.63. Now $u - u_2$ is a thermic majorant of w_1 on E, so that $u - u_2 \geq u_1$ on E, and hence $u \geq u_1 + u_2$. Since u is the least thermic majorant of $w_1 + w_2$ on E, equality follows. □

3.8. Notes and Comments

The strong maximum principle for temperatures is due to Nirenberg [**56**]. The form of the boundary maximum principle given here comes from Watson [**78**].

The material in sections 3.2-3.5 mostly follows Watson [**89**], except that general hypotemperatures and the volume means $\mathcal{V}_m$, for $m > 0$, were not discussed there. The behaviour of the mean values $\mathcal{M}$, given in Theorem 3.59 for hypotemperatures, was studied by Pini [**60**] for real continuous subtemperatures with $n = 1$, and by Watson [**69**] for general subtemperatures and all n. The last part of Theorem 3.59 is given, for subtemperatures, in Watson [**72**].

It follows from Remark 3.40 that the class of temperatures satisfies the Base Axiom of a harmonic space, as in Bauer [**5**] or Constantinescu & Cornea [**12**].

Theorem 3.50 implies that the subtemperatures of Watson [**69**] coincide with the subcaloric functions introduced earlier by Bauer [**5**]. Details can be found in Watson [**89**]. The result was obtained earlier by Bauer [**6**], and also follows from the fact that both Watson [**72**] and Doob [**14**] had a Riesz decomposition theorem with the same kernel. The methods of [**89**] are the most natural way to prove this equivalence.

That the condition $\rho'_-(b) > -\infty$ in Theorem 3.39 can be weakened was known to Petrowsky [**58**], who gave an example where the boundary near $\mathbb{R}^n \times \{b\}$ satisfies an iterated logarithm criterion. That example is also given in Doob [**14**]. Theorem 8.52 below gives a less precise, but more widely applicable, criterion for regularity which also shows that Theorem 3.39 is not sharp.

CHAPTER 4

Temperatures on an Infinite Strip

In this chapter, we consider those temperatures defined on an infinite strip $\mathbb{R}^n \times \,]0,a[$ or half-space $\mathbb{R}^n \times \,]0,+\infty[$, that have a representation as the Gauss-Weierstrass integral of a signed measure. In our main theorem, we show that every nonnegative temperature can be represented as the Gauss-Weierstrass integral of a nonnegative measure. We also consider whether the representing measure is unique, an extension of the boundary maximum principle for subtemperatures, the semigroup property of nonnegative temperatures, and the behaviour of the Gauss-Weierstrass integrals of functions at the hyperplane $\mathbb{R}^n \times \{0\}$. Finally, we look at a minimality property of the fundamental temperature W.

4.1. An Extension of the Maximum Principle on an Infinite Strip

We shall establish an extension of the boundary maximum principle, given in Theorem 3.13, for subtemperatures defined on a strip $\mathbb{R}^n \times \,]0,a[$. The extension allows a subtemperature w to grow quite rapidly at infinity, rather than be upper bounded.

THEOREM 4.1. *Suppose that w is a subtemperature on the strip $\mathbb{R}^n \times \,]0,b[$ for some positive real number b, that*
$$\limsup_{(x,t)\to(y,0+)} w(x,t) \leq A$$
for all $y \in \mathbb{R}^n$, and that for some positive number k
$$\int_0^b \int_{\mathbb{R}^n} \exp(-k|x|^2) w^+(x,t)\, dx\, dt < +\infty.$$
Then $w \leq A$ on $\mathbb{R}^n \times \,]0,b[$.

PROOF. Since $w - A$ is a subtemperature on $\mathbb{R}^n \times \,]0,b[$, and
$$\int_0^b \int_{\mathbb{R}^n} \exp(-k|x|^2)\, dx\, dt < +\infty,$$
it suffices to prove the case $A = 0$.

Furthermore, if we prove the result when $b = \epsilon$ is arbitrarily small, the result for any value of b will follow by repeated application of the result for ϵ. So we may assume that $b \leq (24k)^{-1}$.

Let v be defined on $\mathbb{R}^n \times \,]-\infty, b[$ by
$$v(x,t) = \begin{cases} w(x,t) & \text{if } t > 0, \\ 0 & \text{if } t \leq 0. \end{cases}$$

Then v is a subtemperature on $\mathbb{R}^n \times]-\infty, b[$. Given any point $(x_0, t_0) \in \mathbb{R}^n \times]0, b[$, and any $c > 0$, we put $\Omega^+(x_0, t_0; c) = \Omega(x_0, t_0; c) \cap (\mathbb{R}^n \times]0, b[)$. Then, by Theorem 3.51,

$$w(x_0, t_0) = v(x_0, t_0)$$
$$\leq \tau(c) \int\!\!\int_{\Omega(x_0,t_0;c)} \frac{|x_0 - x|^2}{4(t_0 - t)^2} v(x,t)\, dx\, dt$$
(4.1)
$$\leq \tau(c) \int\!\!\int_{\Omega^+(x_0,t_0;c)} \frac{|x_0 - x|^2}{4(t_0 - t)^2} v^+(x,t)\, dx\, dt.$$

We put
$$\psi(y, s) = s^{-\frac{n+8}{4}} \exp\left(-\frac{|y|^2}{8s}\right)$$
for all $(y, s) \in \mathbb{R}^n \times]0, +\infty[$, and
$$\Psi(x_0, t_0) = \{(x, t) : t < t_0,\ \psi(x_0 - x, t_0 - t) \geq 1\} \cup \{(x_0, t_0)\}.$$

We split the integral in (4.1) into two, namely $I_1(c)$ over $\Omega^+(x_0, t_0; c) \cap \Psi(x_0, t_0)$, and $I_2(c)$ over $\Omega^+(x_0, t_0; c) \backslash \Psi(x_0, t_0)$.

Consider first $I_1(c)$. The inequality $\psi(x_0 - x, t_0 - t) \geq 1$ holds if and only if
$$|x_0 - x|^2 \leq -2(n+8)(t_0 - t)\log(t_0 - t) \quad \text{and} \quad 0 < t_0 - t \leq 1.$$

Therefore, because $(t_0 - t)\log(t_0 - t) \to 0$ as $t \to t_0-$, the set $\Psi(x_0, t_0)$ is closed. Since $-(t_0 - t)\log(t_0 - t) \leq e^{-1}$ whenever $0 < t_0 - t \leq 1$, the set is also bounded. It follows that, because v^+ is upper semicontinuous, v^+ has a maximum value M over $\Psi(x_0, t_0)$. Therefore, if χ_r denotes the characteristic function of $\mathbb{R}^n \times]t_0 - r, t_0[$ for any $r > 0$, we have
$$I_1(c) \leq M \mathcal{V}(\chi_{t_0}; x_0, t_0; c).$$

We now make the change of variables $t_0 - t = c(t_0 - \sigma)$, $x_0 - x = \sqrt{c}(x_0 - \eta)$, which takes $\Omega(x_0, t_0; c)$ to $\Omega(x_0, t_0; 1)$, and χ_r to $\chi_{r/c}$. Furthermore, as $c \to \infty$, the functions $\chi_{r/c}$ decrease to zero. It therefore follows from Lebesgue's monotone convergence theorem that
$$\mathcal{V}(\chi_{t_0}; x_0, t_0; c) = \mathcal{V}(\chi_{t_0/c}; x_0, t_0; 1) \to 0$$
as $c \to \infty$. Hence $I_1(c) \to 0$.

Now consider $I_2(c)$. Whenever $(x, t) \in \Omega(x_0, t_0; c)$, we have
$$c^{-\frac{n}{4}} \leq (t_0 - t)^{-\frac{n}{4}} \exp\left(-\frac{|x_0 - x|^2}{8(t_0 - t)}\right),$$
so that
$$c^{-\frac{n}{2}} \frac{|x_0 - x|^2}{(t_0 - t)^2} \leq c^{-\frac{n}{4}} |x_0 - x|^2 \psi(x_0 - x, t_0 - t).$$

Furthermore, whenever $(x, t) \notin \Psi(x_0, t_0)$ we have $0 \leq \psi(x_0 - x, t_0 - t) \leq 1$. Hence, because $t_0 - t < (24k)^{-1}$ for all $(x, t) \in \Omega^+(x_0, t_0; c)$, it follows that on $\Omega^+(x_0, t_0; c) \backslash \Psi(x_0, t_0)$ we have
$$\psi(x_0 - x, t_0 - t) \leq \psi(x_0 - x, t_0 - t)^{24k(t_0 - t)}$$
$$= (t_0 - t)^{-6k(n+8)(t_0 - t)} \exp(3k|x_0 - x|^2).$$

Since the function $t \mapsto (t_0 - t)^{-6k(n+8)(t_0-t)}$ is bounded on $]0, t_0[$, it follows that, for some constant K,

$$I_2(c) \leq Kc^{-\frac{n}{4}} \int\int_{\Omega^+ \backslash \Psi} |x_0 - x|^2 \exp(-3k|x_0 - x|^2) w^+(x,t)\, dx\, dt$$

$$\leq Kc^{-\frac{n}{4}} \int_0^b \int_{\mathbb{R}^n} \exp(-2k|x_0 - x|^2) w^+(x,t)\, dx\, dt.$$

Furthermore, because $|x|^2 \leq 2(|x_0|^2 + |x_0 - x|^2)$, we also have

$$-2k|x_0 - x|^2 \leq 2k|x_0|^2 - k|x|^2,$$

and hence

$$I_2(c) \leq Kc^{-\frac{n}{4}} \int_0^b \int_{\mathbb{R}^n} \exp(-k|x|^2) w^+(x,t)\, dx\, dt \to 0$$

as $c \to \infty$. The result now follows from (4.1). $\square$

COROLLARY 4.2. *Suppose that u is a temperature on the strip $\mathbb{R}^n \times]0, b[$, that*

$$\lim_{(x,t) \to (y, 0+)} u(x,t) = A$$

for all $y \in \mathbb{R}^n$, and that for some positive number k

$$\int_0^b \int_{\mathbb{R}^n} \exp(-k|x|^2) |u(x,t)|\, dx\, dt < +\infty.$$

Then $u = A$ on $\mathbb{R}^n \times]0, b[$.

PROOF. It suffices to prove the case $A = 0$. We apply Theorem 4.1 with $w = |u|$, and deduce that $|u| \leq 0$, hence $u = 0$, on $\mathbb{R}^n \times]0, b[$. $\square$

4.2. Gauss-Weierstrass Integrals

In Section 2.1, we encountered integrals of the form

$$u(x,t) = \int_{\mathbb{R}^n} W(x - y, t - a) f(y)\, dy,$$

where W is the fundamental temperature and f any bounded, measurable function on $\mathbb{R}^n$. The function u, thus defined, is a temperature on $\mathbb{R}^n \times]0, +\infty[$, and is continuous onto $\mathbb{R}^n \times \{0\}$ at any point of continuity of f. In this section, we study generalizations of such integrals. Since $W(z, r)$ tends rapidly to zero as $|z| \to \infty$, the integral can exist for unbounded functions f, and it is desirable to consider it with f as general as possible. In fact, we replace $f(y)\, dy$ with $d\mu(y)$ for suitable signed measures μ on $\mathbb{R}^n$.

A signed measure takes only finite values, but this is too restrictive for the present situation, and so we adopt the following convention. Let ν be a signed measure on $\mathbb{R}^n$ such that, for some point $(x,t) \in \mathbb{R}^{n+1}$ and some positive number α,

$$\int_{\mathbb{R}^n} W(x - y, t) \exp(\alpha|y|^2)\, d|\nu|(y) < +\infty.$$

Then we write

$$\int_{\mathbb{R}^n} W(x - y, t)\, d\mu(y) = \int_{\mathbb{R}^n} W(x - y, t) \exp(\alpha|y|^2)\, d\nu(y),$$

and thus identify $d\mu(y)$ with $\exp(\alpha|y|^2)\,d\nu(y)$. We call μ a *signed measure*, despite the fact that it may take infinite values. For every *bounded* ν-measurable set A, we have $|\nu|(A) < +\infty$, so that $\mu(A) = \int_A \exp(\alpha|y|^2)\,d\nu(y) \in \mathbb{R}$, and so *locally* μ is a true signed measure.

DEFINITION 4.3. Given a signed measure μ such that the inequality
$$\int_{\mathbb{R}^n} W(x_0 - y, t_0)\,d|\mu|(y) < +\infty$$
holds for some point $(x_0, t_0) \in \mathbb{R}^n \times {]0, +\infty[}$, the *Gauss-Weierstrass Integral* u of μ is defined by
$$u(x,t) = \int_{\mathbb{R}^n} W(x - y, t)\,d\mu(y)$$
for all points $(x,t) \in \mathbb{R}^n \times {]0, +\infty[}$ such that the integral exists. If μ is absolutely continuous with respect to Lebesgue measure on $\mathbb{R}^n$, so that $d\mu(y) = f(y)\,dy$, then u may be called the *Gauss-Weierstrass Integral of f*, rather than of μ.

In this context, the fundamental temperature W is often referred to as the *Gauss-Weierstrass Kernel*.

Analogous to the result that, if a complex power series in ζ converges at a point ζ_0, then it is absolutely convergent to an analytic function on $\{\zeta : |\zeta| < |\zeta_0|\}$, we have the following theorem.

THEOREM 4.4. *Let μ be a signed measure on $\mathbb{R}^n$ whose the Gauss-Weierstrass integral u is defined and finite at some point $(x_0, t_0) \in \mathbb{R}^n \times {]0, +\infty[}$. Then u is defined and is a temperature throughout $\mathbb{R}^n \times {]0, t_0[}$.*

PROOF. Since $u(x_0, t_0)$ is defined and finite, we have
$$\text{(4.2)} \qquad \int_{\mathbb{R}^n} W(x_0 - y, t_0)\,d|\mu|(y) < +\infty.$$
Let $(x,t) \in \mathbb{R}^n \times {]0, t_0[}$, and put
$$D = \{y \in \mathbb{R}^n : |x_0 - y|(\sqrt{t_0} - \sqrt{t}) \geq |x_0 - x|\sqrt{t_0}\}.$$
If $y \in D$, then
$$|y - x| \geq |y - x_0| - |x_0 - x| \geq |y - x_0| - |x_0 - y|\frac{(\sqrt{t_0} - \sqrt{t})}{\sqrt{t_0}} = |x_0 - y|\sqrt{\frac{t}{t_0}},$$
so that
$$W(x - y, t) \leq (4\pi t)^{-\frac{n}{2}}\exp\left(-\frac{|x_0 - y|^2}{4t_0}\right) = \left(\frac{t_0}{t}\right)^{\frac{n}{2}} W(x_0 - y, t_0).$$
Therefore, in view of (4.2),
$$\text{(4.3)} \qquad \int_D W(x - y, t)\,d|\mu|(y) \leq \left(\frac{t_0}{t}\right)^{\frac{n}{2}} \int_D W(x_0 - y, t_0)\,d|\mu|(y) < +\infty.$$
On the other hand, if $y \notin D$ we have $|x_0 - y| < |x_0 - x|\sqrt{t_0}/(\sqrt{t_0} - \sqrt{t})$, so that
$$W(x_0 - y, t_0) \geq (4\pi t_0)^{-\frac{n}{2}}\exp\left(-\frac{|x_0 - x|^2}{4(\sqrt{t_0} - \sqrt{t})^2}\right),$$

and hence

$$W(x-y,t) \leq (4\pi t)^{-\frac{n}{2}} \leq \left(\frac{t_0}{t}\right)^{\frac{n}{2}} \exp\left(\frac{|x_0-x|^2}{4(\sqrt{t_0}-\sqrt{t})^2}\right) W(x_0-y,t_0).$$

Therefore, in view of (4.2),

$$\int_{\mathbb{R}^n \setminus D} W(x-y,t)\, d|\mu|(y)$$
$$\leq \left(\frac{t_0}{t}\right)^{\frac{n}{2}} \exp\left(\frac{|x_0-x|^2}{4(\sqrt{t_0}-\sqrt{t})^2}\right) \int_{\mathbb{R}^n \setminus D} W(x_0-y,t_0)\, d|\mu|(y)$$
$$< +\infty.$$

This, combined with (4.3), shows that u is defined throughout $\mathbb{R}^n \times\,]0, t_0[$, and is locally bounded there. We can now use Theorem 1.29. Thus, whenever the modified heat ball $\overline{\Omega}_5(x,t;c) \subseteq \mathbb{R}^n \times\,]0, t_0[$, we have

$$\mathcal{V}_5(u;x,t;c) = \int_{\mathbb{R}^n} \mathcal{V}_5(W(\,\cdot\, - y, \cdot\,); x, t; c)\, d\mu(y)$$
$$= \int_{\mathbb{R}^n} W(x-y,t)\, d\mu(y)$$
$$= u(x,t).$$

so that u is a temperature. □

COROLLARY 4.5. *If f is a measurable function on $\mathbb{R}^n$ such that*

$$\int_{\mathbb{R}^n} \exp(-\alpha|y|^2)|f(y)|\, dy < +\infty$$

for some positive number α, then the Gauss-Weierstrass integral u of f is defined and is a temperature throughout $\mathbb{R}^n \times\,]0, 1/(4\alpha)[$.

PROOF. Since

$$\int_{\mathbb{R}^n} W\left(y, \frac{1}{4\alpha}\right) |f(y)|\, dy = \left(\frac{\alpha}{\pi}\right)^{\frac{n}{2}} \int_{\mathbb{R}^n} \exp(-\alpha|y|^2)|f(y)|\, dy < +\infty,$$

we see that $u(0, 1/(4\alpha))$ is defined and finite. Now Theorem 4.4 gives the result. □

EXAMPLE 4.6. Let $f(y) = \exp(\beta|y|^2)$ for some $\beta > 0$. Corollary 4.5 implies that the Gauss-Weierstrass integral u of f is defined on the strip $\mathbb{R}^n \times\,]0, 1/(4\beta)[$. Since

$$\int_{\mathbb{R}^n} W\left(y, \frac{1}{4\beta}\right) f(y)\, dy = \left(\frac{\beta}{\pi}\right)^{\frac{n}{2}} \int_{\mathbb{R}^n} dy = +\infty,$$

u is not defined at $(0, 1/(4\beta))$, and so the result is sharp. We can evaluate $u(x,t)$ explicitly, as follows. We put $s = \sqrt{1 - 4\beta t}$, and note that

$$\beta|y|^2 - \frac{|x-y|^2}{4t} = \frac{1}{4t}\sum_{i=1}^{n}\left((4\beta t - 1)y_i^2 + 2x_iy_i - x_i^2\right)$$

$$= -\frac{1}{4t}\sum_{i=1}^{n}(s^2y_i^2 - 2x_iy_i + x_i^2)$$

$$= -\frac{1}{4t}\sum_{i=1}^{n}\left(\left(sy_i - \frac{x_i}{s}\right)^2 - \left(\frac{1}{s^2} - 1\right)x_i^2\right).$$

Now if we put $z = sy - (x/s)$, we obtain

$$\beta|y|^2 - \frac{|x-y|^2}{4t} = -\frac{|z|^2}{4t} + \left(\frac{1}{s^2} - 1\right)\frac{|x|^2}{4t} = -\frac{|z|^2}{4t} + \frac{\beta|x|^2}{s^2}.$$

Hence

$$u(x,t) = \left(\frac{1}{4\pi t}\right)^{\frac{n}{2}} \int_{\mathbb{R}^n} \exp\left(-\frac{|z|^2}{4t} + \frac{\beta|x|^2}{s^2}\right)\frac{1}{s^n}\,dz$$

$$= \left(\frac{1}{1-4\beta t}\right)^{\frac{n}{2}} \exp\left(\frac{\beta|x|^2}{1-4\beta t}\right)\int_{\mathbb{R}^n} W(z,t)\,dz$$

$$= \left(\frac{1}{1-4\beta t}\right)^{\frac{n}{2}} \exp\left(\frac{\beta|x|^2}{1-4\beta t}\right)$$

by Lemma 1.1.

We shall study the behaviour of Gauss-Weierstrass integrals at the boundary $\mathbb{R}^n \times \{0\}$. If we want to assume that a Gauss-Weierstrass integral is defined and finite at some point in $\mathbb{R}^n \times\,]0, +\infty[$, Theorem 4.4 shows that there is no real loss of generality if we suppose that the point is of the form $(0, a)$. Our first result is a generalization and strengthening of Lemma 2.1.

LEMMA 4.7. *Let μ be a signed measure on $\mathbb{R}^n$ whose Gauss-Weierstrass integral u is defined and finite at some point $(0,a) \in \mathbb{R}^n \times\,]0, +\infty[$. Then given any point $x_0 \in \mathbb{R}^n$ and any number θ such that $0 < \theta < 1$, for any $\delta > 0$ we have*

$$\int_{|x_0 - y| > \delta} W(x - y, t)\,d\mu(y) \to 0 \quad as \quad t \to 0+$$

uniformly for $x \in \{x : |x_0 - x| < \theta\delta\}$.

PROOF. Suppose that $t < a/2$, $|x_0 - x| < \theta\delta$, and $|x_0 - y| > \delta$. Two applications of the inequality $|\xi|^2 \leq 2(|\xi - \eta|^2 + |\eta|^2)$ give us

$$2|x-y|^2 \geq |y|^2 - 2|x|^2 \geq |y|^2 - 4(|x_0 - x|^2 + |x_0|^2),$$

and this, together with the fact that $|x - y| > \delta - \theta\delta$, implies that

$$\begin{aligned}\frac{|x-y|^2}{4t} &= \frac{(a-2t)|x-y|^2}{4at} + \frac{2|x-y|^2}{4a} \\ &\geq \frac{(a-2t)(1-\theta)^2\delta^2}{4at} + \frac{|y|^2 - 4\theta^2\delta^2 - 4|x_0|^2}{4a}.\end{aligned}$$

It follows that, for some positive constant C,

$$\begin{aligned}&\left|\int_{|x_0-y|>\delta} W(x-y,t)\,d\mu(y)\right| \\ &\leq \int_{|x_0-y|>\delta} W(x-y,t)\,d|\mu|(y) \\ &\leq Ct^{-\frac{n}{2}} \exp\left(-\frac{(a-2t)(1-\theta)^2\delta^2}{4at} + \frac{\theta^2\delta^2 + |x_0|^2}{a}\right) \int_{\mathbb{R}^n} \exp\left(-\frac{|y|^2}{4a}\right) d|\mu|(y) \\ &\leq Ct^{-\frac{n}{2}} \exp\left(-\frac{(1-\theta)^2\delta^2}{4t}\right) \int_{\mathbb{R}^n} W(y,a)\,d|\mu|(y).\end{aligned}$$

By hypothesis, the last integral is finite. Therefore the last expression, which does not depend on x, tends to zero as $t \to 0+$. $\square$

THEOREM 4.8. *Let f be an extended real-valued function on $\mathbb{R}^n$, whose Gauss-Weierstrass integral u is defined and finite at some point $(0,a) \in \mathbb{R}^n \times]0,+\infty[$, and let $\xi \in \mathbb{R}^n$. Then*

$$\liminf_{\eta \to \xi} f(\eta) \leq \liminf_{(x,t) \to (\xi,0+)} u(x,t) \leq \limsup_{(x,t) \to (\xi,0+)} u(x,t) \leq \limsup_{\eta \to \xi} f(\eta).$$

In particular, if f is continuous at ξ, then

$$\lim_{(x,t) \to (\xi,0+)} u(x,t) = f(\xi).$$

PROOF. It suffices to prove the third inequality, because the second is obvious, and the first follows from the third applied to $-f$ and its Gauss-Weierstrass integral $-u$.

Since the third inequality is otherwise obvious, we suppose that $\limsup_{\eta \to \xi} f(\eta)$ is not $+\infty$. Let A be any real number that satisfies

(4.4) $$\limsup_{\eta \to \xi} f(\eta) < A.$$

Then there is $r > 0$ such that $f(y) < A$ whenever $|y - \xi| < r$. By Theorem 4.4, the function u is defined everywhere on $\mathbb{R}^n \times]0,a[$. We write

$$u(x,t) = \int_{|y-\xi|<r} W(x-y,t)f(y)\,dy + \int_{|y-\xi|\geq r} W(x-y,t)f(y)\,dy,$$

and note that the second integral tends to zero as $(x,t) \to (\xi, 0+)$, by Lemma 4.7. Using our choice of r, Lemma 1.1, and Lemma 2.1, we obtain

$$\int_{|y-\xi|<r} W(x-y,t)f(y)\,dy \leq A \int_{|y-\xi|<r} W(x-y,t)\,dy$$

$$= A\left(1 - \int_{|y-\xi|\geq r} W(x-y,t)\,dy\right)$$

$$\to A$$

as $(x,t) \to (\xi, 0+)$. It follows that

$$\limsup_{(x,t)\to(\xi,0+)} u(x,t) \leq A.$$

Since this holds for every A which satisfies (4.4), the third inequality follows. □

DEFINITION 4.9. If u is a temperature on a strip $\mathbb{R}^n \times\,]0,a[$, and the equality

$$u(x,t) = \int_{\mathbb{R}^n} W(x-y, t-s)u(y,s)\,dy$$

holds whenever $x \in \mathbb{R}^n$ and $0 < s < t < a$, we say that u has the *Semigroup Property* on $\mathbb{R}^n \times\,]0,a[$.

If u has the semigroup property, then the values of $u(x,t)$ for all points (x,t) in $\mathbb{R}^n \times\,]s,a[$, can be obtained from those on the hyperplane $\mathbb{R}^n \times \{s\}$ regardless of those on $\mathbb{R}^n \times\,]0,s[$. We show that this is a property possessed by every Gauss-Weierstrass integral.

THEOREM 4.10. *If* $x \in \mathbb{R}^n$ *and* $0 < s < t$, *then*

(4.5) $$W(x,t) = \int_{\mathbb{R}^n} W(x-y, t-s)W(y,s)\,dy,$$

and if u *is the Gauss-Weierstrass integral on* $\mathbb{R}^n \times\,]0,a[$ *of a signed measure* μ, *then*

(4.6) $$u(x,t) = \int_{\mathbb{R}^n} W(x-y, t-s)u(y,s)\,dy$$

whenever $t < a$.

PROOF. There are several ways to prove this theorem, and we choose the most elementary one. If $x, y \in \mathbb{R}^n$ and $0 < s < t$, we have

$$s|x-y|^2 + (t-s)|y|^2 = \sum_{i=1}^{n}(sx_i^2 + ty_i^2 - 2sx_iy_i)$$

$$= \sum_{i=1}^{n}\left(t\left(y_i - \frac{sx_i}{t}\right)^2 + \frac{x_i^2 s}{t}(t-s)\right)$$

$$= t\left|y - \frac{sx}{t}\right|^2 + \frac{s}{t}(t-s)|x|^2,$$

so that
$$\exp\left(-\frac{|x-y|^2}{4(t-s)}\right)\exp\left(-\frac{|y|^2}{4s}\right) = \exp\left(-\frac{s|x-y|^2+(t-s)|y|^2}{4s(t-s)}\right)$$
$$= \exp\left(-\frac{|y-st^{-1}x|^2}{4st^{-1}(t-s)}\right)\exp\left(-\frac{|x|^2}{4t}\right).$$

Hence, putting $z = y - st^{-1}x$, we obtain
$$\int_{\mathbb{R}^n} W(x-y,t-s)W(y,s)\,dy$$
$$= \frac{1}{((4\pi)^2 s(t-s))^{\frac{n}{2}}}\exp\left(-\frac{|x|^2}{4t}\right)\int_{\mathbb{R}^n}\exp\left(-\frac{|z|^2}{4st^{-1}(t-s)}\right)dz.$$

By Lemma 1.1
$$\int_{\mathbb{R}^n}\exp\left(-\frac{|z|^2}{4st^{-1}(t-s)}\right)dz = \left(4\pi st^{-1}(t-s)\right)^{\frac{n}{2}},$$
so that
$$\int_{\mathbb{R}^n} W(x-y,t-s)W(y,s)\,dy = \frac{1}{(4\pi t)^{\frac{n}{2}}}\exp\left(-\frac{|x|^2}{4t}\right),$$
and (4.5) is established.

We obtain (4.6) from (4.5) by using Fubini's theorem on interchanging the order in a repeated integral. Since, by (4.5),
$$\int_{\mathbb{R}^n}\int_{\mathbb{R}^n} W(x-y,t-s)W(y-z,s)\,dy\,d|\mu|(z) = \int_{\mathbb{R}^n} W(x-z,t)\,d|\mu|(z) < +\infty$$
whenever $0 < s < t < a$ and $x \in \mathbb{R}^n$, Fubini's theorem is applicable. Thus
$$u(x,t) = \int_{\mathbb{R}^n}\int_{\mathbb{R}^n} W(x-y,t-s)W(y-z,s)\,dy\,d\mu(z)$$
$$= \int_{\mathbb{R}^n} W(x-y,t-s)\int_{\mathbb{R}^n} W(y-z,s)\,d\mu(z)\,dy$$
$$= \int_{\mathbb{R}^n} W(x-y,t-s)u(y,s)\,dy.$$
□

If u is a temperature which can be written as the Gauss-Weierstrass integral of a signed measure, the question of whether the measure is uniquely determined by u naturally arises. We now address this question.

THEOREM 4.11. *Suppose that u is a temperature on $\mathbb{R}^n \times\,]0,a[$, and that u can be written as the Gauss-Weierstrass integral of two signed measures μ_1 and μ_2. Then $\mu_1 = \mu_2$.*

PROOF. We may assume that $a < +\infty$. Since the Gauss-Weierstrass integrals of μ_1 and μ_2 are defined and finite at the point $(0,\frac{1}{2}a)$, if we put $\gamma = \frac{1}{2a}$ we have
$$\int_{\mathbb{R}^n}\exp(-\gamma|y|^2)\,d|\mu_i|(y) = (2\pi a)^{\frac{n}{2}}\int_{\mathbb{R}^n} W(y,\tfrac{1}{2}a)\,d|\mu_i|(y) < +\infty$$
for each $i \in \{1,2\}$. Therefore, if we put $d\nu_i = \exp(-\gamma|y|^2)\,d\mu_i(y)$ for each i, then each ν_i is a signed measure of finite total variation. Hence, if we put $\lambda = \nu_1 - \nu_2$,

then λ is also a signed measure of finite total variation. To prove the theorem, we show that λ is the zero measure.

We first observe that

$$\int_{\mathbb{R}^n} W(x-y,t)\exp(\gamma|y|^2)\,d\lambda(y)$$
$$= \int_{\mathbb{R}^n} W(x-y,t)\,d\mu_1(y) - \int_{\mathbb{R}^n} W(x-y,t)\,d\mu_2(y)$$
(4.7)
$$= 0$$

for all $(x,t) \in \mathbb{R}^n \times \,]0,a[$.

By the Hahn-Jordan decomposition theorem, there are disjoint λ-measurable sets P and N such that $P \cup N = \mathbb{R}^n$, and nonnegative finite measures λ^+ and λ^- on $\mathbb{R}^n$, such that $\lambda^+(S) = \lambda(P \cap S)$ and $\lambda^-(S) = -\lambda(N \cap S)$ for every λ-measurable set $S \subseteq \mathbb{R}^n$. We suppose that λ is not null, so that either $\lambda^+(P) > 0$ or $\lambda^-(N) > 0$. It suffices to consider the former case, since the latter follows from the former applied to $-\lambda$. If $\lambda^+(P) = \alpha > 0$, then because of the regularity of λ^+ and λ^-, there exist a compact set $K \subseteq P$ such that

(4.8) $$\lambda(K) = \lambda^+(K) > \tfrac{3}{4}\alpha,$$

and a bounded open set $G \supseteq K$ such that

(4.9) $$\lambda^-(G) = \lambda^-(G \backslash K) < \tfrac{1}{4}\alpha.$$

By Urysohn's lemma, there is a continuous function ϕ on $\mathbb{R}^n$ such that $\phi(y) = 1$ for all $y \in K$, $\phi(y) = 0$ for all $y \in \mathbb{R}^n \backslash G$, and $0 \leq \phi(y) \leq 1$ for all $y \in \mathbb{R}^n$. Given such a function ϕ, we put

$$v(x,t) = \int_{\mathbb{R}^n} W(x-y,t)\phi(y)\exp(-\gamma|y|^2)\,dy.$$

Then the function v is a temperature on $\mathbb{R}^n \times \,]0,+\infty[$ such that

(4.10) $$\lim_{t\to 0+} v(x,t) = \phi(x)\exp(-\gamma|x|^2)$$

for all $x \in \mathbb{R}^n$, by Theorem 2.2.

We now choose a positive number r such that $|y| < r$ whenever $y \in G$. Then $\phi(y) = 0$ whenever $|y| \geq r$, and hence, because $0 \leq \phi \leq 1$, we have

$$0 \leq \exp(\gamma|x|^2)v(x,t)$$
(4.11)
$$\leq (4\pi t)^{-\frac{n}{2}} \int_{|y|<r} \exp\left(\gamma|x|^2 - \frac{|x-y|^2}{4t} - \gamma|y|^2\right) dy.$$

If $0 < t < 1/(16\gamma)$, then for any $x,y \in \mathbb{R}^n$ we have

$$\gamma(|x|^2 - 2|y|^2) \leq 2\gamma|x-y|^2 \leq \frac{|x-y|^2}{8t},$$

so that

$$\gamma|x|^2 \leq \frac{|x-y|^2}{8t} + 2\gamma|y|^2,$$

and hence

$$\gamma|x|^2 - \frac{|x-y|^2}{4t} - \gamma|y|^2 \leq -\frac{|x-y|^2}{8t} + \gamma|y|^2.$$

It therefore follows from (4.11) that

$$0 \leq \exp(\gamma|x|^2)v(x,t)$$
$$\leq (4\pi t)^{-\frac{n}{2}} \int_{|y|<r} \exp\left(-\frac{|x-y|^2}{8t} + \gamma|y|^2\right) dy$$
$$\leq \exp(\gamma r^2) 2^{\frac{n}{2}} \int_{|y|<r} W(x-y, 2t)\, dy$$
(4.12)
$$\leq \exp(\gamma r^2) 2^{\frac{n}{2}},$$

and hence

(4.13) $$\int_{\mathbb{R}^n} \exp(\gamma|x|^2)v(x,t)\, d|\lambda|(x) \leq \exp(\gamma r^2) 2^{\frac{n}{2}} |\lambda|(\mathbb{R}^n) < +\infty,$$

whenever $0 < t < 1/(16\gamma)$. By (4.12), the function $(x,t) \mapsto \exp(\gamma|x|^2)v(x,t)$ is bounded on $\mathbb{R}^n \times\,]0, +\infty[$, and so it follows from the fact that $|\lambda|(\mathbb{R}^n) < +\infty$, the Lebesgue dominated convergence theorem, and (4.10), that

$$\lim_{t \to 0+} \int_{\mathbb{R}^n} \exp(\gamma|x|^2)v(x,t)\, d\lambda(x) = \int_{\mathbb{R}^n} \phi(x)\, d\lambda(x).$$

The properties of ϕ, together with (4.8) and (4.9), show that

$$\int_{\mathbb{R}^n} \phi(x)\, d\lambda(x) = \int_K d\lambda(x) + \int_{G\setminus K} \phi(x)\, d\lambda^+(x) - \int_{G\setminus K} \phi(x)\, d\lambda^-(x)$$
$$\geq \lambda(K) + 0 - \lambda^-(G \setminus K)$$
$$> \tfrac{1}{2}\alpha$$
$$> 0,$$

so that

(4.14) $$\lim_{t \to 0+} \int_{\mathbb{R}^n} \exp(\gamma|x|^2)v(x,t)\, d\lambda(x) > 0.$$

However, if $0 < t < 1/(16\gamma)$, it follows from Fubini's theorem and (4.7) that

$$\int_{\mathbb{R}^n} \exp(\gamma|x|^2)v(x,t)\, d\lambda(x)$$
$$= \int_{\mathbb{R}^n} \exp(\gamma|x|^2) \int_{\mathbb{R}^n} W(x-y,t)\phi(y)\exp(-\gamma|y|^2)\, dy\, d\lambda(x)$$
$$= \int_{\mathbb{R}^n} \phi(y) \exp(-\gamma|y|^2) \int_{\mathbb{R}^n} W(x-y,t) \exp(\gamma|x|^2)\, d\lambda(x)\, dy$$
(4.15) $$= 0.$$

The application of Fubini's theorem is justified by (4.13). Since (4.15) contradicts (4.14), our assumption that λ is not null must be false. Therefore $\mu_1 = \mu_2$. □

4.3. Nonnegative Temperatures

The main result of this section shows that every nonnegative temperature on a strip $\mathbb{R}^n \times\,]0, a[$ can be written as the Gauss-Weierstrass integral of a nonnegative measure on $\mathbb{R}^n$.

LEMMA 4.12. *If μ is a nonnegative measure on $\mathbb{R}^n$ with compact support, then its Gauss-Weierstrass integral u is defined and finite on the half-space $\mathbb{R}^n \times {]}0,+\infty[$, and $u(x,t) \to 0$ as $|x| \to +\infty$, uniformly for $t \in {]}0,+\infty[$.*

PROOF. Since
$$\int_{\mathbb{R}^n} W(x-y,t)\,d\mu(y) \leq (4\pi t)^{-\frac{n}{2}}\mu(\mathbb{R}^n) < +\infty$$
for all $(x,t) \in \mathbb{R}^n \times {]}0,+\infty[$, u is defined on that half-space. We choose r such that $|y| < r$ whenever y belongs to the support of μ. Then, whenever $|x| > r$ and y belongs to the support of μ, we have $|x-y| \geq |x|-r$. Also, writing $\alpha = n/2$ and $\beta = (|x|-r)^2/4$, we have
$$t^{-\frac{n}{2}}\exp\left(-\frac{(|x|-r)^2}{4t}\right) = t^{-\alpha}\exp\left(-\frac{\beta}{t}\right) \leq \left(\frac{\alpha}{\beta e}\right)^\alpha = \left(\frac{2n}{(|x|-r)^2 e}\right)^{\frac{n}{2}}.$$
It follows that, whenever $|x| > r$,
$$u(x,t) = \int_{|y|<r} W(x-y,t)\,d\mu(y)$$
$$\leq (4\pi t)^{-\frac{n}{2}}\int_{|y|<r}\exp\left(-\frac{(|x|-r)^2}{4t}\right)d\mu(y)$$
$$\leq \mu(\mathbb{R}^n)\left(\frac{n}{2\pi e}\right)^{\frac{n}{2}}\frac{1}{(|x|-r)^n},$$
Therefore $u(x,t) \to 0$ as $|x| \to +\infty$, uniformly for $t \in {]}0,+\infty[$. □

LEMMA 4.13. *If $0 < s < t < a$, and u is a nonnegative temperature on the strip $\mathbb{R}^n \times {]}0,a[$, then the inequality*

(4.16) $$u(x,t) \geq \int_{\mathbb{R}^n} W(x-y,t-s)u(y,s)\,dy$$

holds for all $x \in \mathbb{R}^n$.

PROOF. Given any $r > 0$, we put
$$u_r(x,t) = \int_{|y|\leq r} W(x-y,t-s)u(y,s)\,dy,$$
and note that u_r is defined on $\mathbb{R}^n \times {]}s,+\infty[$, in view of Lemma 4.12. We consider the function $v_r = u - u_r$, which is defined on $\mathbb{R}^n \times {]}s,a[$. We show that $v_r \geq 0$. Theorem 4.4 shows that u_r is a temperature, and so v_r is too. We put
$$f(y) = \begin{cases} u(y,s) & \text{if } |y| \leq r, \\ 0 & \text{if } |y| > r. \end{cases}$$
Then Theorem 4.8 shows that
$$\limsup_{(x,t)\to(\xi,s+)} u_r(x,t) \leq \limsup_{\eta\to\xi} f(\eta) \leq u(\xi,s)$$
for all $\xi \in \mathbb{R}^n$. Therefore

(4.17) $$\liminf_{(x,t)\to(\xi,s+)} v_r(x,t) = u(\xi,s) - \limsup_{(x,t)\to(\xi,s+)} u_r(x,t) \geq 0$$

for all ξ. By Lemma 4.12, given any $\epsilon > 0$ we can find a number r_0 such that $u_r(x,t) < \epsilon$ whenever $|x| \geq r_0$ and $s < t < a$. Therefore, given any $R \geq r_0$, we have

$v_r(x,t) \geq -u_r(x,t) > -\epsilon$ whenever $|x| = R$ and $s < t < a$. This, together with (4.17), implies that $\liminf(v_r(x,t) + \epsilon) \geq 0$ whenever (x,t) approaches any point (ζ, ρ) of the boundary of the circular cylinder $B(0,R) \times]s,a[$ such that $\rho < a$. It now follows from the boundary minimum principle (Corollary 3.14) that $v_r \geq -\epsilon$ throughout the cylinder. Since ϵ is arbitrary, $v_r \geq 0$ on $B(0,R) \times]s,a[$. This holds for every $R \geq r_0$, and so $v_r \geq 0$ on $\mathbb{R}^n \times]s,a[$. Therefore $u_r \leq u$ on $\mathbb{R}^n \times]s,a[$ for any $r > 0$, and so the inequality (4.16) follows by making $r \to \infty$. $\square$

THEOREM 4.14. *If u is a nonnegative temperature on the strip $\mathbb{R}^n \times]0,a[$, and continuous on $\mathbb{R}^n \times [0,a[$ with $u(y,0) = 0$ for all $y \in \mathbb{R}^n$, then $u(x,t) = 0$ for all $(x,t) \in \mathbb{R}^n \times [0,a[$.*

PROOF. Suppose that $0 < s < c < b < a$. We shall apply Theorem 4.1 on the substrip $\mathbb{R}^n \times]0,c[$. By hypothesis,
$$\lim_{(x,t) \to (\xi,0+)} u(x,t) = 0$$
for all $\xi \in \mathbb{R}^n$. By Lemma 4.13, we have
$$\int_{\mathbb{R}^n} W(y, b-s) u(y,s) \, dy \leq u(0,b),$$
which implies that
$$\int_0^c \int_{\mathbb{R}^n} W(y, b-s) u(y,s) \, dy \, ds \leq cu(0,b).$$
Furthermore,
$$W(y, b-s) \geq (4\pi b)^{-\frac{n}{2}} \exp\left(-\frac{|y|^2}{4(b-c)}\right),$$
and it therefore follows, writing $k = (4(b-c))^{-1}$, that
$$\int_0^c \int_{\mathbb{R}^n} \exp(-k|y|^2) u(y,s) \, dy \, ds \leq cu(0,t)(4\pi b)^{\frac{n}{2}} < +\infty.$$
Since u is a nonnegative temperature, it now follows from Theorem 4.1 that $u \leq 0$, so that $u = 0$, on $\mathbb{R}^n \times [0,c[$. Because c can be arbitrarily close to a, $u = 0$ on $\mathbb{R}^n \times [0,a[$. $\square$

Theorem 4.14 combines with Theorem 4.1 to give the conclusion of Corollary 4.2 under weaker hypotheses. It is a feature of temperatures on a strip $\mathbb{R}^n \times]0,a[$ that constraints on the positive part often suffice when one might expect similar constraints on the modulus to be required.

COROLLARY 4.15. *Suppose that u is a temperature on the strip $\mathbb{R}^n \times]0,a[$, that*
$$\lim_{(x,t) \to (y,0+)} u(x,t) = A$$
for all $y \in \mathbb{R}^n$, and that for some positive number k
$$\int_0^a \int_{\mathbb{R}^n} \exp(-k|x|^2) u^+(x,t) \, dx \, dt < +\infty.$$
Then $u = A$ on $\mathbb{R}^n \times]0,b[$.

PROOF. We apply Theorem 4.1 with $w = u$, and deduce that $u \leq A$. So the function $A - u$ satisfies the hypotheses of Theorem 4.14, and is therefore identically zero. $\square$

THEOREM 4.16. *If u is a nonnegative temperature on the strip $\mathbb{R}^n \times]0, a[$, then u has the semigroup property on $\mathbb{R}^n \times]0, a[$.*

PROOF. Given any number $s \in]0, a[$, we put
$$v(x,t) = \int_{\mathbb{R}^n} W(x-y, t-s) u(y,s)\, dy.$$
Lemma 4.13 shows that $v(x,t)$ is finite for all $(x,t) \in \mathbb{R}^n \times]s, a[$, so that Theorem 4.4 implies that v is a temperature on that substrip. Furthermore, Theorem 4.8 shows that we can extend v to a continuous function on $\mathbb{R}^n \times [s, a[$, by putting $v(y,s) = u(y,s)$ for all $y \in \mathbb{R}^n$. The function $u - v$ is a temperature on $\mathbb{R}^n \times]s, a[$, is continuous on $\mathbb{R}^n \times [s, a[$, is zero at every point of $\mathbb{R}^n \times \{s\}$, and is nonnegative by Lemma 4.13. Therefore $u - v = 0$ by Theorem 4.14, so that
$$u(x,t) = \int_{\mathbb{R}^n} W(x-y, t-s) u(y,s)\, dy$$
whenever $x \in \mathbb{R}^n$ and $0 < s < t < a$. $\square$

THEOREM 4.17. *If u is a nonnegative temperature on the strip $\mathbb{R}^n \times]0, a[$, and there is a point $(x_0, t_0) \in \mathbb{R}^n \times]0, a[$ such that $u(x_0, t_0) = 0$, then $u(x,t) = 0$ for all $(x,t) \in \mathbb{R}^n \times]0, a[$.*

PROOF. By the strong minimum principle (Theorem 3.11), $u(x,t) = 0$ for all $(x,t) \in \mathbb{R}^n \times]0, t_0[$. If $0 < s < t_0$, then Theorem 4.16 shows that
$$u(x,t) = \int_{\mathbb{R}^n} W(x-y, t-s) u(y,s)\, dy = 0$$
for all $(x,t) \in \mathbb{R}^n \times]s, a[$. This completes the proof. $\square$

Our proof of the fact that every nonnegative temperature on $\mathbb{R}^n \times]0, a[$ can be written as the Gauss-Weierstrass integral of a nonnegative measure, relies not only on the theorems above, but also on two results about the convergence of measures. We use the notation $C_c(\mathbb{R}^n)$ to denote the class of continuous functions on $\mathbb{R}^n$ with compact support. The first of the results we refer to as the *weak compactness theorem*. It reads:

Let κ be a fixed real number, and let $\{\mu_i\}$ be a sequence of nonnegative measures on $\mathbb{R}^n$ such that $\mu_i(\mathbb{R}^n) \leq \kappa$ for all i. Then there exist a nonnegative measure μ on $\mathbb{R}^n$ such that $\mu(\mathbb{R}^n) \leq \kappa$, and a subsequence $\{\mu_{i_j}\}$, such that
$$\lim_{j \to \infty} \int_{\mathbb{R}^n} g\, d\mu_{i_j} = \int_{\mathbb{R}^n} g\, d\mu$$
for all $g \in C_c(\mathbb{R}^n)$.

The second result we refer to as the *weak convergence theorem*. It reads:

Let $\{f_i\}$ be a sequence of continuous functions on $\mathbb{R}^n$, which converges locally uniformly to a function f on $\mathbb{R}^n$, and suppose that there is a positive number λ such that $|f_i| \leq \lambda$ for all i. Let κ be a fixed real number, and let $\{\mu_i\}$ be a sequence of nonnegative measures on $\mathbb{R}^n$ such that:

(a) $\mu_i(\mathbb{R}^n) \leq \kappa$ for all i;

(b) $\{\mu_i\}$ converges to a nonnegative measure μ on $\mathbb{R}^n$, in the sense that

$$\lim_{i \to \infty} \int_{\mathbb{R}^n} g \, d\mu_i = \int_{\mathbb{R}^n} g \, d\mu$$

for all $g \in C_c(\mathbb{R}^n)$; and

(c) given any positive number η, we can find a positive number ρ such that $\mu_i(\{x \in \mathbb{R}^n : |x| \geq \rho\}) < \eta$ for all i.

Then

(4.18) $$\int_{\mathbb{R}^n} |f| \, d\mu < +\infty,$$

and

$$\lim_{i \to \infty} \int_{\mathbb{R}^n} f_i \, d\mu_i = \int_{\mathbb{R}^n} f \, d\mu.$$

Detailed proofs of these results can be found in Chapter 2 of Watson [**79**]. We are now ready to prove our main result.

THEOREM 4.18. *Let u be a nonnegative temperature on $\mathbb{R}^n \times \,]0, a[$. Then there is a nonnegative measure μ on $\mathbb{R}^n$ such that u is the Gauss-Weierstrass integral of μ on $\mathbb{R}^n \times \,]0, a[$. Furthermore, if u is also continuous and real-valued on $\mathbb{R}^n \times [0, a[$, then u is the Gauss-Weierstrass integral of $u(\cdot, 0)$.*

PROOF. If $0 < s < b < a$, then because u has the semigroup property (Theorem 4.16), we have

$$\int_{\mathbb{R}^n} W(y, b - s) u(y, s) \, dy = u(0, b).$$

Therefore, whenever $0 < s < \frac{1}{2}b$, the inequality

$$W(y, b - s) \geq (4\pi b)^{-\frac{n}{2}} \exp\left(-\frac{|y|^2}{2b}\right)$$

implies that

(4.19) $$\int_{\mathbb{R}^n} \exp\left(-\frac{|y|^2}{2b}\right) u(y, s) \, dy \leq (4\pi b)^{\frac{n}{2}} u(0, b).$$

For each $s \in \,]0, \frac{1}{2}b[$ and every Borel set $X \subseteq \mathbb{R}^n$, we put

$$\mu_s(X) = \int_X \exp\left(-\frac{|y|^2}{b}\right) u(y, s) \, dy.$$

It then follows from (4.19) that $\mu_s(\mathbb{R}^n) \leq (4\pi b)^{\frac{n}{2}} u(0, b)$ for all s. We can therefore apply the weak compactness theorem. Thus, there is a nonnegative measure ν on $\mathbb{R}^n$ such that $\nu(\mathbb{R}^n) \leq (4\pi b)^{\frac{n}{2}} u(0, b)$, and a decreasing null sequence $\{s_i\}$ in $]0, \frac{1}{2}b[$ such that

(4.20) $$\lim_{i \to \infty} \int_{\mathbb{R}^n} g \, d\mu_{s_i} = \int_{\mathbb{R}^n} g \, d\nu$$

for all $g \in C_c(\mathbb{R}^n)$.

We aim to apply the weak convergence theorem to the sequence of measures $\{\mu_{s_i}\}$. We know that hypotheses (a) and (b) of that theorem hold in the present case. Furthermore, using the inequality (4.19) we see that, for any positive number r and all $s \in]0, \tfrac{1}{2}b[$,

$$\int_{|y| \geq r} \exp\left(-\frac{|y|^2}{b}\right) u(y,s)\, dy \leq \exp\left(-\frac{r^2}{2b}\right) \int_{|y| \geq r} \exp\left(-\frac{|y|^2}{2b}\right) u(y,s)\, dy$$
$$\leq (4\pi b)^{\frac{n}{2}} u(0,b) \exp\left(-\frac{r^2}{2b}\right).$$

It follows that, given any positive number η, we can find a positive number ρ such that

$$\mu_s(\{x \in \mathbb{R}^n : |x| \geq \rho\}) = \int_{|y| \geq \rho} \exp\left(-\frac{|y|^2}{b}\right) u(y,s)\, dy < \eta,$$

uniformly for $s \in]0, \tfrac{1}{2}b[$. In particular, $\mu_{s_i}(\{x \in \mathbb{R}^n : |x| \geq \rho\}) < \eta$ for all i, so that hypothesis (c) of the weak convergence theorem is also satisfied.

We now fix a point $(x_0, t_0) \in \mathbb{R}^n \times]0, \tfrac{1}{8}b[$, and define functions f_i on $\mathbb{R}^n$ by putting

$$f_i(y) = W(x_0 - y, t_0 - s_i) \exp\left(\frac{|y|^2}{b}\right),$$

for every positive integer i such that $s_i < \tfrac{1}{2}t_0$. For such values of i, we have $\tfrac{1}{2}t_0 < t_0 - s_i < t_0 < \tfrac{1}{8}b$. Therefore, because $-2|x_0 - y|^2 \leq 2|x_0|^2 - |y|^2$ for all $y \in \mathbb{R}^n$, we have

$$f_i(y) \leq (2\pi t_0)^{-\frac{n}{2}} \exp\left(-\frac{|x_0 - y|^2}{4t_0} + \frac{|y|^2}{b}\right)$$
$$\leq (2\pi t_0)^{-\frac{n}{2}} \exp\left(\left(\frac{|x_0|^2}{4t_0} - \frac{|y|^2}{8t_0}\right) + \frac{|y|^2}{b}\right)$$
$$\leq (2\pi t_0)^{-\frac{n}{2}} \exp\left(\frac{|x_0|^2}{4t_0}\right).$$

Thus we have found a number $\lambda = (2\pi t_0)^{-\frac{n}{2}} \exp(|x_0|^2/(4t_0))$ such that $0 \leq f_i \leq \lambda$ for all i. Furthermore, since the function $(y,s) \mapsto W(x_0 - y, t_0 - s) \exp(|y|^2/b)$ is uniformly continuous on $K \times [0, \tfrac{1}{2}t_0]$ for each compact set $K \subseteq \mathbb{R}^n$, we see that $f_i(y) \to W(x_0 - y, t_0) \exp(|y|^2/b)$ locally uniformly in $\mathbb{R}^n$. Hence the sequence $\{f_i\}$ satisfies the hypotheses of the weak convergence theorem.

Applying the weak convergence theorem, we find that

$$\int_{\mathbb{R}^n} W(x_0 - y, t_0) \exp\left(\frac{|y|^2}{b}\right) d\nu(y) < +\infty$$

and

$$\lim_{i \to \infty} \int_{\mathbb{R}^n} W(x_0 - y, t_0 - s_i) u(y, s_i)\, dy = \int_{\mathbb{R}^n} W(x_0 - y, t_0) \exp\left(\frac{|y|^2}{b}\right) d\nu(y).$$

Since u has the semigroup property (by Theorem 4.16), the left-hand side of this equation is $u(x_0, t_0)$. Therefore, putting $d\mu(y) = \exp(|y|^2/b)\, d\nu(y)$, we obtain the representation

$$u(x_0, t_0) = \int_{\mathbb{R}^n} W(x_0 - y, t_0)\, d\mu(y)$$

for any $(x_0, t_0) \in \mathbb{R}^n \times]0, \tfrac{1}{8}b[$.

To extend this representation to all $(x_0, t_0) \in \mathbb{R}^n \times]0, a[$, we choose c such that

$0 < c < \frac{1}{8}b$ and use fact that W has the semigroup property on $\mathbb{R}^n \times]0, +\infty[$, by Theorem 4.10. Thus, for any $(x_0, t_0) \in \mathbb{R}^n \times [\frac{1}{8}b, a[$, we have

$$u(x_0, t_0) = \int_{\mathbb{R}^n} W(x_0 - y, t_0 - c) u(y, c) \, dy$$
$$= \int_{\mathbb{R}^n} W(x_0 - y, t_0 - c) \int_{\mathbb{R}^n} W(y - z, c) \, d\mu(z) \, dy$$
$$= \int_{\mathbb{R}^n} \int_{\mathbb{R}^n} W(x_0 - y, t_0 - c) W(y - z, c) \, dy \, d\mu(z)$$
$$= \int_{\mathbb{R}^n} W(x_0 - z, t_0) \, d\mu(z).$$

The change of the order of integration is justified by Tonelli's theorem.

We now consider the case where u is continuous on $\mathbb{R}^n \times [0, a[$. It follows from Fatou's lemma and the inequality (4.19) that

$$\int_{\mathbb{R}^n} \exp\left(-\frac{|y|^2}{2b}\right) u(y, 0) \, dy \leq \liminf_{s \to 0+} \int_{\mathbb{R}^n} \exp\left(-\frac{|y|^2}{2b}\right) u(y, s) \, dy$$
$$\leq (4\pi b)^{\frac{n}{2}} u(0, b)$$
$$< +\infty.$$

Therefore, by Theorem 4.4, the Gauss-Weierstrass integral v of $u(\cdot, 0)$ is defined and is a nonnegative temperature on $\mathbb{R}^n \times]0, \frac{1}{2}b[$. We put $v(y, 0) = u(y, 0)$ for all $y \in \mathbb{R}^n$, so that v becomes continuous on $\mathbb{R}^n \times [0, \frac{1}{2}b[$, by Theorem 4.8. We now put $w = u - v$ on $\mathbb{R}^n \times [0, \frac{1}{2}b[$. Then $w(y, 0) = 0$ for all $y \in \mathbb{R}^n$, and

$$\int_0^{\frac{1}{2}b} \int_{\mathbb{R}^n} \exp\left(-\frac{|y|^2}{2b}\right) w^+(y, s) \, dy \, ds \leq \int_0^{\frac{1}{2}b} \int_{\mathbb{R}^n} \exp\left(-\frac{|y|^2}{2b}\right) u(y, s) \, dy \, ds$$
$$\leq \frac{1}{2} b (4\pi b)^{\frac{n}{2}} u(0, b)$$

by inequality (4.19). Corollary 4.15 now shows that $w \equiv 0$, hence $u \equiv v$, on $\mathbb{R}^n \times]0, \frac{1}{2}b[$. Since u has the semigroup property on $\mathbb{R}^n \times]0, a[$, choosing c such that $0 < c < \frac{1}{2}b$ we obtain

$$u(x, t) = \int_{\mathbb{R}^n} W(x - y, t - c) u(y, c) \, dy$$
$$= \int_{\mathbb{R}^n} W(x - y, t - c) \int_{\mathbb{R}^n} W(y - z, c) u(z, 0) \, dz \, dy$$
$$= \int_{\mathbb{R}^n} u(z, 0) \int_{\mathbb{R}^n} W(x - y, t - c) W(y - z, c) \, dy \, dz$$
$$= \int_{\mathbb{R}^n} W(x - z, t) u(z, 0) \, dz$$

for all $(x, t) \in \mathbb{R}^n \times]c, a[$, and hence on the whole strip $\mathbb{R}^n \times]0, a[$. $\square$

4.4. Minimality of the Fundamental Temperature

DEFINITION 4.19. Let v be a nonnegative temperature on an open set E such that $v \neq 0$. Then v is called *minimal* if, whenever u is a nonnegative temperature majorized by v on E, then u is a constant multiple of v.

EXAMPLE 4.20. Let u be a nonnegative temperature on a strip or half-space $\mathbb{R}^n \times {]}0,a[$, so that u is the Gauss-Weierstrass integral of a nonnegative measure μ, by Theorem 4.18. If $u(x,t) \leq W(x-x_0,t)$ for all $(x,t) \in \mathbb{R}^n \times {]}0,a[$, then the equation
$$w(x,t) = W(x-x_0,t) - u(x,t)$$
defines a nonnegative temperature on $\mathbb{R}^n \times {]}0,a[$, and so w is the Gauss-Weierstrass integral of a nonnegative measure ω. Theorem 4.11 ensures the uniqueness of the representing measures, so that if δ_{x_0} denotes the unit mass at x_0, we have $\omega = \delta_{x_0} - \mu$, and hence $\mu \leq \delta_{x_0}$. It follows that μ is a constant multiple of δ_{x_0}, and hence that $u(x,t)$ is a constant multiple of $W(x-x_0,t)$ for all $(x,t) \in \mathbb{R}^n \times {]}0,a[$. Thus the spatial translations of W are minimal temperatures on a strip or half-space.

The results in this section are a refinement of the above fact. For simplicity, we treat only the case $x_0 = 0$.

THEOREM 4.21. *Let u be a positive temperature on $\mathbb{R}^n \times {]}0,a[$, let $0 < t_0 < a$, and let*
$$M(t_0, r) = \max\{u(x,t_0) : |x| = r\} \tag{4.21}$$
for all $r > 0$. If u is not a constant multiple of W, then
$$\liminf_{r \to \infty} \left(\frac{\log M(t_0,r)}{r} + \frac{r}{4t_0} \right) > 0;$$
otherwise
$$\lim_{r \to \infty} \left(\frac{\log M(t_0,r)}{r} + \frac{r}{4t_0} \right) = 0.$$

PROOF. The positive temperature u is the Gauss-Weierstrass integral of some nonnegative measure μ, by Theorem 4.18. Since $u > 0$, the measure μ is not null. If μ is concentrated at the point 0, then $u = \kappa W$ for some positive constant κ. Therefore
$$M(t_0,r) = \frac{\kappa}{(4\pi t_0)^{\frac{n}{2}}} \exp\left(-\frac{r^2}{4t_0}\right),$$
so that
$$\lim_{r \to \infty} \left(\frac{\log M(t_0,r)}{r} + \frac{r}{4t_0} \right) = \lim_{r \to \infty} \frac{\log\left(\kappa(4\pi t_0)^{-\frac{n}{2}}\right)}{r} = 0.$$
If μ is not concentrated at 0, then there is a closed ball $B \subseteq \mathbb{R}^n$ such that $0 \notin B$ and $\mu(B) > 0$. For all $x \in \mathbb{R}^n$, we have
$$u(x,t_0) \geq \frac{1}{(4\pi t_0)^{\frac{n}{2}}} \int_B \exp\left(-\frac{|x-y|^2}{4t_0}\right) d\mu(y).$$
If B has centre x_0 and radius ρ, then for all $y \in B$ we have
$$|x-y| \leq |x-x_0| + |x_0-y| \leq |x-x_0| + \rho,$$
so that
$$u(x,t_0) \geq \frac{1}{(4\pi t_0)^{\frac{n}{2}}} \exp\left(-\frac{(|x-x_0|+\rho)^2}{4t_0}\right) \mu(B).$$
Furthermore, whenever $|x| = r \geq |x_0| + \rho$, we have
$$|x-x_0| + \rho \geq |x| - |x_0| + \rho = r - |x_0| + \rho,$$

with equality if $x = rx_0/|x_0|$. Therefore
$$\min\{|x - x_0| + \rho : |x| = r\} = r - |x_0| + \rho,$$
and hence
$$M(t_0, r) \geq \frac{1}{(4\pi t_0)^{\frac{n}{2}}} \exp\left(-\frac{(r - |x_0| + \rho)^2}{4t_0}\right) \mu(B).$$
It follows that, whenever $r \geq |x_0| + \rho$,
$$\frac{\log M(t_0, r)}{r} \geq \frac{\log\left(\mu(B)(4\pi t_0)^{-\frac{n}{2}}\right)}{r} - \frac{r^2 + (|x_0| - \rho)^2 - 2r(|x_0| - \rho)}{4t_0 r}.$$
As $r \to \infty$, the first term on the right-hand side tends to zero, and we obtain
$$\liminf_{r \to \infty} \left(\frac{\log M(t_0, r)}{r} + \frac{r}{4t_0}\right) \geq -\lim_{r \to \infty}\left(\frac{(|x_0| - \rho)^2}{4t_0 r}\right) + \frac{|x_0| - \rho}{2t_0}$$
$$= \frac{|x_0| - \rho}{2t_0},$$
which is positive because $0 \notin B$. $\square$

THEOREM 4.22. *Let u be a nonnegative temperature on $\mathbb{R}^n \times]0, a[$, and let $0 < t_0 < a$. If $u(\cdot, t_0)$ is majorized by $W(\cdot, t_0)$, then u is a constant multiple of W on $\mathbb{R}^n \times]0, a[$.*

PROOF. Let $M(t_0, r)$ be defined by (4.21). Since $u(x, t_0) \leq W(x, t_0)$ for all $x \in \mathbb{R}^n$, we have
$$\limsup_{r \to \infty}\left(\frac{\log M(t_0, r)}{r} + \frac{r}{4t_0}\right) \leq \lim_{r \to \infty} \frac{\log\left((4\pi t_0)^{-\frac{n}{2}}\right)}{r} = 0,$$
and so the result follows from Theorem 4.21. $\square$

4.5. Notes and Comments

Theorem 4.1 was obtained by Watson [69], but Corollary 4.2 was proved earlier for more general parabolic equations by Slobodetskij [61] and Friedman [22, 23]. Other special cases of the theorem were also proved earlier, and detailed references can be found in chapter 2 of Watson [79].

Theorem 4.4 was proved by Flett [21] for the case of the Gauss-Weierstrass integral of a function, and his proof works also for the general case.

The proof of Theorem 4.8 is similar to that of a slightly weaker result, which was given by Watson [70]. The theorem itself can also be found in Watson [79].

The proof of Theorem 4.11 is taken from Aronson [4].

The Gauss-Weierstrass integral representation of nonnegative temperatures on $\mathbb{R}^n \times]0, a[$ is due to Widder [91] for $n = 1$, and to Krzyżański [44] for general n. Our approach is modelled on theirs.

The material on the minimality of the fundamental temperature in section 4.4, comes from Watson [74]. Theorem 4.22 raises the following **open question**. If u is the difference of two nonnegative temperatures on $\mathbb{R}^n \times]0, a[$, $0 < t_0 < a$, and $|u(\cdot, t_0)|$ is majorized by $W(\cdot, t_0)$, does this imply that u is a constant multiple of W on $\mathbb{R}^n \times]0, a[$? Related work was given by Gusarov [29, 30].

CHAPTER 5

Classes of Subtemperatures on an Infinite Strip

In this chapter, we investigate the properties of subtemperatures defined on an infinite strip $\mathbb{R}^n \times\,]0, a[$ or half-space $\mathbb{R}^n \times\,]0, +\infty[$. To do this, we define various classes of subtemperatures by imposing conditions on certain well-behaved integral means over hyperplanes of the form $\mathbb{R}^n \times \{t\}$. When applied to the positive parts of subtemperatures in the appropriate classes, these means are decreasing, and when applied to temperatures in the appropriate classes, they are constant. We apply these means to give characterizations of those temperatures that have nonnegative thermic majorants, those that possess the semigroup property, those that can be represented as the Gauss-Weierstrass integrals of signed measures, and those that can be represented as the Gauss-Weierstrass integrals of functions.

5.1. Hyperplane Mean Values and Classes of Subtemperatures

If $0 < t < b$, and v is a measurable function on $\mathbb{R}^n \times \{t\}$, then we define the *Hyperplane Mean* $M_b(v;t)$ by putting

$$M_b(v;t) = \int_{\mathbb{R}^n} W(x, b-t) v(x,t)\, dx,$$

provided that the integral exists. Observe that, if v is a temperature which has the semigroup property on some strip $\mathbb{R}^n \times\,]0, a[$ with $a > b$, then $M_b(v;t) = v(0,b)$.

It is convenient to begin this section by giving a version of Theorem 4.1 in terms of the hyperplane means M_b.

THEOREM 5.1. *Suppose that $0 \leq s < b$, and that w is a subtemperature on $\mathbb{R}^n \times\,]s, b[$. If $M_b(w^+; \cdot)$ is a locally integrable function on the half-closed interval $[s, b[$, and*

$$\limsup_{(x,t) \to (\xi, s+)} w(x,t) \leq A$$

for all $\xi \in \mathbb{R}^n$, then $w \leq A$ on $\mathbb{R}^n \times\,]s, b[$.

PROOF. Whenever $s < r < b$, we have

$$\int_s^r \int_{\mathbb{R}^n} \exp\left(-\frac{|x|^2}{4(b-r)}\right) w^+(x,t)\, dx\, dt$$

$$\leq \int_s^r \int_{\mathbb{R}^n} \exp\left(-\frac{|x|^2}{4(b-t)}\right) w^+(x,t)\, dx\, dt$$

$$= (4\pi)^{\frac{n}{2}} \int_s^r (b-t)^{\frac{n}{2}} \int_{\mathbb{R}^n} W(x,b-t) w^+(x,t)\, dx\, dt$$

$$= (4\pi)^{\frac{n}{2}} \int_s^r (b-t)^{\frac{n}{2}} M_b(w^+;t)\, dt$$

$$\leq (4\pi b)^{\frac{n}{2}} \int_s^r M_b(w^+;t)\, dt$$

$$< +\infty.$$

Therefore, if $v(x,t) = w(x, s+t)$ for all $(x,t) \in \mathbb{R}^n \times\,]0, b-s[$, Theorem 4.1 shows that $v \leq A$ on $\mathbb{R}^n \times\,]0, b-s[$. Hence $w \leq A$ on $\mathbb{R}^n \times\,]s, b[$. □

COROLLARY 5.2. *Suppose that $0 \leq s < b$, and that u is a temperature on $\mathbb{R}^n \times\,]s, b[$. If $M_b(|u|; \cdot)$ is a locally integrable function on the half-closed interval $[s, b[$, and*

$$\lim_{(x,t) \to (\xi, s+)} u(x,t) = A$$

for all $\xi \in \mathbb{R}^n$, then $u = A$ on $\mathbb{R}^n \times\,]s, b[$.

PROOF. Since $M_b(1; t) = 1$ whenever $0 < t < b$, it suffices to prove the case $A = 0$. We apply Theorem 5.1 with $w = |u|$, and deduce that $|u| \leq 0$, hence $u = 0$, on $\mathbb{R}^n \times\,]s, b[$. □

LEMMA 5.3. *Suppose that $0 \leq s < b$, and that μ is a signed measure on $\mathbb{R}^n$ such that*

(5.1) $$\int_{\mathbb{R}^n} W(y, b-s)\, d|\mu|(y) < +\infty.$$

Then the Gauss-Weierstrass integral u, given by

$$u(x,t) = \int_{\mathbb{R}^n} W(x-y, t-s)\, d\mu(y),$$

represents a temperature on $\mathbb{R}^n \times\,]s, b[$. Furthermore, whenever $s < t < b$, we have

(5.2) $$M_b(|u|; t) \leq \int_{\mathbb{R}^n} W(y, b-s)\, d|\mu|(y),$$

with equality if μ is nonnegative. Moreover, if $d\mu(y) = w(y,s)\, dy$ for some function w, then whenever $s < t < b$ we have the inequality

$$M_b(|u|; t) \leq M_b(|w|; s),$$

and the equality

$$M_b(u; t) = M_b(w; s).$$

PROOF. In view of (5.1), we have

$$|u(0,b)| \leq \int_{\mathbb{R}^n} W(y, b-s)\, d|\mu|(y) < +\infty,$$

so that Theorem 4.4 shows that u is defined and is a temperature on the strip $\mathbb{R}^n \times \,]s, b[$. Furthermore, whenever $s < t < b$, it follows from Fubini's theorem and Theorem 4.10 that

$$M_b(|u|;t) \leq \int_{\mathbb{R}^n} W(x, b-t) \int_{\mathbb{R}^n} W(x-y, t-s) \, d|\mu|(y) \, dx$$

$$= \int_{\mathbb{R}^n} \int_{\mathbb{R}^n} W(x, b-t) W(y-x, t-s) \, dx \, d|\mu|(y)$$

$$= \int_{\mathbb{R}^n} W(y, b-s) \, d|\mu|(y),$$

with equality if μ is nonnegative. This proves (5.2).

Next, if $d\mu(y) = w(y,s) \, dy$ then $d|\mu|(y) = |w(y,s)| \, dy$, so that it follows from (5.2) that

$$M_b(|u|;t) \leq \int_{\mathbb{R}^n} W(y, b-s)|w(y,s)| \, dy = M_b(|w|;s),$$

which is the required inequality. Finally, we obtain the equality from the semigroup property for W, by using Fubini's theorem on interchanging the order in a repeated integral. Since, by Theorem 4.10,

$$\int_{\mathbb{R}^n} \int_{\mathbb{R}^n} W(y, b-t) W(y-z, t-s)|w(z,s)| \, dy \, dz = \int_{\mathbb{R}^n} W(z, b-s)|w(z,s)| \, dz < +\infty$$

whenever $0 < s < t < b$, Fubini's theorem is applicable. Thus

$$M_b(u;t) = \int_{\mathbb{R}^n} W(y, b-t) \int_{\mathbb{R}^n} W(y-z, t-s) w(z,s) \, dz \, dy$$

$$= \int_{\mathbb{R}^n} w(z,s) \int_{\mathbb{R}^n} W(y, b-t) W(z-y, t-s) \, dy \, dz$$

$$= \int_{\mathbb{R}^n} W(z, b-s) w(z,s) \, dz$$

$$= M_b(w;s).$$

□

DEFINITION 5.4. Given any $b > 0$, we say that the function w *belongs to the class* Σ_b if w is a subtemperature on the strip $\mathbb{R}^n \times \,]0, b[$, and the function $t \mapsto M_b(w^+; t)$ is locally integrable on the open interval $]0, b[$.

If w is a subtemperature, then so is w^+, by Example 3.9. Therefore $w \in \Sigma_b$ only if $w^+ \in \Sigma_b$. Furthermore, if u and v are both subtemperatures then so is $u+v$, by Corollary 3.57, so that the inequality $(u+v)^+ \leq u^+ + v^+$ implies that the sum of two functions in Σ_b is also in Σ_b.

We can immediately show that any Gauss-Weierstrass integral which is defined and finite at some point, belong to the class Σ_b for some b.

THEOREM 5.5. *Let μ be a signed measure on $\mathbb{R}^n$ whose Gauss-Weierstrass integral u is defined and finite at some point $(x_0, a) \in \mathbb{R}^n \times \,]0, +\infty[$. Then u is a temperature which belongs to Σ_b for all $b < a$, each function $M_b(|u|;\cdot)$ is bounded on $]0, b[$, and $M_b(u;t) = u(0, b)$ whenever $0 < t < b < a$.*

PROOF. By Theorem 4.4, u is a temperature on $\mathbb{R}^n \times\,]0,a[$. Therefore

$$\int_{\mathbb{R}^n} W(y,b)\,d|\mu|(y) < +\infty$$

whenever $0 < b < a$, and

$$M_b(u^+;t) \leq M_b(|u|;t) \leq \int_{\mathbb{R}^n} W(y,b)\,d|\mu|(y)$$

for all $t \in\,]0,b[$, by (5.2). Thus $u \in \Sigma_b$ for all $b < a$, and each function $M_b(|u|;\cdot)$ is bounded on $]0,b[$. Finally, whenever $0 < t < b$ we have

$$u(0,b) = \int_{\mathbb{R}^n} W(y, b-t)u(y,t)\,dy = M_b(u;t),$$

by the semigroup property of Theorem 4.10. □

Theorem 5.5 raises the question of the relationship between the classes Σ_a and $\bigcap_{b<a} \Sigma_b$ of subtemperatures on the strip $\mathbb{R}^n \times\,]0,a[$.

THEOREM 5.6. *The class Σ_a is a proper subclass of $\bigcap_{b<a} \Sigma_b$.*

PROOF. Suppose that $w \in \Sigma_a$. If $0 < t < b < a$, then

$$\int_{\mathbb{R}^n} W(x, b-t) w^+(x,t)\,dx \leq \left(\frac{a-t}{b-t}\right)^{\frac{n}{2}} \int_{\mathbb{R}^n} W(x, a-t) w^+(x,t)\,dx.$$

Therefore the locally integrability of $M_a(w^+;\cdot)$ on $]0,a[$, implies that $M_b(w^+;\cdot)$ is locally integrable on $]0,b[$. Hence $w \in \bigcap_{b<a} \Sigma_b$.

If $f(x) = \exp(|x|^2/(4a))$ for all $x \in \mathbb{R}^n$, then the Gauss-Weierstrass integral u of f is given by

$$u(x,t) = \left(\frac{a}{a-t}\right)^{\frac{n}{2}} \exp\left(\frac{|x|^2}{4(a-t)}\right),$$

in view of Example 4.6. It belongs to $\bigcap_{b<a} \Sigma_b$ by Theorem 5.5, but not to Σ_a because

$$M_a(u;t) = \left(\frac{\sqrt{a}}{\sqrt{4\pi(a-t)}}\right)^n \int_{\mathbb{R}^n} dx = +\infty$$

whenever $0 < t < a$. □

Theorem 5.5 suggests that the more natural class of subtemperatures on the strip $\mathbb{R}^n \times\,]0,a[$ would be $\bigcap_{b<a} \Sigma_b$ rather than Σ_a. This is, indeed, the case. We shall prove that a subtemperature w belongs to $\bigcap_{b<a} \Sigma_b$ if and only if it satisfies an inequality related to the semigroup property, and that a temperature belongs to the same class if and only if it has the semigroup property.

Theorem 5.8 below is a first step towards this goal. In its proof, we use a form of the Vitali-Carathéodory theorem on approximation by semicontinuous functions, which we now present as a lemma.

LEMMA 5.7. *Suppose that μ is a nonnegative measure on $\mathbb{R}^n$, that ϕ is an extended real-valued function which is μ-integrable on $\mathbb{R}^n$, and that $\epsilon > 0$. Then there is a lower semicontinuous function g on $\mathbb{R}^n$, such that $g(x) \geq \phi(x)$ for all $x \in \mathbb{R}^n$ and*

$$\int_{\mathbb{R}^n} \big(g(x) - \phi(x)\big)\,d\mu(x) < \epsilon.$$

5.1. HYPERPLANE MEAN VALUES AND CLASSES OF SUBTEMPERATURES

PROOF. We write ϕ in terms of its positive and negative parts as $\phi^+ - \phi^-$. Each of ϕ^+ and ϕ^- is the pointwise limit of an increasing sequence of simple functions, and so each can be written as the sum of a series of nonnegative simple functions. Since every simple function is a linear combination of characteristic functions of measurable sets, it follows that ϕ^+ and ϕ^- can be written in the form

$$\phi^+ = \sum_{i=1}^{\infty} a_i \chi_{A_i} \quad \text{and} \quad \phi^- = \sum_{i=1}^{\infty} b_i \chi_{B_i},$$

where χ_S denotes the characteristic function of the measurable set S, and a_i, b_i are nonnegative real numbers. Since ϕ is μ-integrable, we have

$$\sum_{i=1}^{\infty} a_i \mu(A_i) = \int_{\mathbb{R}^n} \phi^+(x)\, d\mu(x) < +\infty$$

and

$$\sum_{i=1}^{\infty} b_i \mu(B_i) = \int_{\mathbb{R}^n} \phi^-(x)\, d\mu(x) < +\infty.$$

For each i, we choose a compact set K_i and an open set V_i, such that $K_i \subseteq B_i$, $A_i \subseteq V_i$, $a_i \mu(V_i \backslash A_i) < 2^{-i-2}\epsilon$, and $b_i \mu(B_i \backslash K_i) < 2^{-i-2}\epsilon$. Then we put

$$f(x) = \sum_{i=1}^{N} b_i \chi_{K_i}(x) \leq \phi^-(x) \quad \text{and} \quad h(x) = \sum_{i=1}^{\infty} a_i \chi_{V_i}(x) \geq \phi^+(x)$$

for all $x \in \mathbb{R}^n$, where N is chosen so that

$$\sum_{i=N+1}^{\infty} b_i \mu(B_i) < \frac{\epsilon}{2}.$$

Note that $h - f \geq \phi$. Also, since each function $a_i \chi_{V_i}$ is lower semicontinuous, so is $\sum_{i=1}^{k} a_i \chi_{V_i}$ for any k, and hence h is lower semicontinuous by Lemma 3.5. Since each function $b_i \chi_{K_i}$ is upper semicontinuous, so is f. Hence $h - f$ is lower semicontinuous. Furthermore,

$$h - f - \phi = \sum_{i=1}^{\infty} a_i \chi_{V_i} - \sum_{i=1}^{N} b_i \chi_{K_i} - \sum_{i=1}^{\infty} a_i \chi_{A_i} + \sum_{i=1}^{\infty} b_i \chi_{B_i}$$

$$= \sum_{i=1}^{\infty} a_i (\chi_{V_i} - \chi_{A_i}) + \sum_{i=1}^{N} b_i (\chi_{B_i} - \chi_{K_i}) + \sum_{i=N+1}^{\infty} b_i \chi_{B_i}.$$

Putting $g = h - f$, we therefore obtain

$$\int_{\mathbb{R}^n} (g - \phi)\, d\mu = \sum_{i=1}^{\infty} a_i \mu(V_i \backslash A_i) + \sum_{i=1}^{N} b_i \mu(B_i \backslash K_i) + \sum_{i=N+1}^{\infty} b_i \mu(B_i)$$

$$< \sum_{i=1}^{\infty} 2^{-i-2}\epsilon + \sum_{i=1}^{N} 2^{-i-2}\epsilon + \frac{\epsilon}{2}$$

$$< \epsilon.$$

□

THEOREM 5.8. *Suppose that $w \in \Sigma_a$, that $M_a(w^-;t)$ is finite for all $t \in \,]0,a[$, and that $0 < s < a$. Then the inequality*

(5.3) $$w(x,t) \leq \int_{\mathbb{R}^n} W(x-y, t-s) w(y,s)\, dy$$

holds for all $(x,t) \in \mathbb{R}^n \times \,]s, a[$, and the integrals are finite. Furthermore, for any $w \in \Sigma_a$, the mean $M_a(w^+; t)$ is finite for all $t \in \,]0, a[$.

PROOF. Suppose first that s is chosen so that $M_a(w^+; s)$ is finite. Then $M_a(|w|; s)$ is finite, so that the integral in (5.3) is finite for all $(x,t) \in \mathbb{R}^n \times \,]s, a[$, by Lemma 5.3.

A key element in our proof is the use of Theorem 4.8, and for this we need to approximate $w(\cdot, s)$ with a sequence of lower semicontinuous functions. Since $M_a(|w|; s) < +\infty$, Lemma 5.7 shows that, for each positive integer m, we can find a lower semicontinuous function g_m on $\mathbb{R}^n \times \{s\}$ such that $g_m(y,s) \geq w(y,s)$ for all $y \in \mathbb{R}^n$, and $M_a(g_m - w; s) < 1/m$. We now define a sequence $\{f_m\}$ on $\mathbb{R}^n \times \{s\}$ by putting $f_1 = g_1$ and $f_m = g_m \wedge f_{m-1}$ for all $m \geq 2$. Then each function f_m is lower semicontinuous, $f_m(y,s) \geq w(y,s)$ for all $y \in \mathbb{R}^n$, $M_a(f_m - w; s) < 1/m$ for all m, and the sequence $\{f_m\}$ is decreasing.

Since $M_a(|f_m|; s) \leq M_a(f_m - w; s) + M_a(|w|; s) < +\infty$, the function

$$h_m(x,t) = \int_{\mathbb{R}^n} W(x-y, t-s) f_m(y,s)\, dy$$

represents a temperature on $\mathbb{R}^n \times \,]s, a[$, by Lemma 5.3. Furthermore, since the sequence $\{f_m\}$ is decreasing, so is $\{h_m\}$, and Lebesgue's monotone convergence theorem shows that

$$\lim_{m \to \infty} h_m(x,t) = \int_{\mathbb{R}^n} W(x-y, t-s) \lim_{m \to \infty} f_m(y,s)\, dy.$$

Next, because $0 \leq M_a(f_m - w; s) < 1/m$ for all m, the sequence $\{f_m\}$ is convergent to $w(\cdot, s)$ in the Lebesgue space $L(W(y, a-s)dy)$. Therefore $\{f_m\}$ has a subsequence which converges almost everywhere to $w(\cdot, s)$, and hence, since $\{f_m\}$ is a decreasing sequence, $\{f_m\}$ itself converges a.e. to $w(\cdot, s)$. Therefore

(5.4) $$\lim_{m \to \infty} h_m(x,t) = \int_{\mathbb{R}^n} W(x-y, t-s) w(y,s)\, dy.$$

The inequality (5.3) will therefore follow if we show that $w \leq h_m$ on $\mathbb{R}^n \times \,]s, a[$ for all m.

To achieve this, we use Theorem 5.1. Since f_m is lower semicontinuous on $\mathbb{R}^n \times \{s\}$, and w is upper semicontinuous on $\mathbb{R}^n \times [s, a[$, for any point $\xi \in \mathbb{R}^n$ we have

$$\liminf_{(x,t) \to (\xi, s+)} h_m(x,t) \geq \liminf_{\eta \to \xi} f_m(\eta, s) \geq f_m(\xi, s) \geq w(\xi, s) \geq \limsup_{(x,t) \to (\xi, s+)} w(x,t),$$

by Theorem 4.8. Therefore

$$\limsup_{(x,t) \to (\xi, s+)} \bigl(w(x,t) - h_m(x,t)\bigr) \leq \limsup_{(x,t) \to (\xi, s+)} w(x,t) - \liminf_{(x,t) \to (\xi, s+)} h_m(x,t) \leq 0$$

for all $\xi \in \mathbb{R}^n$. Furthermore, whenever $s < r < a$, we have

$$\int_s^r \left(M_a((w-h_m)^+;t)\right) dt \le \int_s^r \left(M_a(w^+;t) + M_a(|h_m|;t)\right) dt$$
$$\le \int_s^r M_a(w^+;t)\, dt + (r-s)M_a(|f_m|;s)$$
$$< +\infty,$$

in view of Lemma 5.3. It now follows from Theorem 5.1 that $w - h_m \le 0$ on $\mathbb{R}^n \times {]s,a[}$ for all m. Hence the inequality (5.3) follows from (5.4), in the case where $M_a(w^+;s)$ is finite.

We can now show that $M_a(w^+;s)$ is finite for all $s \in {]0,a[}$. Given any $s \in {]0,a[}$, we choose $s' \in {]0,s[}$ such that $M_a(w^+;s') < +\infty$. Then the inequality (5.3) holds with s' in place of s, so that if we put

$$v(x,t) = \int_{\mathbb{R}^n} W(x-y, t-s') w(y, s')\, dy$$

for all $(x,t) \in \mathbb{R}^n \times {]s', a[}$, then Lemma 5.3 shows that

$$M_a(w^+;s) \le M_a(|v|;s) \le M_a(|w|;s') < +\infty.$$

This completes the proof. $\square$

We can now give a characterization of the subtemperatures in the class $\bigcap_{b<a} \Sigma_b$.

THEOREM 5.9. *Let $0 < a \le +\infty$, and let w be a subtemperature on $\mathbb{R}^n \times {]0, a[}$. Then w belongs to the class $\bigcap_{b<a} \Sigma_b$ if and only if the inequality*

(5.5) $$w^+(x,t) \le \int_{\mathbb{R}^n} W(x-y, t-s) w^+(y, s)\, dy$$

holds whenever $x \in \mathbb{R}^n$, $0 < s < t < a$, and the integrals are finite.

PROOF. If $w \in \bigcap_{b<a} \Sigma_b$ then $w^+ \in \bigcap_{b<a} \Sigma_b$, and therefore Theorem 5.8 shows that (5.5) holds and the integrals are finite.

To prove the converse, we first take any b such that $0 < b < a$. Then, whenever $0 < s < b$, the integral

$$\int_{\mathbb{R}^n} W(y, b-s) w^+(y, s)\, dy = M_b(w^+;s)$$

is finite by hypothesis. Therefore, by Lemma 5.3, if $u(x,t)$ denotes the integral in (5.5) we have $M_b(w^+;t) \le M_b(u;t) = M_b(w^+;s)$ whenever $s < t < b$. Hence the function $M_b(w^+;\cdot)$ is locally bounded on ${]0,b[}$. Thus $w \in \bigcap_{b<a} \Sigma_b$. $\square$

We now turn our attention to temperatures and the semigroup property.

LEMMA 5.10. *Suppose that $w \in \Sigma_a$, and that there are positive numbers b and c with $b < c \le a$, such that w is a temperature on $\mathbb{R}^n \times {]b, c[}$, and w is continuous and real-valued on $\mathbb{R}^n \times {[b, c[}$. Then the equality*

$$w(x,t) = \int_{\mathbb{R}^n} W(x-y, t-b) w(y, b)\, dy$$

holds whenever $(x,t) \in \mathbb{R}^n \times {]b, c[}$.

PROOF. Observe that $w^+ \in \Sigma_a$. Therefore, by Theorem 5.8, the inequality

$$(5.6) \qquad w^+(x,t) \leq \int_{\mathbb{R}^n} W(x-y, t-b)w^+(y,b)\, dy$$

holds whenever $(x,t) \in \mathbb{R}^n \times\,]b, a[$, and the integrals are finite. If $v(x,t)$ denotes the integral in (5.6), then v is a temperature on $\mathbb{R}^n \times\,]b, a[$, and has a continuous extension by $w^+(\cdot, b)$ to $\mathbb{R}^n \times [b, a[$, in view of Theorem 4.8. Therefore $v - w$ is a nonnegative temperature on $\mathbb{R}^n \times\,]b, c[$, with a continuous extension by $w^-(\cdot, b)$ $\mathbb{R}^n \times [b, c[$. So Theorem 4.18 gives the representation

$$(v-w)(x,t) = \int_{\mathbb{R}^n} W(x-y, t-b)w^-(y,b)\, dy$$

for all $(x,t) \in \mathbb{R}^n \times\,]b, c[$. Hence

$$w(x,t) = v(x,t) - \int_{\mathbb{R}^n} W(x-y, t-b)w^-(y,b)\, dy,$$

as we asserted. □

THEOREM 5.11. *Let $0 < a \leq +\infty$, and let u be a temperature on $\mathbb{R}^n \times\,]0, a[$. Then u belongs to the class $\bigcap_{b<a} \Sigma_b$ if and only if the equality*

$$u(x,t) = \int_{\mathbb{R}^n} W(x-y, t-s)u(y,s)\, dy$$

holds whenever $x \in \mathbb{R}^n$ and $0 < s < t < a$.

PROOF. If $u \in \bigcap_{b<a} \Sigma_b$, then Lemma 5.10 shows that the required identity holds.

To prove the converse, we first take any b such that $0 < b < a$. Then, whenever $0 < s < b$, the integral

$$\int_{\mathbb{R}^n} W(y, b-s)u(y,s)\, dy = u(0, b)$$

is finite, so that

$$M_b(|u|; s) = \int_{\mathbb{R}^n} W(y, b-s)|u(y,s)|\, dy$$

is also finite. Therefore, by Lemma 5.3, $M_b(|u|; t) \leq M_b(|u|; s)$ whenever $s < t < b$. Thus the function $M_b(|u|; \cdot)$ is decreasing and real valued, which implies that the function $M_b(u^+; \cdot)$ is locally integrable on $]0, b[$. Hence $u \in \bigcap_{b<a} \Sigma_b$. □

Theorem 5.11 has some interesting consequences.

COROLLARY 5.12. *Let $0 < a \leq +\infty$, and let u be a temperature on $\mathbb{R}^n \times\,]0, a[$. Then the equality*

$$u(x,t) = \int_{\mathbb{R}^n} W(x-y, t-s)u(y,s)\, dy$$

holds whenever $x \in \mathbb{R}^n$ and $0 < s < t < a$, if and only if the inequality

$$u^+(x,t) \leq \int_{\mathbb{R}^n} W(x-y, t-s)u^+(y,s)\, dy$$

holds for all $x \in \mathbb{R}^n$ and $0 < s < t < a$ and the integrals are finite.

PROOF. The equality holds if and only if $u \in \bigcap_{b<a} \Sigma_b$, by Theorem 5.11. The inequality holds and the integrals are finite if and only if $u \in \bigcap_{b<a} \Sigma_b$, by Theorem 5.9. □

COROLLARY 5.13. *Let $0 < a \leq +\infty$, and let u be a temperature on $\mathbb{R}^n \times\,]0, a[$. Then $u \in \bigcap_{b<a} \Sigma_b$ if and only if $-u \in \bigcap_{b<a} \Sigma_b$.*

PROOF. We have $u \in \bigcap_{b<a} \Sigma_b$ if and only if it has the semigroup property, which occurs if and only if $-u$ has the semigroup property, which occurs if and only if $-u \in \bigcap_{b<a} \Sigma_b$. □

Corollary 5.13 is remarkable because the condition $u \in \bigcap_{b<a} \Sigma_b$ apparently does not constrain $u^- = (-u)^+$.

COROLLARY 5.14. *Let $0 < a \leq +\infty$, and let w be a function on $\mathbb{R}^n \times\,]0, a[$. Then w is a temperature in the class $\bigcap_{b<a} \Sigma_b$ if and only if both w and $-w$ are subtemperatures in the class $\bigcap_{b<a} \Sigma_b$.*

PROOF. If w is a temperature in the class $\bigcap_{b<a} \Sigma_b$, then both w and $-w$ are subtemperatures that belong to the class $\bigcap_{b<a} \Sigma_b$, in view of Corollary 5.13.

The converse uses only the fact that, if both w and $-w$ are subtemperatures, then w is a temperature. □

Theorem 5.11 combines with Corollary 5.2 to improve that latter result for temperatures in the class $\bigcap_{b<a} \Sigma_b$. It is a version of Corollary 4.15 in terms of the means M_b.

COROLLARY 5.15. *Let $0 < a \leq +\infty$, and let u be a temperature on $\mathbb{R}^n \times\,]0, a[$ such that, for every $b < a$, the function $M_b(u^+;\cdot)$ is locally integrable on the half-closed interval $[0, b[$. If*

$$\lim_{(x,t) \to (\xi, 0+)} u(x, t) = A$$

for all $\xi \in \mathbb{R}^n$, then $u = A$ on $\mathbb{R}^n \times\,]0, a[$.

PROOF. Since $M_b(1;t) = 1$ whenever $0 < t < b$, it suffices to prove the case $A = 0$. The condition on $M_b(u^+;\cdot)$ implies that $u \in \bigcap_{b<a} \Sigma_b$, so that u has the semigroup property, by Theorem 5.11. Therefore, whenever $0 < t < b < a$, we have

$$M_b(|u|;t) = 2M_b(u^+;t) - \int_{\mathbb{R}^n} W(x, b-t) u(x,t)\, dt = 2M_b(u^+;t) - u(0,b).$$

It follows that, for every $b < a$, the function $M_b(|u|;\cdot)$ is locally integrable on the half-closed interval $[0, b[$. Now Corollary 5.2 shows that $u = 0$ on $\mathbb{R}^n \times\,]0, b[$, and the result follows. □

EXAMPLE 5.16. Let $g(x) = -\exp(|x|^2/(4a))$ for all $x \in \mathbb{R}^n$, and let v be the Gauss-Weierstrass integral of g, so that

$$v(x,t) = -\left(\frac{a}{a-t}\right)^{\frac{n}{2}} \exp\left(\frac{|x|^2}{4(a-t)}\right),$$

in view of Example 4.6. Then v is a negative temperature on $\mathbb{R}^n\times]0,a[$, so that obviously $v \in \Sigma_a$. But

$$M_a(-v;t) = \left(\frac{\sqrt{a}}{\sqrt{4\pi(a-t)}}\right)^n \int_{\mathbb{R}^n} dx = +\infty$$

whenever $0 < t < a$, so that $-v \notin \Sigma_a$. Hence the result of Corollary 5.13, has no counterpart for the class Σ_a.

5.2. Behaviour of the Hyperplane Mean Values of Subtemperatures

The properties of the integral means M_a have already been used in some of our proofs. In this section, we investigate these properties further and show how well behaved the means are.

THEOREM 5.17. *Suppose that $w \in \Sigma_a$, and that $M_a(w^-;t)$ is finite for all $t \in]0,a[$. Then the function $M_a(w;\cdot)$ is real valued, decreasing, and left continuous on $]0,a[$.*

PROOF. Suppose that $0 < s < t < a$, and let u be defined by

$$u(x,t) = \int_{\mathbb{R}^n} W(x-y, t-s)w(y,s)\, dy$$

for all $x \in \mathbb{R}^n$. Theorem 5.8 shows that $w(x,t) \leq u(x,t)$ for all x, and that the mean $M_a(w;\cdot)$ is real-valued. Therefore, by Lemma 5.3,

$$M_a(w;t) \leq M_a(u;t) = M_a(w;s).$$

Thus the mean $M_a(w;\cdot)$ is decreasing on $]0,a[$.

To prove left continuity for a decreasing function, it suffices to prove upper semicontinuity. We take any $r \in]0,a[$, and any sequence $\{r_k\}$ in $]\tfrac{1}{2}r, a[$ with limit r. We put

$$v(x,t) = \int_{\mathbb{R}^n} W\left(x-y, t-\tfrac{1}{2}r\right) w\left(y, \tfrac{1}{2}r\right) dy$$

for all $(x,t) \in \mathbb{R}^n\times]\tfrac{1}{2}r, a[$. Then v is a temperature, so that the function $v - w$ is lower semicontinuous on $\mathbb{R}^n\times]\tfrac{1}{2}r, a[$. Furthermore, by Theorem 5.8, we have $w(x, r_k) \leq v(x, r_k)$ for all $x \in \mathbb{R}^n$ and all k. It therefore follows from Fatou's lemma and Lemma 5.3 that

$$\begin{aligned}
M_a(w;\tfrac{1}{2}r) - M_a(w;r) &= M_a(v;r) - M_a(w;r) \\
&= M_a(v-w;r) \\
&\leq \int_{\mathbb{R}^n} \liminf_{k\to\infty} \Big(W(x, a-r_k)\big(v(x,r_k) - w(x,r_k)\big)\Big) dx \\
&\leq \liminf_{k\to\infty} \big(M_a(v;r_k) - M_a(w;r_k)\big) \\
&= M_a(w;\tfrac{1}{2}r) - \limsup_{k\to\infty} M_a(w;r_k),
\end{aligned}$$

so that $\limsup_{k\to\infty} M_a(w;r_k) \leq M_a(w;r)$. Hence $M_a(w;\cdot)$ is left continuous on $]0,a[$. $\square$

5.2. BEHAVIOUR OF THE HYPERPLANE MEAN VALUES OF SUBTEMPERATURES

EXAMPLE 5.18. Let f be a real-valued, decreasing, left continuous function on a bounded interval $]0, a[$, and put $w(x, t) = f(t)$ for all $(x, t) \in \mathbb{R}^n \times]0, a[$. We have already remarked that w is a subtemperature on $\mathbb{R}^n \times]0, a[$. Furthermore, whenever $0 < t < a$, we have $M_a(w^+; t) = f^+(t)$, so that $w \in \Sigma_a$. This shows that, for an arbitrary subtemperature v on $\mathbb{R}^n \times]0, a[$, left continuity is the most that we can expect from the mean $M_a(v^+; \cdot)$. It also emphasizes the fact that $M_a(v^+; \cdot)$ is a one-dimensional subtemperature.

THEOREM 5.19. *Suppose that $0 < b < c \leq a$ and that $w \in \Sigma_a$. If w is a temperature on the substrip $\mathbb{R}^n \times]b, c[$, then*

$$-\infty \leq M_a(w; r) = M_a(w; s) < +\infty$$

whenever $b < r < s < c$. Conversely, if $M_a(w; \cdot)$ is finite on $]0, a[$ and constant on $]b, c[$, then w is a temperature with the semigroup property on $\mathbb{R}^n \times]b, c[$.

PROOF. Since $w \in \Sigma_a$, we also have $w^+ \in \Sigma_a$, and so Theorem 5.17 shows that $M_a(w^+; \cdot)$ is real-valued on $]0, a[$. Hence $M_a(w; t)$ exists and is not $+\infty$, whenever $0 < t < a$.

Suppose that w is a temperature on $\mathbb{R}^n \times]b, c[$, and that $b < r < s < c$. Then, by Lemma 5.10, the equality

$$w(x, t) = \int_{\mathbb{R}^n} W(x - y, t - r) w(y, r) \, dy$$

holds whenever $(x, t) \in \mathbb{R}^n \times]r, c[$. Therefore

$$M_a(w; s) = \int_{\mathbb{R}^n} W(x, a - s) \int_{\mathbb{R}^n} W(x - y, s - r) w(y, r) \, dy \, dx.$$

Also, because the fundamental temperature W has the semigroup property by Theorem 4.10,

$$M_a(w; r) = \int_{\mathbb{R}^n} w(y, r) \int_{\mathbb{R}^n} W(x, a - s) W(x - y, s - r) \, dx \, dy.$$

It now follows from Fubini's theorem that, if either $M_a(w; s)$ or $M_a(w; r)$ is finite, then both are finite and the two are equal. This is equivalent to our assertion.

For the converse, suppose that $M_a(w; \cdot)$ is finite on $]0, a[$ and constant on $]b, c[$, and that $b < r < c$. Then $M_a(w^-; \cdot) < +\infty$ on $]0, a[$, so that

$$w(x, t) \leq \int_{\mathbb{R}^n} W(x - y, t - r) w(y, r) \, dy$$

for all $(x, t) \in \mathbb{R}^n \times]r, a[$ and the integrals are finite, by Theorem 5.8. We put

$$v(x, t) = \int_{\mathbb{R}^n} W(x - y, t - r) w(y, r) \, dy$$

whenever $(x, t) \in \mathbb{R}^n \times]r, c[$, so that $w(x, t) \leq v(x, t)$ there. Furthermore, by our hypothesis and Lemma 5.3, whenever $r < t < c$ we have

$$M_a(w; r) = M_a(w; t) \leq M_a(v; t) = M_a(w; r),$$

so that $M_a(v - w; t) = 0$. The function $(x, t) \mapsto W(x, a - t)\big(v(x, t) - w(x, t)\big)$ is nonnegative and lower semicontinuous, and is zero at a point if and only if $v - w$ is zero at that point. Therefore its integral $M_a(v-w;t)$ can be zero only if $v-w=0$ throughout $\mathbb{R}^n \times \{t\}$. Hence $v=w$ on $\mathbb{R}^n \times]r,c[$, which gives us the equality

$$w(x,t) = \int_{\mathbb{R}^n} W(x-y, t-r) w(y,r) \, dy$$

for all $(x,t) \in \mathbb{R}^n \times]r,c[$, whenever $b < r < c$. Thus w has the semigroup property on $\mathbb{R}^n \times]b,c[$, and so is a temperature there. □

REMARK 5.20. We must allow the value $-\infty$ in Theorem 5.19. If v is the function in Example 5.16, then v is a temperature on $\mathbb{R}^n \times]0,a[$, it belongs to the class Σ_a, and has $M_a(v;t) = -\infty$ for all $t \in]0,a[$.

REMARK 5.21. If w is defined as in Example 5.18, then $w + v \in \Sigma_a$ and $M_a(w+v;t) = -\infty$ for all $t \in]0,a[$, but $w+v$ may not be a temperature on $\mathbb{R}^n \times]0,a[$. So the converse part of Theorem 5.19 does not follow if we allow M_a to be identically $-\infty$.

Theorem 5.19 gives a necessary and sufficient condition for a function in the class $\bigcap_{b<a} \Sigma_b$ to be a temperature on $\mathbb{R}^n \times]0,a[$.

THEOREM 5.22. *Suppose that $0 < a \leq +\infty$, and that $w \in \bigcap_{b<a} \Sigma_b$. Then w is a temperature on $\mathbb{R}^n \times]0,a[$ if and only if, for every $b < a$, the function $M_b(w;\cdot)$ is constant and finite on $]0,b[$.*

PROOF. If w is a temperature on $\mathbb{R}^n \times]0,a[$ and $b<a$, then Theorem 5.19 shows that $M_b(w;\cdot)$ is constant, possibly $-\infty$, on $]0,b[$. By Corollary 5.13, $-w \in \Sigma_b$, which excludes the possibility that $M_b(w;\cdot) = -\infty$.

Conversely if, for every $b < a$, the function $M_b(w;\cdot)$ is constant and finite on $]0,b[$, then Theorem 5.19 shows that w is a temperature on $\mathbb{R}^n \times]0,b[$ for every $b < a$. □

EXAMPLE 5.23. Clearly Example 5.18 implies that, whenever $0 < a \leq +\infty$ and $w \in \bigcap_{b<a} \Sigma_b$, the means $M_b(w^+;\cdot)$ may all be unbounded. Here we show that this may happen if w is a temperature.

Let w be defined on $\mathbb{R}^n \times]0,+\infty[$ by the formula

$$w(x,t) = (|x|^2 - 2nt) t^{-\frac{n}{2}-2} \exp\left(-\frac{|x|^2}{4t}\right).$$

Then w is a temperature on $\mathbb{R}^n \times]0,+\infty[$. If $0 < t < b$, we have

$$M_b(w^+;t) = t^{-\frac{n}{2}-2} (4\pi(b-t))^{-\frac{n}{2}} \int_{|x|^2 \geq 2nt} (|x|^2 - 2nt) \exp\left(-\frac{b|x|^2}{4t(b-t)}\right) dx$$

$$= t^{-\frac{n}{2}-2} (4\pi(b-t))^{-\frac{n}{2}} \int_{\sqrt{2nt}}^{\infty} (\rho^2 - 2nt) \exp\left(-\frac{b\rho^2}{4t(b-t)}\right) \omega_n \rho^{n-1} \, d\rho,$$

where ω_n denotes the $(n-1)$-dimensional surface area of the unit sphere in $\mathbb{R}^n$. We now put

$$\eta = \frac{b\rho^2}{4t(b-t)} \qquad \text{and} \qquad s = \frac{bn}{2(b-t)},$$

to obtain

$$M_b(w^+;t) = A_{n,b} \left(\frac{b-t}{t}\right) \int_s^{\infty} (\eta-s) \eta^{\frac{n}{2}-1} e^{-\eta} \, d\eta,$$

where $A_{n,b}$ depends only on n and b. Hence, because s decreases to $\frac{n}{2}$ as t decreases to 0,

$$M_b(w^+;t) \leq A_{n,b}\left(\frac{b-t}{t}\right)\int_{\frac{n}{2}}^{\infty}\left(\eta - \frac{n}{2}\right)\eta^{\frac{n}{2}-1}e^{-\eta}\,d\eta,$$

which implies that $w \in \Sigma_b$. Finally, as $t \to 0+$,

$$\int_s^{\infty}(\eta - s)\eta^{\frac{n}{2}-1}e^{-\eta}\,d\eta \to \int_{\frac{n}{2}}^{\infty}\left(\eta - \frac{n}{2}\right)\eta^{\frac{n}{2}-1}e^{-\eta}\,d\eta > 0,$$

so that $M_b(w^+;t) \to +\infty$.

We conclude this section by demonstrating that temperatures in $\bigcap_{b<a}\Sigma_b$ have a locally uniform growth rate as the spatial variables tend to infinity.

THEOREM 5.24. *Let $0 < a \leq +\infty$, and let u be a temperature on $\mathbb{R}^n \times\,]0,a[$. If $u \in \bigcap_{c<a}\Sigma_c$ and $0 < b < a$, then*

$$W(x, b-t)u(x,t) \to 0 \qquad as \qquad |x| \to \infty,$$

uniformly for t in any compact subset of $]0,b[$.

Conversely if, given any positive number $b < a$ and compact subset K of $]0,b[$, we can find a constant $C(b,K)$ such that

$$W(x, b-t)u^+(x,t) \leq C(b,K)$$

for all $(x,t) \in \mathbb{R}^n \times K$, then $u \in \bigcap_{c<a}\Sigma_c$.

PROOF. Suppose that $u \in \bigcap_{c<a}\Sigma_c$. Given b such that $0 < b < a$, we put

$$v(x,t) = W(x, t+1)u\left(\frac{x\sqrt{b}}{t+1}, \frac{tb}{t+1}\right)$$

for all $(x,t) \in \mathbb{R}^n \times\,]0,+\infty[$. Then v is a temperature on $\mathbb{R}^n \times\,]0,+\infty[$, and if we put

$$x' = \frac{x\sqrt{b}}{t+1}, \qquad t' = \frac{tb}{t+1},$$

then

(5.7) $$v(x,t) = \left(\frac{b-t'}{\sqrt{b}}\right)^n W(x', b-t')u(x',t').$$

Therefore, for any $t > 0$, we have

$$\int_{\mathbb{R}^n}|v(x,t)|\,dx = M_b(|u|;t').$$

If $0 < \alpha < \beta < b$, then the inequalities $\alpha \leq t' \leq \beta$ hold if and only if

$$\gamma \equiv \frac{\alpha}{b-\alpha} \leq t \leq \frac{\beta}{b-\beta} \equiv \delta.$$

We choose t_0 such that $0 < t_0 < \gamma$, and put $c = \gamma - t_0$. For all $t \geq t_0$, we have

$$t'_0 = \frac{t_0 b}{t_0 + 1} \leq \frac{tb}{t+1} = t'.$$

It follows from Corollary 5.13 that $|u| \in \bigcap_{c<a} \Sigma_c$, and so Theorem 5.17 shows that the function $M_b(|u|;\cdot)$ is decreasing and real-valued on $]0,b[$. Therefore, for all $t \geq t_0$, we have
$$\int_{\mathbb{R}^n} |v(x,t)|\, dx = M_b(|u|;t') \leq M_b(|u|;t'_0),$$
and hence
(5.8) $$\int_{t_0}^{\delta} \int_{\mathbb{R}^n} |v(x,t)|\, dx\, dt \leq M_b(|u|;t'_0)(\delta - t_0) < +\infty.$$

Let m be an integer, $m \geq 3$, put $l = 2(m+n)/e$, and let κ be the positive number in Corollary 1.26. It follows from (5.8) that, given any $\epsilon > 0$, we can find $\rho > 0$ such that
$$\kappa \int_{t_0}^{\delta} \int_{|x| \geq \rho} |v(x,t)|\, dx\, dt < \left(\frac{b-\beta}{\sqrt{b}}\right)^n \epsilon.$$

Hence, if $\gamma \leq r \leq \delta$ and $|z| \geq \rho + l\sqrt{c}$, Corollary 1.26 shows that
$$|v(z,r)| \leq \kappa \int_{r-c}^{r} \int_{|x-z|<l\sqrt{c}} |v(x,t)|\, dx\, dt$$
$$\leq \kappa \int_{t_0}^{\delta} \int_{|x|\geq\rho} |v(x,t)|\, dx\, dt$$
$$< \left(\frac{b-\beta}{\sqrt{b}}\right)^n \epsilon.$$

In terms of u, this inequality is
$$\left(\frac{b-r'}{\sqrt{b}}\right)^n W(z',b-r')|u(z',r')| < \left(\frac{b-\beta}{\sqrt{b}}\right)^n \epsilon.$$

Because $\alpha \leq r' \leq \beta$, we have $b - r' \geq b - \beta$, and so this implies that
$$W(z',b-r')|u(z',r')| < \epsilon,$$
provided that $|z| \geq \rho + l\sqrt{c}$. Since
$$z' = \frac{zr'}{r\sqrt{b}}$$
with $\gamma \leq r \leq \delta$, we have
$$|z'| \leq \frac{|z|\beta}{\gamma\sqrt{b}},$$
so that we obtain $|z| \geq \rho + l\sqrt{c}$ whenever
$$|z'| \geq \frac{\beta(\rho + l\sqrt{c})}{\gamma\sqrt{b}}.$$

This completes the proof of the first statement.

To prove the converse, we let $0 < c < b < a$, and choose any compact interval

$[\alpha, \beta] \subseteq]0, c[$. Then for all $t \in [\alpha, \beta]$, we have

$$M_c(u^+; t) = \int_{\mathbb{R}^n} W(x, c-t) u^+(x, t) \, dx$$

$$= \left(\frac{b-t}{c-t}\right)^{\frac{n}{2}} \int_{\mathbb{R}^n} \exp\left(-\frac{|x|^2}{4}\left(\frac{1}{c-t} - \frac{1}{b-t}\right)\right) W(x, b-t) u^+(x, t) \, dx$$

$$= \left(\frac{b-t}{c-t}\right)^{\frac{n}{2}} \int_{\mathbb{R}^n} \exp\left(-\frac{(b-c)|x|^2}{4(c-t)(b-t)}\right) W(x, b-t) u^+(x, t) \, dx.$$

Therefore the inequality $W(x, b-t) u^+(x, t) \leq C(b, K)$, for $K = [\alpha, \beta]$, implies that

$$M_c(u^+; t) \leq C(b, K) \left(\frac{b-t}{c-t}\right)^{\frac{n}{2}} \int_{\mathbb{R}^n} \exp\left(-\frac{(b-c)|x|^2}{4(c-t)(b-t)}\right) dx$$

$$= C(b, K) \left(\frac{b-t}{c-t}\right)^{\frac{n}{2}} \left(\frac{4\pi(c-t)(b-t)}{b-c}\right)^{\frac{n}{2}}$$

$$\leq C(b, K) \left(\frac{4\pi b^2}{b-c}\right)^{\frac{n}{2}},$$

in view of Lemma 1.1. It follows that $u \in \Sigma_c$ whenever $0 < c < b < a$, so that $u \in \bigcap_{c<a} \Sigma_c$. $\square$

5.3. Classes of Subtemperatures and Nonnegative Thermic Majorants

We define classes of subtemperatures on $\mathbb{R}^n \times]0, a[$ which are subclasses of the classes Σ_a, and a have a close relationship with the thermic majorization of nonnegative subtemperatures. They characterize those temperatures that can be represented by Gauss-Weierstrass integrals.

DEFINITION 5.25. Given any positive real number a, we say that the function w *belongs to the class* Φ_a if w is a subtemperature on the strip $\mathbb{R}^n \times]0, a[$, the function $M_a(w^+; \cdot)$ is locally integrable on $]0, a[$, and $\liminf_{t \to 0+} M_a(w^+; t) < +\infty$.

Obviously $\Phi_a \subseteq \Sigma_a$. Theorem 5.17 shows that the function $M_a(w^+; \cdot)$ is real-valued and decreasing on $]0, a[$, and so the condition $\liminf_{t \to 0+} M_a(w^+; t) < +\infty$ implies that the function is bounded. Example 5.18 implies that Φ_a is a proper subset of Σ_a.

THEOREM 5.26. *The class* Φ_a *is a proper subset of* $\bigcap_{b<a} \Phi_b$.

PROOF. Suppose that $w \in \Phi_a$. If $0 < t < b < a$, then the inequality

$$\int_{\mathbb{R}^n} W(x, b-t) w^+(x, t) \, dx \leq \left(\frac{a-t}{b-t}\right)^{\frac{n}{2}} \int_{\mathbb{R}^n} W(x, a-t) w^+(x, t) \, dx$$

implies that $M_b(w^+; \cdot)$ is locally integrable on $]0, b[$, and that

$$\liminf_{t \to 0+} M_b(w^+; t) \leq \left(\frac{a}{b}\right)^{\frac{n}{2}} \liminf_{t \to 0+} M_a(w^+; t) < +\infty.$$

Hence $w \in \bigcap_{b<a} \Phi_b$.

If $f(x) = \exp(|x|^2/(4a))$ for all $x \in \mathbb{R}^n$, then the Gauss-Weierstrass integral u

of f is given by
$$u(x,t) = \left(\frac{a}{a-t}\right)^{\frac{n}{2}} \exp\left(\frac{|x|^2}{4(a-t)}\right),$$
in view of Example 4.6. It belongs to $\bigcap_{b<a} \Phi_b$ because, by Theorem 4.10,
$$M_b(u;t) = \int_{\mathbb{R}^n} W(x,b-t) u(x,t)\, dx = u(0,b)$$
whenever $0 < t < b < a$. But it does not belong to Φ_a, because
$$M_a(u;t) = \left(\frac{\sqrt{a}}{\sqrt{4\pi(a-t)}}\right)^n \int_{\mathbb{R}^n} dx = +\infty$$
whenever $0 < t < a$. □

THEOREM 5.27. *Let $0 < a \leq +\infty$, and let u be a temperature on $\mathbb{R}^n \times\,]0, a[$. Then $u \in \bigcap_{b<a} \Phi_b$ if and only if $-u \in \bigcap_{b<a} \Phi_b$.*

PROOF. If $u \in \bigcap_{b<a} \Phi_b$, then $u \in \bigcap_{b<a} \Sigma_b$, and hence u has the semigroup property, by Theorem 5.11. Therefore, whenever $0 < t < b < a$, we have
$$M_b(u^-;t) = M_b(u^+;t) - M_b(u;t) = M_b(u^+;t) - u(0,b),$$
which implies the result. □

COROLLARY 5.28. *Let $0 < a \leq +\infty$, and let w be a function on $\mathbb{R}^n \times\,]0, a[$. Then w is a temperature in the class $\bigcap_{b<a} \Phi_b$ if and only if both w and $-w$ are subtemperatures in the class $\bigcap_{b<a} \Phi_b$.*

PROOF. If w is a temperature in the class $\bigcap_{b<a} \Phi_b$, then both w and $-w$ are subtemperatures that belong to the class $\bigcap_{b<a} \Phi_b$, by Theorem 5.27.

The converse uses only the fact that, if both w and $-w$ are subtemperatures, then w is a temperature. □

THEOREM 5.29. *Let $0 < a \leq +\infty$, and let w be a subtemperature on $\mathbb{R}^n \times\,]0, a[$. Then w^+ has a thermic majorant on $\mathbb{R}^n \times\,]0, a[$ if and only if $w \in \bigcap_{b<a} \Phi_b$. In this case, the least thermic majorant of w^+ is the function u, defined by*
$$u(x,t) = \lim_{r \to 0+} \int_{\mathbb{R}^n} W(x-y, t-r) w^+(y,r)\, dy$$
for all $(x,t) \in \mathbb{R}^n \times\,]0, a[$.

PROOF. If w^+ has a thermic majorant v on $\mathbb{R}^n \times\,]0, a[$, then v is the Gauss-Weierstrass integral of a measure, by Theorem 4.18. Therefore $v \in \bigcap_{b<a} \Phi_b$ by Theorem 5.5, and it follows that $w \in \bigcap_{b<a} \Phi_b$.

Conversely, suppose that $w \in \bigcap_{b<a} \Phi_b$. For each r such that $0 < r < a$, we define the function u_r on $\mathbb{R}^n \times\,]r, a[$ by putting
$$u_r(x,t) = \int_{\mathbb{R}^n} W(x-y, t-r) w^+(y,r)\, dy.$$
By Theorem 5.9, u_r is a thermic majorant of w^+ on $\mathbb{R}^n \times\,]r, a[$. Therefore, if $0 < r < s < a$ and $(x,t) \in \mathbb{R}^n \times\,]s, a[$, we have
$$u_r(x,t) = \int_{\mathbb{R}^n} W(x-y, t-s) u_r(y,s)\, dy \geq \int_{\mathbb{R}^n} W(x-y, t-s) w^+(y,s)\, dy = u_s(x,t),$$

by the semigroup property (Theorem 4.16). Hence the function $r \mapsto u_r$ is decreasing on $]0, a[$, and $u = \lim_{r \to 0+} u_r$ exists on $\mathbb{R}^n \times]0, a[$. Also, whenever $0 < r < b < a$ we have
$$\lim_{r \to 0+} u_r(0, b) = \lim_{r \to 0+} M_b(w^+; r) < +\infty.$$
It therefore follows from the Harnack monotone convergence theorem (Theorem 1.31) that $u = \lim_{r \to 0+} u_r$ is a temperature on $\mathbb{R}^n \times]0, b[$ for every $b < a$, and hence on $\mathbb{R}^n \times]0, a[$. Since u_r is a majorant of w^+ on $\mathbb{R}^n \times]r, a[$ whenever $0 < r < a$, u is a majorant of w^+ on $\mathbb{R}^n \times]0, a[$.

It remains to prove that u is the least thermic majorant of w^+. Let h be any thermic majorant of w^+ on $\mathbb{R}^n \times]0, a[$. Then, whenever $0 < r < a$, the semigroup property of h (Theorem 4.16) gives us
$$h(x, t) = \int_{\mathbb{R}^n} W(x - y, t - r) h(y, r)\, dy \geq \int_{\mathbb{R}^n} W(x - y, t - r) w^+(y, r)\, dy = u_r(x, t)$$
for all $(x, t) \in \mathbb{R}^n \times]r, a[$. Therefore $h \geq \lim_{r \to 0+} u_r = u$, as required. $\square$

COROLLARY 5.30. *Suppose that $0 < s < a$ and $w \in \Sigma_a$. Then the least thermic majorant of w^+ on $\mathbb{R}^n \times]s, a[$ is the function u, defined by*
$$u(x, t) = \lim_{r \to s+} \int_{\mathbb{R}^n} W(x - y, t - r) w^+(y, r)\, dy$$
for all $(x, t) \in \mathbb{R}^n \times]s, a[$.

PROOF. Define a function v on $\mathbb{R}^n \times]0, a - s[$ by putting $v(x, t) = w(x, t + s)$. Then, whenever $0 < t < a - s$, we have
$$M_{a-s}(v^+; t) = \int_{\mathbb{R}^n} W(x, a - t - s) w^+(x, t + s)\, dx = M_a(w^+; t + s).$$
It therefore follows from Theorem 5.17 that $v \in \Phi_{a-s}$. Hence $v \in \bigcap_{b < a - s} \Phi_b$ by Theorem 5.26, so that v^+ has a least thermic majorant on $\mathbb{R}^n \times]0, a - s[$, given for all points $(x, t) \in \mathbb{R}^n \times]0, a - s[$ by
$$\lim_{\rho \to 0+} \int_{\mathbb{R}^n} W(x - y, t - \rho) v^+(y, \rho)\, dy = \lim_{\rho \to 0+} \int_{\mathbb{R}^n} W(x - y, t - \rho) w^+(y, \rho + s)\, dy.$$
It follows that the least thermic majorant of w^+ on $\mathbb{R}^n \times]s, a[$ is given at every point $(x, t) \in \mathbb{R}^n \times]s, a[$ by
$$\lim_{\rho \to 0+} \int_{\mathbb{R}^n} W(x - y, t - s - \rho) w^+(y, \rho + s)\, dy = \lim_{r \to s+} \int_{\mathbb{R}^n} W(x - y, t - r) w^+(y, r)\, dy$$
as required. $\square$

REMARK 5.31. If u is the function in the above corollary, then $u(x, t)$ is not generally equal to $\int_{\mathbb{R}^n} W(x - y, t - s) w^+(y, s)\, dy$. This can be seen by taking w as in Example 5.18, with s a discontinuity of a positive function f.

We now give some characterizations of the temperatures in the class $\bigcap_{b < a} \Phi_b$.

THEOREM 5.32. *Let $0 < a \leq +\infty$, and let u be a temperature on $\mathbb{R}^n \times]0, a[$. Then the following statements are equivalent:*
  (a) $u \in \bigcap_{b < a} \Phi_b$.
  (b) u^+ *has a thermic majorant on $\mathbb{R}^n \times]0, a[$.*

(c) u can be written as a difference of nonnegative temperatures on $\mathbb{R}^n \times \,]0,a[$.

(d) u has a representation as the Gauss-Weierstrass integral of a signed measure on $\mathbb{R}^n \times \,]0,a[$.

PROOF. If (a) holds, then Theorem 5.29 shows that (b) holds also.

Suppose that (b) holds, and that v is a thermic majorant of u^+ on $\mathbb{R}^n \times \,]0,a[$. Then $u = v - (v-u)$ expresses u as a difference of two nonnegative temperatures on $\mathbb{R}^n \times \,]0,a[$.

If (c) holds, then it follows from Theorem 4.18 that (d) holds also.

Finally, if (d) holds then Theorem 5.5 shows that (a) holds. □

In the context of Theorem 5.32, it is natural to ask what the connection is between the temperature u and the least thermic majorant of u^+. Theorem 5.29 gives one answer, but our next result gives a neater one.

THEOREM 5.33. *If u is the Gauss-Weierstrass integral on $\mathbb{R}^n \times \,]0,a[$ of a signed measure μ, then the least thermic majorant of u^+ on $\mathbb{R}^n \times \,]0,a[$ is the Gauss-Weierstrass integral of μ^+.*

PROOF. Let w be any thermic majorant of u^+ on $\mathbb{R}^n \times \,]0,a[$. By Theorem 4.18, w is the Gauss-Weierstrass integral of a nonnegative measure ν. We put $\alpha = (2a)^{-1}$, and define the measures λ and ω on $\mathbb{R}^n$ by putting $d\lambda(y) = \exp(-\alpha|y|^2)\,d\mu(y)$ and $d\omega(y) = \exp(-\alpha|y|^2)\,d\nu(y)$. Then λ and ω have finite total variation, because

$$\omega(\mathbb{R}^n) = \int_{\mathbb{R}^n} \exp(-\alpha|y|^2)\,d\nu(y) = (2\pi a)^{n/2} w(0, a/2) < +\infty,$$

and

$$|\lambda|(\mathbb{R}^n) = \int_{\mathbb{R}^n} \exp(-\alpha|y|^2)\,d|\mu|(y) = (2\pi a)^{n/2} \int_{\mathbb{R}^n} W(0-y, a/2)\,d|\mu|(y) < +\infty$$

since $u(0, \tfrac{1}{2}a)$ is defined and finite. We write $\lambda = \omega - (\omega - \lambda)$, and will show that $\omega - \lambda$ is a nonnegative measure. Since $w \geq u$, Theorem 4.18 shows that there is a nonnegative measure η such that $w - u$ is the Gauss-Weierstrass integral of η on $\mathbb{R}^n \times \,]0,a[$. However, we also have the representation

$$(w-u)(x,t) = \int_{\mathbb{R}^n} W(x-y,t)\exp(\alpha|y|^2)\,d(\omega - \lambda)(y)$$

for all $(x,t) \in \mathbb{R}^n \times \,]0,a[$. Theorem 4.11 now shows that

$$d\eta(y) = \exp(\alpha|y|^2)\,d(\omega - \lambda)(y),$$

so that $\omega - \lambda$ is nonnegative. It now follows from the minimality of the Hahn-Jordan decomposition that $\lambda^+ \leq \omega$ (and $\lambda^- \leq \omega - \lambda$). Therefore

$$w(x,t) = \int_{\mathbb{R}^n} W(x-y,t)\exp(\alpha|y|^2)\,d\omega(y)$$
$$\geq \int_{\mathbb{R}^n} W(x-y,t)\exp(\alpha|y|^2)\,d\lambda^+(y)$$
$$= \int_{\mathbb{R}^n} W(x-y,t)\,d\mu^+(y)$$
$$\geq u(x,t)$$

for all $(x,t) \in \mathbb{R}^n \times {]}0,a[$. Thus the Gauss-Weierstrass integral v of μ^+ satisfies $w \geq v \geq u$ on $\mathbb{R}^n \times {]}0,a[$, and is a temperature there by Theorem 4.4. Hence v, being nonnegative, is the least thermic majorant of u^+. □

5.4. Characterizations of the Gauss-Weierstrass Integrals of Functions

Theorem 5.32 gives some characterizations of the temperatures which can be represented as the Gauss-Weierstrass integrals of measures, but characterizing those that are the Gauss-Weierstrass integrals of *functions* is a different matter. Theorem 5.34 gives such a characterization.

THEOREM 5.34. *Let $0 < a \leq +\infty$, and let u be a temperature on $\mathbb{R}^n \times {]}0,a[$. Then the following statements are equivalent:*

(a) There is a strictly increasing, nonnegative, convex function ϕ on $[0,+\infty[$ such that $r^{-1}\phi(r) \to +\infty$ as $r \to +\infty$ and $\phi \circ |u| \in \bigcap_{b<a} \Phi_b$.

(b) There is a measurable function f on $\mathbb{R}^n$ such that u is the Gauss-Weierstrass integral of f on $\mathbb{R}^n \times {]}0,a[$.

PROOF. Suppose that statement (a) holds. We first consider the case where $u \geq 0$. Since $\phi \circ u \in \bigcap_{b<a} \Phi_b$, it is a subtemperature that has a thermic majorant v on $\mathbb{R}^n \times {]}0,a[$, by Theorem 5.29. For each positive integer k, we put

$$a_k = \sup\left\{\frac{r}{\phi(r)} : r \geq k\right\}.$$

Then $r \leq a_k \phi(r) + k$ for all $r \geq 0$, so that $u \leq a_k \phi \circ u + k$, and hence $u \leq a_k v + k$, on $\mathbb{R}^n \times {]}0,a[$. The nonnegative temperatures u, v and $w_k = a_k v + k - u$ are the Gauss-Weierstrass integrals of nonnegative measures μ, ν and ω_k, respectively, by Theorem 4.18. Furthermore, by the uniqueness of the representing measures (Theorem 4.11), we have $d\omega_k(y) = a_k d\nu(y) + kdy - d\mu(y)$. Therefore, denoting the singular part of a measure λ by λ^*, we obtain $0 \leq \omega_k^* = a_k \nu^* - \mu^*$, and hence $\mu^* \leq a_k \nu^*$. The growth condition on ϕ implies that $a_k \to 0$ as $k \to \infty$, and so $\mu^* = 0$. Statement (b) now follows from the Radon-Nikodým theorem.

We now consider the case where u may take negative values at some points. Since $\phi \circ |u|$ belongs to $\bigcap_{b<a} \Phi_b$, it is a subtemperature that has a thermic majorant v on $\mathbb{R}^n \times {]}0,a[$, by Theorem 5.29. Therefore the subtemperature $|u|$ is majorized by $\phi^{-1} \circ v$. Because the function $r \mapsto -\phi^{-1}(-r)$ is convex, and $-v$ is a temperature, the function $p \mapsto -\phi^{-1}(-(-v(p)))$ is a subtemperature (Section 3.2), and hence $\phi^{-1} \circ v$ is a supertemperature. Therefore, by Theorem 3.64, $|u|$ has a least thermic majorant w which satisfies $w \leq \phi^{-1} \circ v$, and hence $\phi \circ w \leq v$. Furthermore $w - u \leq 2w$, and so the monotonicity of ϕ ensures that $\phi \circ (\frac{1}{2}(w-u)) \leq \phi \circ w \leq v$. Thus the nonnegative temperatures w and $\frac{1}{2}(w-u)$ are such that the subtemperatures $\phi \circ w$ and $\phi \circ (\frac{1}{2}(w-u))$ belong to $\bigcap_{b<a} \Phi_b$, by Theorem 5.29. We can now apply the case of nonnegative temperatures, proved above, to deduce that w and $\frac{1}{2}(w-u)$ are both Gauss-Weierstrass integrals of functions. Hence the same is true of $u = w - (w-u)$.

Now we suppose, conversely, that statement (b) holds. For each positive integer k, we put

$$S_k = \{y \in \mathbb{R}^n : k-1 \leq |f(y)| < k\}.$$

For each $b < a$, we define a measure σ_b on $\mathbb{R}^n$ by putting $d\sigma_b(y) = W(y,b)\,dy$. Then, whenever $0 < b < a$, we have

$$+\infty > \int_{\mathbb{R}^n} W(y,b)(|f(y)| + 1)\,dy$$

$$= \sum_{k=1}^{\infty} \int_{S_k} W(y,b)(|f(y)| + 1)\,dy$$

$$\geq \sum_{k=1}^{\infty} k \int_{S_k} W(y,b)\,dy$$

$$= \sum_{k=1}^{\infty} k\sigma_b(S_k).$$

We can therefore find a positive, increasing sequence $\{c_k\}$, such that $c_k \to +\infty$ as $k \to \infty$ and

(5.9) $$\sum_{k=1}^{\infty} kc_k\sigma_b(S_k) < +\infty.$$

Putting $c_0 = 0$, we define the continuous, piecewise affine function ϕ by

$$\phi(r) = c_j(r - j + 1) + \sum_{i=0}^{j-1} c_i \quad \text{whenever} \quad j - 1 \leq r < j,$$

for every positive integer j. Because the sequence $\{c_k\}$ is positive and increasing, the function ϕ is nonnegative, strictly increasing and convex. Furthermore, because $c_k \to +\infty$ as $k \to \infty$, we have $r^{-1}\phi(r) \to +\infty$ as $r \to +\infty$. Moreover, whenever $0 < b < a$ we have

$$\int_{\mathbb{R}^n} (\phi \circ |f|)\,d\sigma_b = \sum_{k=1}^{\infty} \int_{S_k} (\phi \circ |f|)\,d\sigma_b$$

$$\leq \sum_{k=1}^{\infty} \phi(k)\sigma_b(S_k)$$

$$= \sum_{k=1}^{\infty} \left(\sum_{i=1}^{k} c_i\right) \sigma_b(S_k)$$

$$\leq \sum_{k=1}^{\infty} kc_k\sigma_b(S_k)$$

(5.10) $$< +\infty,$$

by (5.9). Since ϕ is convex, and in view of Lemma 1.1, it follows from Jensen's inequality that

$$\phi\left(\int_{\mathbb{R}^n} W(x-y,t)|f(y)|\,dy\right) \leq \int_{\mathbb{R}^n} W(x-y,t)\phi(|f(y)|)\,dy,$$

for all $(x,t) \in \mathbb{R}^n \times\,]0,a[$. Therefore, whenever $0 < t < b < a$, we have

$$\begin{aligned}
M_b(\phi \circ |u|;t) &= \int_{\mathbb{R}^n} W(x,b-t)\phi\left(\left|\int_{\mathbb{R}^n} W(x-y,t)f(y)\,dy\right|\right) dx \\
&\leq \int_{\mathbb{R}^n} W(x,b-t)\int_{\mathbb{R}^n} W(x-y,t)\phi(|f(y)|)\,dy\,dx \\
&= \int_{\mathbb{R}^n} \phi(|f(y)|)\int_{\mathbb{R}^n} W(y-x,t)W(x,b-t)\,dx\,dy \\
&= \int_{\mathbb{R}^n} W(y,b)\phi(|f(y)|)\,dy \\
&= \int_{\mathbb{R}^n} (\phi \circ |f|)\,d\sigma_b \\
&< +\infty,
\end{aligned}$$

by (5.10). Since $\phi \circ |u|$ is a subtemperature on $\mathbb{R}^n \times\,]0,a[$, by Example 3.9, it follows that $\phi \circ |u| \in \bigcap_{b<a} \Phi_b$. Thus statement (a) holds. $\square$

COROLLARY 5.35. *If u is a bounded temperature on $\mathbb{R}^n \times\,]0,a[$, then there is a measurable function f on $\mathbb{R}^n$ such that u is the Gauss-Weierstrass integral of f on $\mathbb{R}^n \times\,]0,a[$, and*

$$\sup_{\mathbb{R}^n \times]0,a[} |u| = \sup_{\mathbb{R}^n} |f|.$$

PROOF. The function u^2 is a subtemperature on $\mathbb{R}^n \times\,]0,a[$, and

$$M_b(u^2;t) = \int_{\mathbb{R}^n} W(x,b-t)u(x,t)^2\,dx \leq \sup u^2$$

whenever $0 < t < b < a$. Therefore $u^2 \in \bigcap_{b<a} \Phi_b$. Hence, by Theorem 5.34, there is a measurable function f on $\mathbb{R}^n$ such that u is the Gauss-Weierstrass integral of f on $\mathbb{R}^n \times\,]0,a[$. Furthermore, the nonnegative temperatures $\sup|u| - u$ and $u + \sup|u|$ are the Gauss-Weierstrass integrals of $\sup|u| - f$ and $f + \sup|u|$, respectively, in view of Theorem 4.11, and so those latter functions are nonnegative almost everywhere. Re-defining f on a set of measure zero, we get $\sup|u| \geq f \geq -\sup|u|$ everywhere, and hence $\sup|f| \leq \sup|u|$. Finally,

$$|u(x,t)| \leq \int_{\mathbb{R}^n} W(x-y,t)|f(y)|\,dy \leq \sup|f|$$

for all $(x,t) \in \mathbb{R}^n \times\,]0,a[$, so that $\sup|u| \leq \sup|f|$, and equality holds. $\square$

For *nonnegative* temperatures to be Gauss-Weierstrass integrals of functions, there is another criterion not included in Theorem 5.34.

THEOREM 5.36. *Let $0 < a \leq +\infty$, and let u be a nonnegative temperature on $\mathbb{R}^n \times\,]0,a[$. Then the following statements are equivalent:*

(a) There is an increasing sequence of bounded, nonnegative temperatures with limit u on $\mathbb{R}^n \times\,]0,a[$.

(b) There is a measurable function f on $\mathbb{R}^n$ such that u is the Gauss-Weierstrass integral of f on $\mathbb{R}^n \times\,]0,a[$.

PROOF. Suppose that u is the limit of an increasing sequence $\{v_k\}$ of bounded, nonnegative temperatures on $\mathbb{R}^n \times\,]0,a[$. By Corollary 5.35, each function v_k is the Gauss-Weierstrass integral of a function g_k on $\mathbb{R}^n \times\,]0,a[$. Since the sequence $\{v_k\}$ is increasing, we have

$$0 \leq (v_{k+1} - v_k)(x,t) = \int_{\mathbb{R}^n} W(x-y,t)(g_{k+1} - g_k)(y)\, dy$$

for all $(x,t) \in \mathbb{R}^n \times\,]0,a[$, so that $g_{k+1} - g_k \geq 0$ almost everywhere. Thus the sequence $\{g_k\}$ is increasing a.e., and therefore tends a.e. to a measurable function g. Lebesgue's monotone convergence theorem now shows that u is the Gauss-Weierstrass integral of g on $\mathbb{R}^n \times\,]0,a[$.

Conversely, suppose that u is the Gauss-Weierstrass integral of f on $\mathbb{R}^n \times\,]0,a[$. Then $f \geq 0$ a.e. on $\mathbb{R}^n$, and we can re-define f on a set of measure zero to ensure that $f \geq 0$ everywhere. For each positive integer k we put $f_k = f \wedge k$, and denote by u_k the Gauss-Weierstrass integral of f_k. Then each function u_k is a nonnegative, bounded temperature on $\mathbb{R}^n \times\,]0,+\infty[$, by Theorem 2.2. Moreover, $\{f_k\}$ is an increasing sequence of bounded functions that converges to f on $\mathbb{R}^n$, so that Lebesgue's monotone convergence theorem shows that the sequence $\{u_k\}$ increases to u on $\mathbb{R}^n \times\,]0,a[$. $\square$

DEFINITION 5.37. A nonnegative temperatures which can be written as the limit of an increasing sequence of bounded, nonnegative temperatures, is called *quasi-bounded*.

Thus Theorem 5.36 shows that a nonnegative temperature on $\mathbb{R}^n \times\,]0,a[$ is quasi-bounded if and only if it is the Gauss-Weierstrass integral of a function.

5.5. Notes and Comments

Sections 5.1-5.3 mostly follow Watson [**70**]. The main exceptions are Theorem 5.24, which essentially comes from Watson [**77**], although the hypotheses for the converse part are weaker here, and Theorem 5.33 which comes from Watson [**76**].

The results in section 5.4 are modelled on Theorem 1.3.9, Corollary 1.3.10, and Theorem 9.4.8 in Armitage & Gardiner [**3**], and their proofs are adaptations of the proofs in [**3**] to the present situation. Very similar results can be found in section 1.XVI.6 of Doob [**14**].

CHAPTER 6

Green Functions and Heat Potentials

For any point $q \in \mathbb{R}^{n+1}$, we let $G(\cdot;q)$ denote the fundamental supertemperture with pole at q. In this chapter, we define the corresponding notion $G_E(\cdot;q)$ relative to an arbitrary open subset E of $\mathbb{R}^{n+1}$, and call it the Green function for E with pole at q. Such functions are important because, if μ is a nonnegative measure on E, then the equation

$$G_E\mu = \int_E G_E(\cdot;q)\,d\mu(q)$$

defines a nonnegative supertemperature on E, provided that $G_E\mu < +\infty$ on a dense subset of E. A supertemperature of this form is called a heat potential. The importance of heat potentials is due to the Riesz decomposition theorem, which shows that every supertemperature is locally the sum of a heat potential and a temperature. If $E = \mathbb{R}^n \times\,]a,b[$, where $-\infty \leq a < b \leq +\infty$, then $G_E(\cdot;q)$ is just the restriction of $G(\cdot;q)$ to E, and so is known explicitly. In this way we obtain a representation theorem from the Riesz decomposition theorem. It has interesting consequences for the class $\bigcap_{b<a} \Sigma_b$ of subtemperatures on $\mathbb{R}^n \times\,]0,a[$, defined in Chapter 5. Further consequences of the Riesz decomposition theorem include the fact that the fundamental mean values of subtemperatures over heat spheres are always finite and have the convexity property of Theorem 1.11, that locally every supertemperature is the limit of an increasing sequence of smooth ones, and that supertemperatures can be characterized in terms of the mean values over modified heat balls.

6.1. Green Functions

Let E be an open set in $\mathbb{R}^{n+1}$. Let $q \in E$, and consider the restriction to E of $G(\cdot;q)$, the fundamental supertemperature with pole at q (cf. Section 3.2). Since $G(\cdot;q) \geq 0$, it has a nonnegative greatest thermic minorant $h_E(\cdot;q)$ (cf. Section 3.7).

DEFINITION 6.1. The *Green function* G_E for E is the nonnegative, real-valued function defined on $E \times E$ by putting

$$G_E(p;q) = G(p;q) - h_E(p;q),$$

where for each $q \in E$ the function $h_E(\cdot;q)$ is the greatest thermic minorant of $G(\cdot;q)$ on E. We may refer to $G_E(\cdot;q)$ as the *Green function for E with pole at q*.

Observe that, if D is an open subset of E and $q \in D$, then $h_D(\cdot;q) \geq h_E(\cdot;q)$ on D. Therefore $G_D \leq G_E$ on $D \times D$.

EXAMPLE 6.2. If $-\infty \leq a < b \leq +\infty$ and $E = \mathbb{R}^n \times\,]a,b[$, then G_E is the restriction of G to $E \times E$. To see this, we take any point $q = (y,s) \in E$. Since $G(\cdot;q) = 0$ throughout $\mathbb{R}^n \times\,]-\infty,s]$, the same is true of the greatest thermic minorant $h_E(\cdot;q)$ of $G(\cdot;q)$ on E. Therefore, by Theorem 4.14, we have $h_E(\cdot;q) = 0$ on the strip or half-space $\mathbb{R}^n \times\,]s,b[$. So $G_E(\cdot;q)$ is the restriction to E of $G(\cdot;q)$ for all $q \in E$.

We now present two simple characterizations of the Green function.

THEOREM 6.3. *Let G_E be the Green function for an open set E, let $q \in E$, and let*
$$\mathcal{F} = \{w \geq 0 : w = G(\cdot;q) + u \text{ for some supertemperature } u \text{ on } E\}.$$
Then $G_E(\cdot;q) = \inf \mathcal{F}$ on E.

PROOF. The definition of G_E shows that $G_E(\cdot;q) \in \mathcal{F}$, and so $G_E(\cdot;q) \geq \inf \mathcal{F}$. Suppose that $w = G(\cdot;q) + u \geq 0$ for some supertemperature u on E, and let $h_E(\cdot;q)$ denote the greatest thermic minorant of $G(\cdot;q)$ on E. Since $w \geq 0$, we have $G(\cdot;q) \geq -u$ on E, and so $-u \leq h_E(\cdot;q)$ on E. Therefore
$$G_E(\cdot;q) = G(\cdot;q) - h_E(\cdot;q) \leq G(\cdot;q) + u = w$$
for every $w \in \mathcal{F}$, and hence $G_E(\cdot;q) \leq \inf \mathcal{F}$. □

COROLLARY 6.4. *Let G_E be the Green function for an open set E, let $q \in E$, and let*
$$\mathcal{G} = \{w \geq 0 : w = G(\cdot;q) + u \text{ for some temperature } u \text{ on } E\}.$$
Then $G_E(\cdot;q) = \inf \mathcal{G}$ on E.

PROOF. The definition of G_E shows that $G_E(\cdot;q) \in \mathcal{G}$, and so $G_E(\cdot;q) \geq \inf \mathcal{G}$. By the theorem, $G_E(\cdot;q) = \inf \mathcal{F} \leq \inf \mathcal{G}$ on E. □

For convex domains of revolution, we can now give another characterization of the Green function. This will be extended to other open sets in Theorem 8.53(a), when we have treated the generalized Dirichlet problem.

THEOREM 6.5. *Let R be a convex domain of revolution, and let $q \in R$. If f denotes the restriction of $G(\cdot;q)$ to $\partial_n R$, then*
$$G_R(\cdot;q) = G(\cdot;q) - S_f$$
on R, where S_f denotes the PWB solution of the Dirichlet problem for f on R.

PROOF. Note that $f \in C(\partial_n R)$, so that f is resolutive by Theorem 3.38. By definition, $G_R(\cdot;q) = G(\cdot;q) - h_R(\cdot;q)$ on R. Since $0 \leq h_R(\cdot;q) \leq G(\cdot;q)$, we have $h_R(\cdot;q) \in \mathfrak{L}_f$ (the lower class determined by f), so that $h_R(\cdot;q) \leq S_f$. Therefore
$$G(\cdot;q) - S_f \leq G(\cdot;q) - h_R(\cdot;q) = G_R(\cdot;q).$$
But also $G(\cdot;q) \geq G(\cdot;q) \wedge (\max f) \in \mathfrak{U}_f$ (the upper class determined by f), which implies that $G(\cdot;q) - S_f \geq 0$. Now Theorem 6.3 shows that $G_R(\cdot;q) \leq G(\cdot;q) - S_f$, and hence equality holds. □

EXAMPLE 6.6. Suppose that $n = 1$, and let R be the open rectangle $]a, b[\times]\alpha, \beta[$. Put $c = b - a$. The Green function for R can be found by using reflections in the lateral boundary. For any two points $p = (x, t)$ and $q = (y, s)$, this results in the difference of the sum functions of two series:

$$\sum_{j \in \mathbb{Z}} W(2jc - x + y, t - s) - \sum_{j \in \mathbb{Z}} W(2jc + 2a - x - y, t - s),$$

where $\mathbb{Z}$ denotes the set of all integers.

Consider the first series. Let k be a positive integer. If $|x - y| < kc$ and $2|j| > k$, then we have $|2jc - (x - y)| \geq |2jc| - |x - y| > 2|j|c - kc > 0$, and therefore $(2jc - x + y)^2 \geq (2|j| - k)^2 c^2$. Hence, if also $t > s$, we have

$$W(2jc - x + y, t - s) \leq [4\pi(t - s)]^{-1/2} \exp\left(-\frac{(2|j| - k)^2 c^2}{4(t - s)}\right).$$

Since also $|2jc - (x - y)| \leq |2jc| + |x - y| < 2|j|c + kc < 4|j|c$, it follows that

$$|D_x W(2jc - x + y, t - s)| = \left|\frac{2(2jc - x + y)}{4(t - s)}\right| W(2jc - x + y, t - s)$$

$$\leq \frac{8|j|c}{\sqrt{\pi}[4(t - s)]^{3/2}} \exp\left(-\frac{(2|j| - k)^2 c^2}{4(t - s)}\right).$$

If we also have $t - s < k^2 c$, then $t - s < 4j^2 c$, and hence

$$|D_{xx} W(2jc - x + y, t - s)| = |D_t W(2jc - x + y, t - s)|$$

$$= \left|\frac{(2jc - x + y)^2 - 2(t - s)}{4(t - s)^2}\right| W(2jc - x + y, t - s)$$

$$\leq \frac{96 j^2 c}{\sqrt{\pi}[4(t - s)]^{5/2}} \exp\left(-\frac{(2|j| - k)^2 c^2}{4(t - s)}\right).$$

Given any $\alpha > 0$, we have

$$r^{-5/2} \exp\left(-\frac{\alpha^2}{r}\right) \leq \left(\frac{5}{2e}\right)^{5/2} \alpha^{-5}$$

for all $r > 0$. Therefore, taking $r = 4(t - s)$ and $\alpha = (2|j| - k)c$, we obtain

$$\exp\left(-\frac{(2|j| - k)^2 c^2}{4(t - s)}\right) \leq \left(\frac{5}{2e}\right)^{5/2} \frac{[4(t - s)]^{5/2}}{(2|j| - k)^5 e^5} \leq C \frac{(t - s)^{5/2}}{|j|^5}$$

for some constant C, provided that $|j| \geq k$. It follows that

$$W(2jc - x + y, t - s) \leq C(t - s)^2 |j|^{-5},$$

$$|D_x W(2jc - x + y, t - s)| \leq C(t - s)|j|^{-4},$$

and

$$|D_{xx} W(2jc - x + y, t - s)| = |D_t W(2jc - x + y, t - s)| \leq C|j|^{-3},$$

where the constant C is different on each occurence. It follows that, given given $k \in \mathbb{N}$ and $(y, s) \in \mathbb{R}^{n+1}$, the series

$$\sum_{|j| \geq k} W(2jc - x + y, t - s), \qquad \sum_{|j| \geq k} D_x W(2jc - x + y, t - s),$$

and
$$\sum_{|j|\geq k} D_{xx}W(2jc - x + y, t - s) = \sum_{|j|\geq k} D_t W(2jc - x + y, t - s),$$
all converge uniformly on the set $]y - kc, y + kc[\times]s, s + k^2c[$. Therefore the sum function of the series
$$\sum_{j\in\mathbb{Z}} W(2jc - x + y, t - s)$$
is partial differentiable twice with respect to x and once with respect to t, and these partial derivatives can be calculated by termwise differentiation of the series.

Similar calculations produce similar results for the second series.

For each fixed $q \in R$, each term in both series represents a temperature on R, except $W(-x + y, t - s)$, the term of the first series with $j = 0$; the exceptional term is, of course, a temperature on $R\setminus\{q\}$. We now define the function H_R on $R \times R$ by putting
$$H_R(p, q) = \sum_{j\in\mathbb{Z}} W(2jc - x + y, t - s) - \sum_{j\in\mathbb{Z}} W(2jc + 2a - x - y, t - s).$$

The function $H_R(\cdot, q)$ is, for fixed q, a function which differs from $G(\cdot, q)$ by a temperature on R, has limit 0 at every point of $\partial_n S$, and is identically zero when $t \leq s$.

Therefore, if f denotes the restriction of $G(\cdot; q)$ to $\partial_n R$, there is a temperature u on R such that $H_R(\cdot; q) = G(\cdot; q) + u$ and $\lim_{p\to r} u(p) = f(r)$ for all $r \in \partial_n R$. It follows from Theorem 3.32 that u is the PWB solution for f on R, and so Theorem 6.5 shows that $H_R(\cdot; q) = G_R(\cdot; q)$, the Green function for R with pole at q.

We now investigate the set of zeroes of the Green function of an arbitrary open set E. For this, we require a notation for the dual for the adjoint heat operator of the set $\Lambda(p; E)$, where p is an arbitrary point of E. If $q = (y, s) \in E$, we denote by $\Lambda^*(q; E) = \Lambda^*(y, s; E)$ the set of points $p \in E$ for which there is a polygonal path $\gamma \subseteq E$ joining q to p, along which the temporal variable is strictly *increasing*.

THEOREM 6.7. *Let G_E be the Green function for an open set E, let S be any subset of E, and let*
$$\Lambda^*(S; E) = \bigcup_{q\in S} \Lambda^*(q; E).$$
Then:

(a) $G_E(\cdot; q) > 0$ on $\Lambda^*(q; E)$, and $G_E(\cdot; q) = 0$ on $E\setminus\Lambda^*(q; E)$, for any $q \in E$.

(b) *The Green function for $\Lambda^*(S; E)$ is the restriction to $\Lambda^*(S; E) \times \Lambda^*(S; E)$ of G_E.*

PROOF. If $G_E(\cdot; q)$ had a zero in $\Lambda^*(q; E)$, then the minimum principle would imply that $G_E(\cdot; q) = 0$ in a neighbourhood of q, contrary to the definition of G_E.

We now put $D = \Lambda^*(S; E)$, and note that $\Lambda^*(p; E) \subseteq D$ for all $p \in D$. Let G_D denote the Green function of the open set D, so that $G_D \leq G_E$ on $D \times D$. We take any point $p \in D$, and put
$$v(\cdot; p) = \begin{cases} G_D(\cdot; p) & \text{on} \quad \Lambda^*(p; E), \\ 0 & \text{on} \quad E\setminus\Lambda^*(p; E). \end{cases}$$

Then $v(\cdot;p) \leq G_E(\cdot;p)$ on E. To prove the reverse inequality, we first note that, if $h_D(\cdot;p)$ denotes the greatest thermic minorant of $G(\cdot;p)$ on D, then

$$v(\cdot;p) - G(\cdot;p) = \begin{cases} -h_D(\cdot;p) & \text{on} \quad \Lambda^*(p;E), \\ -G(\cdot;p) & \text{on} \quad E\backslash\Lambda^*(p;E). \end{cases}$$

We show that this function is a supertemperature on E, so that we can apply Theorem 6.3. It is clearly finite-valued, and is lower semicontinuous except possibly at p. Choose $\epsilon > 0$ such that the open ball $B(p,\epsilon)$ is contained in D. The restriction of $G(\cdot;p)$ to $\mathbb{R}^{n+1}\backslash\Lambda^*(p;B(p,\epsilon))$ is a continuous function, and $h_D(\cdot;p)$ is continuous on the superset D of $\Lambda^*(p;E) \cup B(p,\epsilon)$, with $-h_D(\cdot;p) \geq -G(\cdot;p)$ on D, so that $v(\cdot;p) - G(\cdot;p)$ is lower semicontinuous at p, and therefore on E. Furthermore, condition (δ_4) is satisfied, except possibly at points of $E \cap \partial\Lambda^*(p;E)$. For each point $(x',t') \in E \cap \partial\Lambda^*(p;E)$ we can find a neighbourhood N of (x',t') such that $N \cap \partial\Lambda^*(p;E) \subseteq \mathbb{R}^n \times \{t'\}$. If the closure of the heat cylinder $\Delta = \Delta(x',t';c)$ is contained in N, we have $v(\cdot;p) - G(\cdot;p) = -G(\cdot;p)$ on $\partial_n\Delta$, so that the condition (δ_4) is satisfied at (x',t'). It follows that $v(\cdot;p) - G(\cdot;p)$ is a supertemperature on E. Therefore $v(\cdot;p)$ is the sum of $G(\cdot;p)$ and a supertemperature on E, and so Theorem 6.3 implies that $G_E(\cdot;p) \leq v(\cdot;p)$ on E. Hence $G_E(\cdot;p) = v(\cdot;p)$ on E.

If we take $S = E$, then $D = E$ and we deduce that $G_E(\cdot;p) = 0$ on $E\backslash\Lambda^*(p;E)$.

For any choice of D, we have $\Lambda^*(p;D) = \Lambda^*(p;E)$, and so $G_D(\cdot;p) = 0$ on $D\backslash\Lambda^*(p;D)$. Since

$$G_E(\cdot;p) = \begin{cases} G_D(\cdot;p) & \text{on} \quad \Lambda^*(p;E), \\ 0 & \text{on} \quad E\backslash\Lambda^*(p;E), \end{cases}$$

it follows that $G_D(\cdot;p)$ is the restriction of $G_E(\cdot;p)$ to D. $\square$

6.2. Green Functions and the Adjoint Heat Equation

Every result about the heat equation $\Theta u = 0$ has a dual for the adjoint heat equation $\Theta^* u = 0$, obtained by reversing the temporal variable. However, with Green functions we find a nontrivial interaction between the concepts related to the two equations. The Green function G for $\mathbb{R}^{n+1}$ not only satisfies $\Theta G(\cdot;q) = 0$ on $\mathbb{R}^{n+1}\backslash\{q\}$, but also satisfies $\Theta^* G(p;\cdot) = 0$ on $\mathbb{R}^{n+1}\backslash\{p\}$. We shall show that the Green function for any open set E has a similar property. To prove this, we shall require a particular method for the construction of the greatest thermic minorant of an appropriate supertemperature. This is the subject of our next theorem, in which, given a supertemperature w that is minorized by a subtemperature on an open set E, and any circular cylinder D such that $\overline{D} \subseteq E$, we denote by $\pi_D w$ the supertemperature on E defined in the dual for supertemperatures of Theorem 3.21.

THEOREM 6.8. *Let E be an open set, and let w be a supertemperature which is minorized by a subtemperature on E. Let $\{D_k\}$ be a sequence of open circular cylinders such that $\overline{D_k} \subseteq E$ for all k, such that $\bigcup_{k=1}^{\infty} D_k = E$, and such that each cylinder in the collection $\{D_k : k \in \mathbb{N}\}$ occurs infinitely often in the sequence. For each k, let $w_k = \pi_{D_k} \cdots \pi_{D_1} w$. Then the sequence $\{w_k\}$ is a decreasing sequence of supertemperatures whose limit is the greatest thermic minorant of w on E.*

PROOF. Theorem 3.64 shows that w has a greatest thermic minorant v on E. It follows from Theorem 3.21 that the sequence $\{w_k\}$ is decreasing on E, that each function w_k is a supertemperature on E and a temperature on D_k, and that $w \geq w_k \geq v$ on E. We put $w_\infty = \lim_{k\to\infty} w_k$ on E. Given a circular cylinder D in the collection $\{D_k : k \in \mathbb{N}\}$, the fact that it occurs infinitely often in the sequence implies that there is a subsequence $\{w_{k_j}\}$ of $\{w_k\}$ such that the restriction of w_{k_j} to D is a temperature for all j. Since the sequence $\{w_k\}$ is decreasing, it follows that $w_\infty = \lim_{j\to\infty} w_{k_j}$, so that on D the function w_∞ is the limit of a decreasing sequence of temperatures, and so is itself a temperature, by the Harnack monotone convergence theorem (Theorem 1.31). Since $\bigcup_{k=1}^\infty D_k = E$, it follows that w_∞ is a temperature on E. The inequalities $w \geq w_k \geq v$ for all k, imply that $w \geq w_\infty \geq v$, so that $w_\infty = v$ on E. □

For any points $p, q \in \mathbb{R}^{n+1}$, we put
$$G^*(p;q) = G(q;p),$$
and define the Green function relative to Θ^* for any open set E as follows.

DEFINITION 6.9. The *Green cofunction* G_E^* for E, is the nonnegative, real-valued function defined on $E \times E$ by putting
$$G_E^*(p;q) = G^*(p;q) - h_E^*(p;q),$$
where for each $q \in E$ the function $h_E^*(\cdot;q)$ is the greatest *cothermic* minorant of $G^*(\cdot;q)$ on E.

Thus the definition of the Green cofunction G_E^* is just the dual of that of the Green function G_E.

THEOREM 6.10. *For any open set E, we have the identities*
$$h_E(p;q) = h_E^*(q;p) \qquad \text{and} \qquad G_E(p;q) = G_E^*(q;p)$$
for all $p, q \in E$. In particular, for each point $p \in E$, the function $G_E(p;\cdot)$ is a cotemperature on $E\backslash\{p\}$ and a cosupertemperature on E.

PROOF. We choose an arbitrary open circular cylinder D_0 such that $\overline{D}_0 \subseteq E$. We then choose a sequence $\{D_k\}$ of open circular cylinders, with closures in E, such that $\overline{D}_0 \subseteq D_1$, $\bigcup_{k=1}^\infty D_k = E$, $\partial D_k \cap D_0 = \emptyset$ for each k, and such that each cylinder in the collection $\{D_k : k \in \mathbb{N}\}$ occurs infinitely often in the sequence.

We use the notation π_D of Theorem 3.21. Given any point $q \in E$, we define
$$u_0(\cdot;q) = \pi_{D_0} G(\cdot;q)$$
on E. Then, by Theorem 3.21, the function $u_0(\cdot;q)$ is a supertemperature on E, is a temperature on $\overline{D}_0 \backslash \partial_n D_0$, and satisfies $h_E(\cdot;q) \leq u_0(\cdot;q) \leq G(\cdot;q)$ on E. We now define a sequence $\{u_k(\cdot;q)\}$ on E, as in Theorem 6.8 with $w = u_0(\cdot;q)$, that decreases to $h_E(\cdot;q)$ as $k \to \infty$.

For any point p', the function $G(p';\cdot)$ is a cotemperature on $\mathbb{R}^{n+1}\backslash\{p'\}$. Hence, because $u_0(p;q) = G(p;q)$ whenever $q \in E$ and $p \in (E\backslash\overline{D}_0) \cup \partial_n D_0$, and
$$u_0(p;q) = \int_{\partial_n D_0} G(\cdot;q) \, d\mu_p$$
for all $q \in E$ and $p \in \overline{D}_0 \backslash \partial_n D_0$, where μ_p denotes the caloric measure at p for D_0, it follows that $u_0(p;\cdot)$ is a cotemperature on $E\backslash\partial D_0$ for any $p \in E$ (using

differentiation under the integral sign if $p \in D_0$). For $k \geq 0$, suppose that $u_k(p;\cdot)$ is a cotemperature on $E\backslash\bigcup_{i=0}^{k}\partial D_i$. Then, because $u_{k+1}(p;q) = u_k(p;q)$ whenever $q \in E$ and $p \in (E\backslash\overline{D}_{k+1}) \cup \partial_n D_{k+1}$, and because

$$u_{k+1}(p;q) = \int_{\partial_n D_{k+1}} u_k(\cdot;q)\,d\mu_p$$

for all $q \in E$ and $p \in \overline{D}_{k+1}\backslash\partial_n D_{k+1}$, where μ_p denotes the caloric measure at p for D_{k+1}, it follows as before that $u_{k+1}(p;\cdot)$ is a cotemperature on $E\backslash\bigcup_{i=0}^{k+1}\partial D_i$. In particular, for any $p \in E$, the function $u_{k+1}(p;\cdot)$ is a cotemperature on D_0, because we chose the sequence $\{D_k\}$ to satisfy $\partial D_k \cap D_0 = \emptyset$ for each k. Since the sequence $\{u_k(p;\cdot)\}$ is decreasing, it follows from the dual for cotemperatures of the Harnack monotone convergence theorem that its limit is a cotemperature on D_0, for each $p \in E$. Thus $h_E(p;\cdot)$ is a cotemperature on the arbitrary open cylinder D_0 with closure in E, and so $h_E(p;\cdot)$ is a cotemperature on E, for each $p \in E$.

Because $h_E(p;\cdot)$ is a cothermic minorant of $G(p;\cdot) = G^*(\cdot;p)$, it follows that $h_E(p;\cdot) \leq h_E^*(\cdot;p)$, and hence that $h_E(p;q) \leq h_E^*(q;p)$ for all $p,q \in E$. The dual of this argument, with the roles of Θ and Θ^* interchanged, gives $h_E^*(p;q) \leq h_E(q;p)$ for all $p,q \in E$. Hence $h_E(p;q) = h_E^*(q;p)$, and therefore

$$G_E(p;q) = G(p;q) - h_E(p;q) = G^*(q;p) - h_E^*(q;p) = G_E^*(q;p)$$

whenever $p,q \in E$. □

Theorem 6.10 combines with Theorem 6.7 to give more information about the zeroes and restrictions of Green functions, as follows.

THEOREM 6.11. *Let G_E be the Green function for an open set E, let S be any subset of E, and let*

$$\Lambda(S;E) = \bigcup_{p \in S} \Lambda(p;E).$$

Then:
(a) $G_E(p;\cdot) > 0$ on $\Lambda(p;E)$, and $G_E(p;\cdot) = 0$ on $E\backslash\Lambda(p;E)$, for any $p \in E$.
(b) The Green function for $\Lambda(S;E)$ is the restriction to $\Lambda(S;E) \times \Lambda(S;E)$ of G_E.

PROOF. By Theorem 6.10, we have $G_E(p;q) = G_E^*(q;p)$ for all $p,q \in E$. So if we apply the dual for G_E^* of Theorem 6.7, we obtain the result. □

THEOREM 6.12. *The Green function G_E for E is lower semicontinuous at every point of $\{(p,p) : p \in E\}$, and is continuous at all other points of $E \times E$.*

PROOF. The assertion of lower semicontinuity at all points of $\{(p,p) : p \in E\}$ follows from the facts that G_E is zero at such points and $G_E \geq 0$ everywhere.

We take any point $(p_0,q_0) \in E \times E$ such that $p_0 \neq q_0$, and let U, U_1 and V be euclidean balls in $\mathbb{R}^{n+1}$, with centres p_0, p_0 and q_0 respectively, such that

$$\overline{U} \subseteq U_1, \quad \overline{U}_1 \cup \overline{V} \subseteq E, \quad \overline{U}_1 \cap \overline{V} = \emptyset.$$

We choose a point $p_1 \in U_1\backslash\overline{U}$ such that $\Lambda(p_1;U_1) \supseteq U$. Since $G_E(p_1;q) \leq G(p_1;q)$ for all $q \in E$, and $G(p_1;\cdot)$ is bounded on the set $E\backslash\Lambda(p_1;U_1) \supseteq V$, we can find a positive number l such that $G_E(p_1;q) \leq l$ for all $q \in V$. Since $G_E(\cdot;q)$ is a nonnegative temperature on U_1 for each $q \in V$, it follows from the Harnack inequality for temperatures (Corollary 1.33) that there is a positive number κ such that $G_E(p;q) \leq \kappa G_E(p_1;q)$ for all $(p,q) \in U \times V$. Therefore $G_E \leq \kappa l$ on $U \times V$, so that the family $\mathcal{G} = \{G_E(\cdot;q) : q \in V\}$ is uniformly bounded on U. Each member of $\mathcal{G}$ is a temperature on U, and hence $\mathcal{G}$ is uniformly equicontinuous on any compact subset of U, by Theorem 1.40. Therefore, given $\epsilon > 0$, we can find a neighbourhood M of p_0 such that

$$|G_E(p;q) - G_E(p_0;q)| < \frac{\epsilon}{2} \quad \text{whenever} \quad q \in V \quad \text{and} \quad p \in M.$$

Furthermore, because $G_E(p_0;\cdot)$ is continuous on V in view of Theorem 6.10, we can find a neighbourhood N of q_0 such that

$$|G_E(p_0;q) - G_E(p_0;q_0)| < \frac{\epsilon}{2} \quad \text{whenever} \quad q \in N.$$

Hence

$$|G_E(p;q) - G_E(p_0;q_0)| < \epsilon \quad \text{whenever} \quad (p,q) \in M \times N,$$

which proves that G_E is continuous at $(p_0;q_0)$. $\square$

6.3. Heat Potentials

DEFINITION 6.13. Let E be an open set, let G_E be its Green function, and let μ be a nonnegative measure on E. We define a nonnegative, extended real-valued function $G_E\mu$ by putting

$$G_E\mu(p) = \int_E G_E(p;q)\,d\mu(q)$$

for all $p \in E$. If $G_E\mu$ is a supertemperature on E, then it is called a *Heat Potential*. Dually, we define $G_E^*\mu$ by putting

$$G_E^*\mu(p) = \int_E G_E^*(p;q)\,d\mu(q) = \int_E G_E(q;p)\,d\mu(q)$$

(see Theorem 6.10) for all $p \in E$. If $G_E^*\mu$ is a cosupertemperature on E, then it is called a *Coheat Potential*.

EXAMPLE 6.14. Let μ be a nonnegative measure on $\mathbb{R}^n$ such that the Gauss-Weierstrass integral u of μ is finite at some point $(x_0,t_0) \in \mathbb{R}^n \times]0,+\infty[$, so that u is defined and is a temperature throughout $\mathbb{R}^n \times]0,t_0[$, by Theorem 4.4. If we put $E = \mathbb{R}^n \times]-\infty,t_0[$, then G_E is the restriction of G to $E \times E$, by Example 6.2. Let the measure μ' be defined on $\mathbb{R}^n \times \{0\}$ by putting $\mu'(S \times \{0\}) = \mu(S)$ for all Borel subsets S of $\mathbb{R}^n$, and the measure μ'' be defined on E by putting $\mu''(T) = \mu'(T \cap (\mathbb{R}^n \times \{0\}))$ for all Borel subsets T of E. Then for all $(x,t) \in E$ we have

$$G_E\mu''(x,t) = \int_E G((x,t);(y,s))\,d\mu''(y,s)$$
$$= \int_{\mathbb{R}^n \times \{0\}} G((x,t);(y,0))\,d\mu'(y,0)$$
$$= \int_{\mathbb{R}^n} W(x-y,t)\,d\mu(y),$$

so that

$$G_E\mu''(x,t) = \begin{cases} u(x,t) & \text{whenever} \quad 0 < t < t_0, \\ 0 & \text{whenever} \quad t \leq 0. \end{cases}$$

EXAMPLE 6.15. We take a fixed heat sphere $\partial\Omega(x_0,t_0;c)$, and let μ denote the measure on it that occurs in the definition of the fundamental means $\mathcal{M}$. Thus $d\mu(x,t) = \tau(c)Q(x_0 - x, t_0 - t)\,d\sigma(x,t)$, where σ denotes surface area measure. Then the coheat potential $G^*\mu$ can be evaluated explicitly. By Theorem 6.10 and Example 1.12, we have

$$G^*\mu(y,s) = \int_{\mathbb{R}^{n+1}} G(x,t;y,s)\,d\mu(x,t)$$
$$= \mathcal{M}(G(\cdot,\cdot;y,s);x_0,t_0;c)$$
$$= G(x_0,t_0;y,s) \wedge \tau(c).$$

Note that $G^*\mu$ is a cotemperature outside the support of μ, and that its greatest cothermic minorant is the zero function.

We proceed to give some criteria for $G_E\mu$ to be a heat potential. In view of Example 6.14, the next result can be seen as complementary to Theorem 4.4.

THEOREM 6.16. *If $G_E\mu(p_0) < +\infty$ for some point $p_0 \in E$, then $G_E\mu$ is a supertemperature on $\Lambda(p_0;E)$.*

PROOF. Let $\{U_j\}$ be an expanding sequence of bounded open sets such that $\overline{U}_j \subseteq E$ for all j, and $\bigcup_{j=1}^\infty U_j = E$. For each j, we put

$$u_j(p) = \int_{U_j} \bigl(G_E(p;q) \wedge j\bigr)\,d\mu(q)$$

whenever $p \in E$. We show that each function u_j is a supertemperature on E, with a view to using Theorem 3.60. Certainly the finiteness conditions are satisfied, because $0 \le u_j(p) \le j\mu(\overline{U}_j) < +\infty$ for all $p \in E$, due to the compactness of $\overline{U}_j$. To show that u_j is lower semicontinuous, we use Fatou's lemma and the lower semicontinuity of $G_E(\cdot;q) \wedge j$ for every $q \in E$. Thus, for any point $r \in E$, we have

$$\liminf_{p \to r} u_j(p) \ge \int_{U_j} \liminf_{p \to r} \bigl(G_E(p;q) \wedge j\bigr)\,d\mu(q)$$
$$\ge \int_{U_j} \bigl(G_E(r;q) \wedge j\bigr)\,d\mu(q)$$
$$= u_j(r),$$

as required. To show that u_j has the appropriate mean value property, we use the fact that $G_E(\cdot;q) \wedge j$ is a supertemperature on E for each $q \in E$, which follows from Corollary 3.18. Thus, whenever the closed heat cylinder $\overline{\Delta}(p;c)$ is contained in E, we have

$$\mathcal{L}(u_j;p;c) = \int_{U_j} \mathcal{L}(G_E(\cdot;q) \wedge j;p;c)\,d\mu(q) \le \int_{U_j} \bigl(G_E(p;q) \wedge j\bigr)\,d\mu(q) = u_j(p).$$

Therefore u_j is a supertemperature on E.

The sequence of sets $\{U_j\}$ expands to E, and the sequence of functions $\{G_E \wedge j\}$ increases to G_E, so that the sequence $\{u_j\}$ is increasing to the limit $G_E\mu$ as $j \to \infty$, by Lebesgue's monotone convergence theorem. It now follows from Theorem 3.60 that $G_E\mu$ is a supertemperature on $\Lambda(p_0;E)$. □

LEMMA 6.17. *Let E be an open set, let G_E be its Green function, and let μ be a nonnegative measure on E such that $\mu(E) < +\infty$ and $\mu(B) = 0$ for some open euclidean ball B such that $\overline{B} \subseteq E$. Then $G_E\mu$ is a subtemperature on the set $\Lambda(B;E) = \bigcup_{p \in B} \Lambda(p;E)$.*

PROOF. Let $p \in B$. Since $0 \leq G_E \leq G$ on $E \times E$, the function $G_E(p;\cdot)$ is bounded on $E\backslash B$. Therefore our hypotheses on μ show that

$$G_E\mu(p) = \int_{E\backslash B} G_E(p;q)\, d\mu(q) < +\infty.$$

Now Theorem 6.16 gives the result. $\square$

THEOREM 6.18. *Let E be an open set, let G_E be its Green function, and let μ be a nonnegative measure on E such that $\mu(E) < +\infty$. Then $G_E\mu$ is a heat potential.*

PROOF. Let $p \in E$, and let B be an open euclidean ball such that $p \in B$ and $\overline{B} \subseteq E$. We put $M = E\backslash B$, and denote by μ_S the restriction of μ to a Borel set S. Then $G_E\mu = G_E\mu_B + G_E\mu_M$. Since $\mu_M(B) = 0$ and $\mu_M(E) < +\infty$, it follows from Lemma 6.17 that $G_E\mu_M$ is a supertemperature on $\Lambda(p;E)$. Since $\overline{B} \subseteq E$, we can find a point $p' \in E\backslash\overline{B}$ such that $p \in \Lambda(p';E)$. We choose an open euclidean ball V such that $p' \in V$ and $\overline{V} \subseteq E\backslash\overline{B}$. Then $\mu_B(V) = 0$ and $\mu_B(E) < +\infty$, so that $G_E\mu_B$ is a supertemperature on $\Lambda(p';E)$, by Lemma 6.17. Since $p \in \Lambda(p';E)$ we have $\Lambda(p;E) \subseteq \Lambda(p';E)$, and so $G_E\mu_B$ is a supertemperature on $\Lambda(p;E)$. It follows that $G_E\mu$ is a supertemperature on $\Lambda(p;E)$. Since p is arbitrary, $G_E\mu$ is a supertemperature on E. $\square$

Our next result shows that the greatest thermic minorant of any heat potential is the zero function. This property will later be seen to characterize heat potentials in the class of all nonnegative supertemperatures (in Corollary 6.39).

THEOREM 6.19. *Let E be an open set, let G_E be its Green function, and let μ be a nonnegative measure on E such that $G_E\mu$ is a heat potential. Then the greatest thermic minorant of $G_E\mu$ on E is zero.*

PROOF. We employ the method of constructing greatest thermic minorants in Theorem 6.8. Let $\{D_k\}$ be a sequence of open circular cylinders satisfying the conditions in Theorem 6.8. Let $u = G_E\mu$, let $q \in E$, and for each k let $w_k = \pi_{D_k} \cdots \pi_{D_1} w$ for both $w = u$ and $w = G_E(\cdot;q)$. Theorem 6.8 tells us that each sequence $\{w_k\}$ is a decreasing sequence of supertemperatures that converges to the greatest thermic minorant of w on E. So $\{G_E(\cdot;q)_k\}$ decreases to zero on E, and we need to prove that $\{u_k\}$ does the same.

We claim that

(6.1) $$u_k(p) = \int_E G_E(p;q)_k\, d\mu(q)$$

for all $p \in E$. This can be proved by induction on k. If we put $u_0 = u$ and $G_E(\cdot;q)_0 = G_E(\cdot;q)$, then the case $k = 0$ is just the definition of u. Suppose that (6.1) holds when $k = j$, for some $j \geq 0$. By definition of the π_D operator, for both $w = u$ and $w = G_E(\cdot;q)$, each function w_{j+1} is equal to the Poisson integral of the restriction of w_j to $\partial_n D_{j+1}$ on $\overline{D}_{j+1}\backslash\partial_n D_{j+1}$, and to w_j elsewhere on E. Thus, if p does not belong to $\overline{D}_{j+1}\backslash\partial_n D_{j+1}$, we have

$$u_{j+1}(p) = u_j(p) = \int_E G_E(p;q)_j\, d\mu(q) = \int_E G_E(p;q)_{j+1}\, d\mu(q),$$

so that (6.1) holds with $k = j+1$ in this case. On the other hand, if p does belong to $\overline{D}_{j+1}\backslash\partial_n D_{j+1}$, then

$$u_{j+1}(p) = \int_{\partial_n D_{j+1}} u_j(r)\, d\mu_p(r),$$

where μ_p denotes the caloric measure at p for D_{j+1}. Therefore, by the induction hypothesis,

$$u_{j+1}(p) = \int_{\partial_n D_{j+1}} \int_E G_E(r;q)_j\, d\mu(q)\, d\mu_p(r)$$
$$= \int_E \int_{\partial_n D_{j+1}} G_E(r;q)_j\, d\mu_p(r)\, d\mu(q)$$
$$= \int_E G_E(p;q)_{j+1}\, d\mu(q),$$

so that (6.1) holds with $k = j+1$ in this case also. Thus (6.1) is true for all $k \geq 0$.

Since u is a supertemperature on E, it is finite on a dense subset of E. If p_0 is chosen such that $u(p_0) < +\infty$, then the function $G_E(p_0;\cdot)$ is μ-integrable on E. Therefore the inequalities $0 \leq G_E(p_0;q)_k \leq G_E(p_0;q)$ show that we can use the Lebesgue dominated convergence theorem to obtain

$$u_\infty(p_0) = \lim_{k\to\infty} u_k(p_0) = \int_E \lim_{k\to\infty} G_E(p_0;q)_k\, d\mu(q) = 0.$$

Thus u_∞, the greatest thermic minorant of u on E, is zero on a dense subset of E, and hence everywhere on E. $\square$

Theorem 6.19 implies that the heat potential of a non-null measure can never be a temperature, as follows.

COROLLARY 6.20. *If $G_E\mu$ is a temperature on E, then μ is null.*

PROOF. If $G_E\mu$ is a temperature on E, then its greatest thermic minorant on E is $G_E\mu$ itself, so that

$$\int_E G_E(p;q)\, d\mu(q) = 0$$

for all $p \in E$, by Theorem 6.19. Since $G_E(p;\cdot) > 0$ on $\Lambda(p;E)$ for all p, by Theorem 6.11, we have $\mu(\Lambda(p;E)) = 0$ for all p. By Lindelöf's theorem, there is a sequence $\{p_i\}$ of points in E such that $E = \bigcup_{i=1}^\infty \Lambda(p_i;E)$, and hence

$$\mu(E) \leq \sum_{i=1}^\infty \mu(\Lambda(p_i;E)) = 0.$$

$\square$

We proceed to investigate the question, raised by the above corollary, of the relationship between open subsets of E where μ is null, and those subsets where $G_E\mu$ is a temperature.

THEOREM 6.21. *Let E be an open set with Green function G_E, let $p_0 \in E$, let μ be a nonnegative measure on E such that $G_E\mu(p_0) < +\infty$, and let F denote the support of μ. Then $G_E\mu$ is a temperature on $\Lambda(p_0; E)\backslash F$.*

PROOF. We suppose that $E\backslash F \neq \emptyset$, because otherwise the result is vacuous. Let m be an integer with $m \geq 5$, and for each point $p \in E$ let $\Omega_m(p; c)$ denote the corresponding modified heat ball with centre p and radius c. For each point $q \in E \cap F$, the Green function with pole at q is a temperature on $E\backslash F$, so that $G_E(p; q) = \mathcal{V}_m(G_E(\cdot; q); p; c)$ whenever $\overline{\Omega}_m(p; c) \subseteq E\backslash F$, by Theorem 1.25. It follows that

$$G_E\mu(p) = \int_{E \cap F} G_E(p; q)\, d\mu(q) = \int_{E \cap F} \mathcal{V}_m(G_E(\cdot; q); p; c)\, d\mu(q) = \mathcal{V}_m(G_E\mu; p; c)$$

whenever $\overline{\Omega}_m(p; c) \subseteq E\backslash F$. Since $G_E\mu(p_0) < +\infty$, $G_E\mu$ is a supertemperature on $\Lambda(p_0; E)$, by Theorem 6.16. Therefore $G_E\mu$ is locally integrable on $\Lambda(p_0; E)$, by Theorem 3.56. Now Theorem 1.29 shows that $G_E\mu$ is a temperature on $\Lambda(p_0; E)\backslash F$. □

COROLLARY 6.22. *If $G_E\mu$ is a heat potential, then $G_E\mu$ is a temperature on $E\backslash F$.*

PROOF. If $G_E\mu$ is a heat potential, then it is finite on a dense subset S of E. Therefore, by the theorem, $G_E\mu$ is a temperature on the set

$$\bigcup_{p \in S} \Lambda(p; E)\backslash F = E\backslash F.$$

□

LEMMA 6.23. *Let U and V be open sets, let μ be a nonnegative measure on V, let J be a nonnegative function on $U \times V$, and let u be a function defined on U by putting*

$$u(p) = \int_V J(p; q)\, d\mu(q).$$

If

  (a) $J(p; \cdot)$ is continuous on V for each $p \in U$,
  (b) $J(\cdot; q)$ is a temperature on U for each $q \in V$, and
  (c) $u(p)$ is finite for all p in a dense subset of U,
then u is a temperature on U.

PROOF. The function J is measurable on $U \times V$, and so Tonelli's theorem is applicable. Let m be an integer, $m \geq 5$, and let $\Omega_m(p; c)$ denote the corresponding modified heat ball with centre p and radius c. Then, whenever $\overline{\Omega}_m(p; c) \subseteq U$, we have

$$\mathcal{V}_m(u;p;c) = \mathcal{V}_m\left(\int_V J(\cdot;q)\,d\mu(q);p;c\right)$$
$$= \int_V \mathcal{V}_m(J(\cdot;q);p;c)\,d\mu(q)$$
$$= \int_V J(p;q)\,d\mu(q)$$
$$= u(p),$$

using Theorem 1.25 applied to $J(\cdot;q)$ for each $q \in V$. The function $w = -u$ is upper bounded, finite-valued on a dense subset of U, and satisfies $w(p) = \mathcal{V}_m(w;p;c)$ whenever $\overline{\Omega}_m(p;c) \subseteq U$, so that it is locally integrable on U by Lemma 3.54. Therefore u is a temperature on U, by Theorem 1.29. $\square$

THEOREM 6.24. *Let D and E be open sets such that $D \subseteq E$, and let G_D and G_E denote their Green functions. Let μ be a nonnegative measure on E such that $\mu(E\backslash D) = 0$ and $G_E\mu$ is a heat potential, and let μ_D denote the restriction of μ to D. Then there is a nonnegative temperature u on D such that $G_E\mu = G_D\mu_D + u$ on D.*

PROOF. Since $D \subseteq E$, we have $G_D \leq G_E$ on $D \times D$. Therefore, whenever $p \in D$, we have

$$G_D\mu_D(p) = \int_D G_D(p;q)\,d\mu(q) \leq \int_E G_E(p;q)\,d\mu(q) = G_E\mu(p).$$

Since $G_E\mu$ is a heat potential, it is finite on a subset S of E such that the Lebesgue measure of $E\backslash S$ is zero, by Theorem 3.56. Hence $G_D\mu_D(p) < +\infty$ for all $p \in S \cap D$, so that $G_D\mu_D$ is a supertemperature on $\bigcup_{p \in S \cap D} \Lambda(p;D) = D$, by Theorem 6.16. Thus $G_D\mu_D$ is a heat potential, and we can write

$$G_E\mu(p) - G_D\mu_D(p) = \int_D \left(G_E(p;q) - G_D(p;q)\right)d\mu(q)$$

for all $p \in S \cap D$. Given any point $q \in D$, we let $h_D(\cdot;q)$ and $h_E(\cdot;q)$ denote the greatest thermic minorants of $G(\cdot;q)$ on D and E respectively. Then for all $p, q \in D$, we have

$$0 \leq G_E(p;q) - G_D(p;q) = h_D(p;q) - h_E(p;q) = J(p;q),$$

say, where $J(\cdot;q)$ is a temperature on D for each q, and $J(p;\cdot)$ is a cotemperature on D for each p, by Theorem 6.10. Therefore, whenever $\overline{\Omega}(p;c) \subseteq D$, we have

$$\mathcal{V}(G_E\mu - G_D\mu_D;p;c) = \int_D \mathcal{V}(G_E(\cdot;q) - G_D(\cdot;q);p;c)\,d\mu(q)$$
$$= \int_D \mathcal{V}(J(\cdot;q);p;c)\,d\mu(q)$$
(6.2)
$$= \int_D J(p;q)\,d\mu(q),$$

by Theorem 1.16 applied to $J(\cdot;q)$ for each $q \in D$. Since $G_E\mu$ and $G_D\mu_D$ are nonnegative supertemperatures on D, we have

$$0 \leq \mathcal{V}(G_E\mu - G_D\mu_D;p;c) \leq \mathcal{V}(G_E\mu;p;c) \leq G_E\mu(p) < +\infty$$

for all $p \in S \cap D$, by Theorem 3.56. Hence the function u, defined on D by

$$u(p) = \int_D J(p;q)\, d\mu(q),$$

is finite for all $p \in S \cap D$, so that Lemma 6.23 implies that u is a temperature on D. Equation (6.2) implies that $\mathcal{V}(G_E\mu;p;c) = \mathcal{V}(G_D\mu_D;p;c) + u(p)$ almost everywhere on D. Therefore, if we make $c \to 0+$ we obtain $G_E\mu(p) = G_D\mu_D(p) + u(p)$ for almost all $p \in D$, by Theorem 3.59, so that $G_E\mu(p) = G_D\mu_D(p) + u(p)$ for *all* $p \in D$, again by Theorem 3.59. □

THEOREM 6.25. *Let $G_E\mu$ be a heat potential on an open set E, and let D be an open subset of E. Then $G_E\mu$ is a temperature on D if and only if $\mu(D) = 0$.*

PROOF. If $\mu(D) = 0$, then $G_E\mu$ is a temperature on D, by Corollary 6.22.

Conversely, suppose that $G_E\mu$ is a temperature on D. For any Borel set $S \subseteq E$, we denote by μ_S the restriction of μ to S. Now $G_E\mu = G_E\mu_D + G_E\mu_{E\setminus D}$, and both $G_E\mu_D$ and $G_E\mu_{E\setminus D}$ are heat potentials. Corollary 6.22 shows that $G_E\mu_{E\setminus D}$ is a temperature on D, so that the same is true of $G_E\mu_D$, because of our hypothesis on $G_E\mu$. Furthermore, Theorem 6.24 shows that there is a nonnegative temperature u on D such that $G_E\mu_D = G_D\mu_D + u$ on D, which implies that $G_D\mu_D$ is also a temperature on D. Hence μ_D is null, by Corollary 6.20. □

6.4. The Distributional Heat Operator

In this section, we present preparatory material for the proof of the Riesz Decomposition Theorem, which is extremely important and has many interesting consequences.

We use the following notation. We denote by $C_c(E)$ the linear space of all real-valued continuous functions on $\mathbb{R}^{n+1}$ whose support is a compact subset of E. We denote by $C_c^{2,1}(E)$ the subspace $C_c(E) \cap C^{2,1}(\mathbb{R}^{n+1})$.

DEFINITION 6.26. Let u be a locally integrable, extended real-valued function on an open set E. We define a linear functional L_u on $C_c^{2,1}(E)$ by putting

$$L_u(\phi) = \int\!\!\int_E u\Theta^*\phi \, dx\, dt.$$

We call L_u the *Distributional Heat Operator* of u. Clearly $L_{u+v} = L_u + L_v$ on $C_c^{2,1}(E)$.

THEOREM 6.27. *(a) If $u \in C^{2,1}(E)$ and $\phi \in C_c^{2,1}(E)$, then*

$$L_u(\phi) = \int\!\!\int_E \phi\Theta u\, dx\, dt.$$

(b) If u is a temperature on E, then $L_u = 0$ on $C_c^{2,1}(E)$.
(c) If u is a subtemperature on E, then $L_u \geq 0$ on $C_c^{2,1}(E)$.

PROOF. Let $u \in C^{2,1}(E)$, let $\phi \in C_c^{2,1}(E)$, let K_ϕ denote the support of ϕ, and let D be a bounded open superset of K_ϕ such that $\overline{D} \subseteq E$. We choose a function $\psi \in C_c^{2,1}(D)$ such that $\psi = 1$ on an open superset of K_ϕ, and an open ball B such that $D \subseteq B$. The function $u\psi$ belongs to $C^{2,1}(E)$, and is 0 on $E\backslash D$. We define $u\psi$ to be zero on $\mathbb{R}^{n+1}\backslash E$. By Green's formula for the heat equation, we have

$$\int\int_B (u\psi\Theta^*\phi - \phi\Theta(u\psi))\,dx\,dt = \int_{\partial B} 0\,d\sigma,$$

so that

(6.3) $$\int\int_{K_\phi} u\Theta^*\phi\,dx\,dt = \int\int_{K_\phi} \phi\Theta u\,dx\,dt.$$

Hence

$$\int\int_E u\Theta^*\phi\,dx\,dt = \int\int_E \phi\Theta u\,dx\,dt,$$

which proves (a) and (b).

Now suppose that u is an arbitrary subtemperature on E, and let ϕ, K_ϕ, D, and ψ be as above. By Theorems 3.56 and 3.51, u is locally integrable on E and the inequality $u(p) \leq \mathcal{V}(u;p;c)$ holds whenever $\overline{\Omega}(p;c) \subseteq E$. It therefore follows from Theorem 1.28 that there is a sequence $\{w_k\}$ of smooth subtemperatures on D such that

$$\lim_{k\to\infty} \int\int_{K_\phi} |w_k(x,t) - u(x,t)|\,dx\,dt = 0.$$

It follows that

$$\left|\int\int_{K_\phi} w_k\Theta^*\phi\,dx\,dt - \int\int_{K_\phi} u\Theta^*\phi\,dx\,dt\right| \leq \int\int_{K_\phi} |w_k - u||\Theta^*\phi|\,dx\,dt$$

$$\leq \max|\Theta^*\phi| \int\int_{K_\phi} |w_k - u|\,dx\,dt$$

$$\to 0$$

as $k \to \infty$. In view of (6.3), we have

$$\int\int_{K_\phi} w_k\Theta^*\phi\,dx\,dt = \int\int_{K_\phi} \phi\Theta w_k\,dx\,dt \geq 0$$

whenever $\phi \geq 0$. It follows that

$$L_u(\phi) = \int\int_{K_\phi} u\Theta^*\phi\,dx\,dt = \lim_{k\to\infty} \int\int_{K_\phi} w_k\Theta^*\phi\,dx\,dt \geq 0$$

whenever $\phi \geq 0$. $\square$

THEOREM 6.28. *If u is a subtemperature on E, then there is a nonnegative measure μ_u on E such that*

$$L_u(\phi) = \int_E \phi\,d\mu_u$$

for all $\phi \in C_c^{2,1}(E)$. Moreover, this measure is unique.

PROOF. We shall extend L_u to a nonnegative linear functional on $C_c(E)$, in order to apply the Riesz representation theorem. Let $\psi \in C_c(E)$, and let A denote the support of ψ. Let B be a bounded open set such that $A \subseteq B$ and $\overline{B} \subseteq E$. Then we can find a sequence $\{\psi_k\}$ in $C_c^{2,1}(B)$ such that $\psi_k \to \psi$ uniformly on $\mathbb{R}^{n+1}$ as $k \to \infty$. Moreover, because the standard mollifiers are nonnegative, if $\psi \geq 0$ we can have $\psi_k \geq 0$ for all k. Let D be a bounded open set such that $\overline{B} \subseteq D$ and $\overline{D} \subseteq E$. Then we can find a function $\chi \in C_c^{2,1}(D)$ such that $\chi = 1$ on B and

$0 \leq \chi \leq 1$ on $\mathbb{R}^{n+1}$. For any function $\phi \in C_c^{2,1}(B)$ we have $\phi = \phi\chi$ on $\mathbb{R}^{n+1}$, so that $|\phi| \leq \sup|\phi|\chi$. Therefore, because Theorem 6.27 shows that L_u is a nonnegative linear functional on $C_c^{2,1}(E)$, we have $|L_u(\phi)| \leq L_u(|\phi|) \leq \sup|\phi|L_u(\chi)$. Hence, for all positive integers i, j,

$$|L_u(\psi_i) - L_u(\psi_j)| = |L_u(\psi_i - \psi_j)| \leq L_u(\chi)\sup|\psi_i - \psi_j|.$$

Since $\psi_k \to \psi$ uniformly on $\mathbb{R}^{n+1}$, it follows that $\{L_u(\psi_k)\}$ is a Cauchy sequence in $\mathbb{R}$, and hence convergent. Moreover, $\lim_{k\to\infty} L_u(\psi_k)$ is independent of the choice of $\{\psi_k\}$. For suppose that $\{\psi_k^*\}$ is another sequence in $C_c^{2,1}(B)$ such that $\psi_k^* \to \psi$ uniformly on $\mathbb{R}^{n+1}$ as $k \to \infty$. Then, as before, $\{L_u(\psi_k^*)\}$ is a Cauchy sequence in $\mathbb{R}$, and hence convergent. Put $\lambda = \lim_{k\to\infty} L_u(\psi_k)$ and $\lambda^* = \lim_{k\to\infty} L_u(\psi_k^*)$. Then for all k we have

$$\begin{aligned}|\lambda - \lambda^*| &\leq |\lambda - L_u(\psi_k)| + |L_u(\psi_k) - L_u(\psi_k^*)| + |L_u(\psi_k^*) - \lambda^*| \\ &\leq |\lambda - L_u(\psi_k)| + L_u(\chi)\sup|\psi_k - \psi_k^*| + |L_u(\psi_k^*) - \lambda^*| \\ &\leq |\lambda - L_u(\psi_k)| + L_u(\chi)\big(\sup|\psi_k - \psi| + \sup|\psi_k^* - \psi|\big) + |L_u(\psi_k^*) - \lambda^*| \\ &\to 0\end{aligned}$$

as $k \to \infty$. Therefore $\lambda = \lambda^*$, and we can define

$$\lambda_u(\psi) = \lim_{k\to\infty} L_u(\psi_k)$$

unambiguously. Since we can have $\psi_k \geq 0$ for all k if $\psi \geq 0$, it follows that $\lambda_u(\psi) \geq 0$ if $\psi \geq 0$. Hence λ_u a nonnegative linear functional on $C_c(E)$ that coincides with L_u on $C_c^{2,1}(E)$, and as such is unique. By the Riesz representation theorem, there is a unique nonnegative measure μ_u on E such that $\lambda_u(\psi) = \int_E \psi\, d\mu_u$ for all $\psi \in C_c(E)$. Since $\lambda_u = L_u$ on $C_c^{2,1}(E)$, the result follows. $\square$

DEFINITION 6.29. Let u be a subtemperature on E. The nonnegative measure μ_u which represents L_u as in Theorem 6.28, is called the *Riesz Measure* associated with u. If v is a supertemperature on E, then the *Riesz Measure* associated with v is defined to be the Riesz measure associated with the subtemperature $-v$.

Observe that, if D is an open subset of E and $\phi \in C_c^{2,1}(D)$, then

$$\int\int_D u\Theta^*\phi\, dx\, dt = \int\int_E u\Theta^*\phi\, dx\, dt = \int_E \phi\, d\mu_u = \int_D \phi\, d\mu_u.$$

Hence, by the uniqueness part of Theorem 6.28, the restriction to D of the Riesz measure associated with u on E, is the Riesz measure associated with u on D.

We shall prove that the Riesz measure associated with a heat potential $G_E\mu$ is μ itself. This requires the following result, which of independent interest.

THEOREM 6.30. *If $q_0 = (y_0, s_0) \in \mathbb{R}^{n+1}$ and w is the restriction to E of $G(\cdot; q_0)$, then $L_w(\phi) = -\phi(q_0)$ for all $\phi \in C_c^{2,1}(E)$.*

PROOF. Let $\phi \in C_c^{2,1}(E)$, and let B be an open ball which contains both q_0 and the support of ϕ. For each $a > 0$, we put $B_a = \{(x, t) \in B : t > s_0 + a\}$. We apply Green's formula for the heat equation (1.1), with $D = B_a$, $v = \phi$, and $w = G(\cdot; q_0)$. Since $\Theta G(\cdot; q_0) = 0$ on B_a, and $\phi = 0$ on a neighbourhood of ∂B, we thus obtain

$$\iint_{B_a} -G(\cdot; q_0)\Theta^*\phi \, dx \, dt$$

$$= \int_{\partial B_a} \left(\langle \phi \nabla_x G(\cdot; q_0) - G(\cdot; q_0)\nabla_x \phi, \nu_x \rangle - \phi G(\cdot; q_0)\nu_t\right) d\sigma,$$

(6.4)
$$= \int_{H_a} \phi G(\cdot; q_0) \, dx,$$

where $H_a = \{(x,t) \in \partial B_a : t = s_0 + a\}$ and we have used the facts that $\nu_x = 0$, $\nu_t = -1$, $d\sigma = dx$ on H_a. We shall make $a \to 0+$ to obtain the result.

We write $p = (x,t)$ and

(6.5) $\int_{H_a} \phi(p) G(p; q_0) \, dx = \int_{H_a} \left(\phi(p) - \phi(q_0)\right) G(p; q_0) \, dx + \phi(q_0) \int_{H_a} G(p; q_0) \, dx.$

Given $\epsilon > 0$, we choose a closed circular cylinder

$$S = \{(x,t) : |x - y_0| \leq \delta, \, s_0 \leq t \leq s_0 + \delta\} \subseteq B$$

such that $|\phi(p) - \phi(q_0)| < \epsilon$ whenever $p \in S$. Then if $a < \delta$, we have

$$\int_{H_a} |\phi(p) - \phi(q_0)| G(p; q_0) \, dx$$

$$= \left(\int_{H_a \cap S} + \int_{H_a \setminus S}\right) |\phi(p) - \phi(q_0)| G(p; q_0) \, dx$$

$$\leq \epsilon \int_{\mathbb{R}^n} G(x, s_0 + a; y_0, s_0) \, dx + 2 \max|\phi| \int_{|x - y_0| > \delta} G(x, s_0 + a; y_0, s_0) \, dx$$

$$= \epsilon \int_{\mathbb{R}^n} W(x - y_0, a) \, dx + 2 \max|\phi| \int_{|x - y_0| > \delta} W(x - y_0, a) \, dx.$$

The first integral is 1, by Lemma 1.1. Furthermore, for any fixed $\eta > 0$, we have

$$\int_{|x-y_0| \leq \eta} W(x - y_0, a) \, dx = (4\pi a)^{-\frac{n}{2}} \int_0^\eta \exp\left(-\frac{r^2}{4a}\right) \omega_n r^{n-1} \, dr$$

$$= \left(\frac{\omega_n}{2\pi^{\frac{n}{2}}}\right) \int_0^{\eta^2/4a} e^{-z} z^{\frac{n}{2}-1} \, dz$$

$$\to 1$$

as $a \to 0+$. Therefore

$$\int_{|x-y_0| > \delta} W(x - y_0, a) \, dx \to 0$$

as $a \to 0+$, and hence

$$\int_{H_a} \left(\phi(p) - \phi(q_0)\right) G(p; q_0) \, dx \to 0$$

as $a \to 0+$. Furthermore, $H_a \supseteq \{(x, s_0 + a) : |x - y_0| \leq \alpha\}$ for some $\alpha > 0$, and therefore

$$1 \geq \int_{H_a} G(p; q_0) \, dx \geq \int_{|x-y_0| \leq \alpha} W(x - y_0, a) \, dx \to 1$$

as $a \to 0+$. It now follows from (6.5) that
$$\int_{H_a} \phi(p) G(p;q_0)\, dx \to \phi(q_0)$$
as $a \to 0+$. Making $a \to 0+$ in (6.4), we now obtain
$$\int\int_B G(p;q_0) \Theta^* \phi(p)\, dx\, dt = -\phi(q_0),$$
which implies the result of the theorem because B contains the support of ϕ. □

THEOREM 6.31. *The Riesz measure associated with a heat potential $G_E \mu$ on E is μ itself.*

PROOF. By definition, $G_E(p;q) = G(p;q) - h_E(p;q)$, where $h_E(\cdot;q)$ is the greatest thermic minorant of $G(\cdot;q)$ on E. By Theorem 6.27, we have
$$L_{G_E(\cdot;q)}(\phi) = L_{G(\cdot;q)}(\phi) - L_{h_E(\cdot;q)}(\phi) = L_{G(\cdot;q)}(\phi)$$
for all $\phi \in C_c^{2,1}(E)$. It now follows from Fubini's theorem and Theorem 6.30 that
$$L_{G_E\mu}(\phi) = \int_E \left(\int_E G_E(p;q)\, d\mu(q) \right) \Theta^* \phi(p)\, dp$$
$$= \int_E \left(\int_E G_E(p;q) \Theta^* \phi(p)\, dp \right) d\mu(q)$$
$$= \int_E L_{G_E(\cdot;q)}(\phi)\, d\mu(q)$$
$$= \int_E L_{G(\cdot;q)}(\phi)\, d\mu(q)$$
$$= -\int_E \phi(q)\, d\mu(q)$$
for all $\phi \in C_c^{2,1}(E)$. Theorem 6.28 now gives the result. □

We need one more result on the distributional heat operator, before we can prove the Riesz decomposition theorem.

THEOREM 6.32. *Let u and v be subtemperatures on E such that $L_u(\phi) = L_v(\phi)$ for all $\phi \in C_c^{2,1}(E)$. Then there is a temperature h such that $u = v + h$ on E.*

PROOF. Given a bounded open set D such that $\overline{D} \subseteq E$, we choose two other bounded open sets U and V such that $\overline{D} \subseteq U$, $\overline{U} \subseteq V$, and $\overline{V} \subseteq E$. We fix an integer $m \geq 5$, and for each $c > 0$ put
$$V_c = \{p \in \mathbb{R}^{n+1} : \overline{\Omega}_m(p;c) \subseteq V\}.$$
The set V_c clearly depends on m, but because m does not vary in this proof, we do not indicate that dependence in the notation. Since V_c expands to V as c decreases to 0, we can find a number $c_0 > 0$ such that $\overline{U} \subseteq V_c$ whenever $0 < c \leq c_0$.

Given any subtemperature w on E, and any positive number $c \leq c_0$, we define a function w_c on V_c by putting
$$w_c(p) = \mathcal{V}_m(w;p;c).$$

6.4. THE DISTRIBUTIONAL HEAT OPERATOR

By Theorems 3.56 and 3.51, w is locally integrable on E, and satisfies the inequality $w(p) \leq \mathcal{V}(w; p; c)$ whenever $\overline{\Omega}(p; c) \subseteq E$. Therefore, by Theorem 1.28, each function w_c is a smooth subtemperature on V_c, and for each compact subset K of V_c we have

$$(6.6) \qquad \lim_{c \to 0+} \int\int_K |w_c(y,s) - w(y,s)|\, dy\, ds = 0.$$

In view of Lemma 1.27 and formula (1.23), we have

$$w_c(x,t) = \int\int_{\mathbb{R}^{n+1}} \lambda_{m,c}(|x-y|, t-s) w(y,s)\, dy\, ds,$$

and the function $\Phi_c^{x,t}$, defined by $\Phi_c^{x,t}(y,s) = \lambda_{m,c}(|x-y|, t-s)$, belongs to $C_c^{2,1}(E)$. Therefore

$$\Theta w_c(x,t) = \int\int_{\mathbb{R}^{n+1}} \Theta_{x,t} \lambda_{m,c}(|x-y|, t-s) w(y,s)\, dy\, ds$$
$$= \int\int_{\mathbb{R}^{n+1}} \left(\Theta_{y,s}^* \Phi_c^{x,t}(y,s)\right) w(y,s)\, dy\, ds$$
$$(6.7) \qquad = L_w\left(\Phi_c^{x,t}\right).$$

For each $d > 0$, we now put $U_d = \{p \in \mathbb{R}^{n+1} : \overline{\Omega}_m(p;d) \subseteq U\}$, and choose $d_0 > 0$ such that $\overline{D} \subseteq U_d$ whenever $0 < d \leq d_0$. Then, if $0 < d \leq d_0$ and $(x,t) \in U_d$, the support of $\Phi_d^{x,t}$ is a subset of U. Therefore

$$|\mathcal{V}_m(w_c; x, t; d) - \mathcal{V}_m(w; x, t; d)| = \left|\int\int_{\mathbb{R}^{n+1}} \Phi_d^{x,t}(y,s)\left(w_c(y,s) - w(y,s)\right) dy\, ds\right|$$
$$\leq \max \Phi_d^{x,t} \int\int_{\overline{U}} |w_c(y,s) - w(y,s)|\, dy\, ds.$$

It now follows from (6.6) with $K = \overline{U}$ that

$$(6.8) \qquad \lim_{c \to 0+} \mathcal{V}_m(w_c; x, t; d) = \mathcal{V}_m(w; x, t; d).$$

Now let u and v be subtemperatures on E such that $L_u(\phi) = L_v(\phi)$ for all $\phi \in C_c^{2,1}(E)$. For each positive number c such that $c \leq c_0$, we define a function h_c by putting

$$h_c(p) = u_c(p) - v_c(p)$$

whenever $p \in V_c$. Then each h_c is a temperature on U_c, because

$$\Theta h_c(x,t) = \Theta u_c(x,t) - \Theta v_c(x,t) = L_u\left(\Phi_c^{x,t}\right) - L_v\left(\Phi_c^{x,t}\right) = 0$$

by (6.7) and our hypotheses on u and v, because $\Phi_c^{x,t} \in C_c^{2,1}(E)$. It now follows from Theorem 1.25 and (6.8) that, whenever $(x,t) \in D$ and $0 < d \leq d_0$,

$$h_c(x,t) = \mathcal{V}_m(h_c; x, t; d)$$
$$= \mathcal{V}_m(u_c; x, t; d) - \mathcal{V}_m(v_c; x, t; d)$$
$$\to \mathcal{V}_m(u; x, t; d) - \mathcal{V}_m(v; x, t; d)$$

as $c \to 0+$. Thus h_c has a pointwise limit h on D as $c \to 0+$, where

$$h(x,t) = \mathcal{V}_m(u - v; x, t; d)$$

whenever $0 < d \leq d_0$. Furthermore, (6.6) with $K = \overline{D}$ implies that

$$\lim_{c \to 0+} \int\int_{\overline{D}} |h_c(y,s) - (u-v)(y,s)|\, dy\, ds = 0,$$

and therefore there is a sequence $\{h_{c_k}\}$ which converges to $u-v$ pointwise almost everywhere on D. Thus h is defined everywhere on D, and is equal to $u-v$ almost everywhere on D. Therefore h is locally integrable on D, and satisfies

$$h(x,t) = \mathcal{V}_m(u-v;x,t;d) = \mathcal{V}_m(h;x,t;d)$$

whenever $0 < d \leq d_0$ and $\overline{\Omega}_m(x,t;d) \subseteq D$. It now follows from Theorem 1.29 that h is a temperature on D. Hence $u = v + h$ a.e. on D, with both u and $v+h$ subtemperatures on D. Now Theorem 3.59 shows that $u = v + h$ everywhere on D. In view of the arbitrary nature of D, the result is established. □

COROLLARY 6.33. *If u is a subtemperature on E such that L_u is the zero functional on $C_c^{2,1}(E)$, then u is a temperature on E.*

PROOF. Take $v = 0$ in the Theorem. □

6.5. The Riesz Decomposition Theorem

THEOREM 6.34. *Let u be a supertemperature on an open set E, and let μ_u be its associated Riesz measure. If u is minorized by a subtemperature on E, then $G_E \mu_u$ is a heat potential, and u has the representation*

$$u = G_E \mu_u + h$$

on E, where h is the greatest thermic minorant of u on E.

PROOF. Let $\{K_j\}$ be a sequence of compact sets such that $K_j \subseteq K_{j+1}^\circ$ for all j, and $\bigcup_{j=1}^\infty K_j = E$. For each j, we let μ_{u,K_j} denote the restriction of μ_u to K_j. Since K_j is compact, we have $\mu_{u,K_j}(E) < +\infty$, and so $G_E \mu_{u,K_j}$ is a heat potential, by Theorem 6.18. Furthermore, by Theorem 6.31, the Riesz measure associated with $G_E \mu_{u,K_j}$ on E is μ_{u,K_j}. Therefore the distributional heat operator of $G_E \mu_{u,K_j}$ is equal to that of u on $C_c^{2,1}(K_j^\circ)$. Theorem 6.32 now implies that there is a temperature v_j on K_j° such that $u = G_E \mu_{u,K_j} + v_j$ on K_j°. Naturally, a similar identity holds with j replaced by $j+1$, and it follows that

$$G_E \mu_{u,K_{j+1}} + v_{j+1} = G_E \mu_{u,K_j} + v_j$$

on K_j°. We now put

$$v_j^*(p) = v_{j+1}(p) + \int_{K_{j+1} \setminus K_j} G_E(p;q) \, d\mu_u(q)$$

for all $p \in K_{j+1}^\circ$. Since heat potentials are finite almost everywhere, by Theorem 3.56, we have $v_j = v_j^*$ a.e. on K_j°. It follows from Corollary 6.22 that v_j^* is a temperature on K_j°, and hence that $v_j = v_j^*$ everywhere on K_j°. Therefore, on K_{j+1}° we have

$$u = G_E \mu_{u,K_{j+1}} + v_{j+1} = G_E \mu_{u,K_j} + v_j^*,$$

with v_j^* a supertemperature on K_{j+1}° and a temperature on K_j°. It follows that the function v_j^{**}, defined on E by putting

$$v_j^{**}(p) = \begin{cases} v_j^*(p) & \text{if } p \in K_{j+1}^\circ, \\ u(p) - G_E \mu_{u,K_j}(p) & \text{if } p \in E \setminus K_j, \end{cases}$$

is a well-defined supertemperature on E (in view of Corollary 6.22), is a temperature on K_j°, and satisfies $u = v_j^{**} + G_E \mu_{u,K_j}$ on E.

Since u is minorized by a subtemperature on E, it has a greatest thermic minorant h on E, by Theorem 3.64. Since $u \geq h$, we have $v_j^{**} - h \geq -G_E\mu_{u,K_j}$ on E, so that the supertemperature $v_j^{**} - h$ has a greatest thermic minorant h^{**} on E, and $G_E\mu_{u,K_j} \geq -h^{**}$. Theorem 6.19 now shows that $h^{**} \geq 0$. Hence

$$u - h = v_j^{**} - h + G_E\mu_{u,K_j} \geq h^{**} + G_E\mu_{u,K_j} \geq G_E\mu_{u,K_j}.$$

Making $j \to \infty$ and using Lebesgue's monotone convergence theorem, we deduce that $u - h \geq G_E\mu_u$ which, together with Theorem 6.16, implies that $G_E\mu_u$ is a heat potential. By Theorem 6.31, the Riesz measure associated with $G_E\mu_u$ is μ_u. Therefore Theorem 6.32 shows that there is a temperature h^* on E such that $u = G_E\mu_u + h^*$. Taking greatest thermic minorants, we obtain $h = 0 + h^*$, by Theorems 6.19 and 3.66. This completes the proof. □

The Riesz decomposition theorem combines with results in Chapter 4 to give representation theorems for nonnegative supertemperatures on a strip or half-space.

COROLLARY 6.35. *Let u be a nonnegative supertemperature on a strip or half-space $S = \mathbb{R}^n \times\,]0, a[$. Then u has the representation*

$$u(x,t) = \int_S W(x-y, t-s)\, d\mu_u(y,s) + \int_{\mathbb{R}^n} W(x-y, t)\, d\nu(y)$$

for some nonnegative measure ν on $\mathbb{R}^n$.

PROOF. By the Riesz decomposition theorem, we can write $u = G_S\mu_u + h$, where h is the greatest thermic minorant of u on S. By Example 6.2, we have

$$G_S(x,t;y,s) = G(x,t;y,s) = W(x-y, t-s)$$

for all $(x,t), (y,s) \in S$. Furthermore, since $u \geq 0$ we have $h \geq 0$, and so h is the Gauss-Weierstrass integral of a nonnegative measure ν on $\mathbb{R}^n$, by Theorem 4.18. The result follows. □

COROLLARY 6.36. *Let u be a nonnegative supertemperature on a strip or half-space $S = \mathbb{R}^n \times\,]0, a[$, and let $0 < t_0 < a$. If $u(\cdot, t_0) \leq W(\cdot, t_0)$ on $\mathbb{R}^n$, then u has the representation*

$$u(x,t) = \int_S W(x-y, t-s)\, d\mu_u(y,s) + \kappa W(x,t),$$

where κ is a nonnegative number.

PROOF. By Corollary 6.35, u has the representation

$$u(x,t) = \int_S W(x-y, t-s)\, d\mu_u(y,s) + h(x,t)$$

for some temperature h such that $0 \leq h \leq u$ on S. Since $0 \leq h(\cdot, t_0) \leq W(\cdot, t_0)$ on $\mathbb{R}^n$, it follows from Theorem 4.22 that h is a constant multiple of W. □

COROLLARY 6.37. *Let u be a supertemperature on an open set E, and let D be a bounded open set such that $\overline{D} \subseteq E$. Then there is a temperature h on D such that*

$$u(p) = \int_D G(p;q)\, d\mu_u(q) + h(p)$$

for all $p \in D$.

PROOF. Since $\overline{D}$ is compact, u is lower bounded on D and $\mu(\overline{D}) < +\infty$. By the Riesz decomposition theorem, u has the representation $u = G_D\mu_{u,D} + v$ on D, where $\mu_{u,D}$ is the restriction of μ_u to D, and v is the greatest thermic minorant of u on D. By Theorem 6.18, $G\mu_{u,D}$ is a heat potential, and therefore Theorem 6.24 shows that there is a temperature w on D such that $G\mu_{u,D} = G_D\mu_{u,D} + w$ on D. Hence $u = G\mu_{u,D} + (v - w)$ on D, as required. □

COROLLARY 6.38. *Let u be a supertemperature on an open set E. Then u has a thermic minorant on E if and only if $G_E\mu_u$ is a heat potential.*

PROOF. If u has a thermic minorant on E, then Theorem 6.34 shows that $G_E\mu_u$ is a heat potential.

Conversely, suppose that the function $v = G_E\mu_u$ is a heat potential. Then the Riesz measure associated with v is μ_u, by Theorem 6.31, so that there is a temperature h such that $u = v + h$ on E, by Theorem 6.32. Since $v \geq 0$, h is a thermic minorant of u on E. □

COROLLARY 6.39. *Let u be a nonnegative supertemperature on an open set E. Then u is a heat potential if and only if the greatest thermic minorant of u on E is the zero function.*

PROOF. By the Riesz decomposition theorem, $u = G_E\mu_u + h$ on E, where h is the greatest thermic minorant of u on E. So if $h = 0$, then u is a heat potential.

Conversely, if u is a heat potential, then the greatest thermic minorant of u on E is the zero function, by Theorem 6.19. □

For the next corollary, we recall the definition of the heat annulus, with centre p_0, inner radius b, and outer radius $c > b$. It is

$$A(p_0; b, c) = \Omega(p_0; c) \backslash \overline{\Omega}(p_0; b).$$

We also recall the notation $\tau(a) = (4\pi a)^{-\frac{n}{2}}$.

COROLLARY 6.40. *Let u be a supertemperature on an open superset E of the closed heat annulus $\overline{A}(p_0; b, c)$. Then the function $\mathcal{M}(u; p_0; \cdot)$ is real-valued on $[b, c]$.*

PROOF. Let D be a bounded open set such that $\overline{A}(p_0; b, c) \subseteq D$ and $\overline{D} \subseteq E$. By Corollary 6.37, there is a temperature h on D such that

$$u(p) = \int_D G(p; q)\, d\mu_u(q) + h(p)$$

for all $p \in D$. The function $\mathcal{M}(h; p_0; \cdot)$ is clearly real-valued on $[b, c]$. Moreover, by Tonelli's theorem and Example 1.12, we have

$$\begin{aligned}\mathcal{M}(G\mu_{u,D}; p_0; a) &= \int_D \mathcal{M}(G(\cdot; q); p_0; a)\, d\mu_u(q) \\ &= \int_D \bigl(G(p_0; q) \wedge \tau(a)\bigr)\, d\mu_u(q) \\ &\leq \tau(a)\mu_u(D) \\ &< +\infty,\end{aligned}$$

because $\overline{D}$ is compact. This completes the proof. □

6.5. THE RIESZ DECOMPOSITION THEOREM

THEOREM 6.41. *Let u be a supertemperature on an open set E, let $\overline{\Omega}(p_0;d) \subseteq E$, let $0 < c < d$, and let $\overline{\Omega}'(p_0;\cdot)$ denote $\overline{\Omega}(p_0;\cdot)\backslash\{p_0\}$. Then*

$$(6.9) \qquad \mathcal{M}(u;p_0;c) = \mathcal{M}(u;p_0;d) - \int_c^d \tau'(l)\mu_u\big(\overline{\Omega}'(p_0;l)\big)\,dl$$

and

$$(6.10) \qquad u(p_0) = \mathcal{M}(u;p_0;d) - \int_0^d \tau'(l)\mu_u\big(\overline{\Omega}'(p_0;l)\big)\,dl.$$

PROOF. Let D be a bounded open set such that $\overline{\Omega}(p_0;d) \subseteq D$ and $\overline{D} \subseteq E$. By Corollary 6.37, there is a temperature h on D such that

$$u(p) = \int_D G(p;q)\,d\mu_u(q) + h(p)$$

for all $p \in D$. We put $\mu = \mu_{u,D}$, so that the above integral becomes $G\mu(p)$. By Corollary 6.40, the mean values in (6.9) are finite, so we can subtract them. Since $h(p_0) = \mathcal{M}(h;p_0;a)$ whenever $0 < a \leq d$, by Theorem 1.6, it follows that

$$\mathcal{M}(u;p_0;c) - \mathcal{M}(u;p_0;d) = \mathcal{M}(G\mu;p_0;c) - \mathcal{M}(G\mu;p_0;d)$$
$$= \int_D \big(\mathcal{M}(G(\cdot;q);p_0;c) - \mathcal{M}(G(\cdot;q);p_0;d)\big)\,d\mu(q)$$
$$= \int_D \big((G(p_0;q) \wedge \tau(c)) - (G(p_0;q) \wedge \tau(d))\big)\,d\mu(q),$$

where we have used Tonelli's theorem and Example 1.12. The definition of $\Omega(p_0;c)$ shows that the equality $G(p_0;q) \wedge \tau(c) = \tau(c)$ holds if and only if $q \in \overline{\Omega}'(p_0;c)$, and hence

$$(G(p_0;q) \wedge \tau(c)) - (G(p_0;q) \wedge \tau(d)) = \begin{cases} \tau(c) - \tau(d) & \text{if } q \in \overline{\Omega}'(p_0;c), \\ G(p_0;q) - \tau(d) & \text{if } q \in A(p_0;c,d), \\ 0 & \text{if } q \notin \Omega(p_0;d). \end{cases}$$

Therefore

$$\mathcal{M}(u;p_0;c) - \mathcal{M}(u;p_0;d) = \int_{\Omega(p_0;d)} \big((G(p_0;q) \wedge \tau(c)) - \tau(d)\big)\,d\mu(q).$$

We now put $\lambda(l) = \mu(\overline{\Omega}'(p_0;l))$ whenever $0 < l \leq d$, and use integration by parts for Stieltjes integrals to obtain

$$\mathcal{M}(u;p_0;c) - \mathcal{M}(u;p_0;d) = \int_0^d \big((\tau(l) \wedge \tau(c)) - \tau(d)\big)\,d\lambda(l)$$
$$= \big((\tau(l) \wedge \tau(c)) - \tau(d)\big)\lambda(l)\Big|_0^d - \int_c^d \tau'(l)\lambda(l)\,dl$$
$$= -\int_c^d \tau'(l)\mu\big(\overline{\Omega}'(p_0;l)\big)\,dl,$$

which proves (6.9). Making $c \to 0+$ in (6.9), and using Theorem 3.59, we obtain (6.10). $\square$

6.6. Monotone Approximation by Smooth Supertemperatures

Let u be a supertemperature on an open set E. In view of Theorems 3.51 and 3.56, u is a locally integrable function such that the inequality $u(p) \leq \mathcal{V}(u;p;c)$ holds whenever $\overline{\Omega}(p;c) \subseteq E$. It therefore follows from Theorem 1.28 that u can be approximated locally in the mean by smooth supertemperatures. We can now show that u can also be approximated locally by an increasing sequence of smooth supertemperatures. An application follows immediately.

THEOREM 6.42. *Let u be a supertemperature on an open set E, and let D be a bounded open set such that $\overline{D} \subseteq E$. Then there exists an increasing sequence $\{u_k\}$ of smooth supertemperatures on D, such that $\lim_{k \to \infty} u_k(p) = u(p)$ for all $p \in D$.*

PROOF. If μ denotes the restriction to D of the Riesz measure associated with u on E, then there is a temperature h on D such that $u = G\mu + h$ on D, by Corollary 6.37. It therefore suffices to prove the theorem with $G\mu$ in place of u.

We choose a continuously differentiable function ψ on $\mathbb{R}$ such that $\psi(t) = 1$ whenever $t \geq 1$, $\psi(t) = 0$ whenever $t \leq \frac{1}{2}$, and $\psi'(t) \geq 0$ for all t. For each positive integer k, we put $\psi_k(t) = \psi(kt)$. Then $\lim_{k \to \infty} \psi_k(t) = 1$ for all $t > 0$. For all $(x,t) \in \mathbb{R}^{n+1}$, we now put

$$u_k(x,t) = \int_{\mathbb{R}^{n+1}} G(x,t;y,s)\psi_k(t-s)\,d\mu(y,s),$$

and note that the integrand, as a function of (x,t) for any fixed (y,s), belongs to $C^{2,1}(\mathbb{R}^{n+1})$. Since $G(x,t;y,s) = 0$ whenever $t \leq s$, the integrand increases to $G(x,t;y,s)$ as $k \to \infty$. Therefore, by Lebesgue's monotone convergence theorem, the sequence $\{u_k\}$ increases to $G\mu$ as $k \to \infty$. Differentiation under the integral sign shows that each u_k belongs to $C^{2,1}(\mathbb{R}^{n+1})$, and that

$$\Theta u_k(x,t) = \int_{\mathbb{R}^{n+1}} \Big((\Theta_{x,t} G(x,t;y,s))\psi_k(t-s) - G(x,t;y,s)\psi_k'(t-s) \Big)\,d\mu(y,s) \leq 0$$

for all $(x,t) \in \mathbb{R}^{n+1}$. Hence each u_k is a smooth supertemperature on $\mathbb{R}^{n+1}$. $\square$

As an application of Theorem 6.42, we extend the result of Theorem 1.11 to arbitrary subtemperatures.

THEOREM 6.43. *Let w be a subtemperature on an open set E, and let the closed heat annulus $\overline{A}(p_0;b,c) \subseteq E$. Then there is a convex function ϕ such that*

$$\mathcal{M}(w;p_0;a) = \phi(\tau(a))$$

whenever $b \leq a \leq c$.

PROOF. We choose a bounded open set D such that $\overline{A}(p_0;b,c) \subseteq D$ and $\overline{D} \subseteq E$. By Theorem 6.42, there is a decreasing sequence $\{w_k\}$ of smooth subtemperatures that converges pointwise to w on D. By Theorem 1.11, for each k there is a convex function ϕ_k such that $\mathcal{M}(w_k;p_0;a) = \phi_k(\tau(a))$ whenever $b \leq a \leq c$. Therefore, by Lebesgue's monotone convergence theorem,

$$\mathcal{M}(w;p_0;a) = \lim_{k \to \infty} \mathcal{M}(w_k;p_0;a) = \lim_{k \to \infty} \phi_k(\tau(a))$$

whenever $b \leq a \leq c$. Furthermore $\lim_{k\to\infty} \phi_k$ is convex, because $\mathcal{M}(w; p_0; \cdot)$ is finite by Corollary 6.40. □

REMARK 6.44. Theorem 6.43 has no counterpart for the means $\mathcal{L}$ with which we defined subtemperatures. To see this, let w be the characteristic function of the half-space $\mathbb{R}^n \times \,]-\infty, 0]$. If $t_0 > 0$, then the mean $\mathcal{L}(w; x_0, t_0; \cdot)$ has a discontinuity at t_0, because $\mathcal{L}(w; x_0, t_0; a) = 0$ whenever $0 < a < t_0$, but

$$\mathcal{L}(w; x_0, t_0; t_0) = \mu_{(x_0, t_0)}(B(x_0, \sqrt{t_0}) \times \{0\}) > 0$$

by Lemma 2.10.

Theorem 6.43 permits the following improvement on Theorem 1.6. In order to have $w(p_0) = \mathcal{M}(w; p_0; c)$ whenever $0 < c \leq c_0$, it is not necessary for w to be a temperature on an open superset of $\overline{\Omega}(p_0; c_0)$, it is sufficient if w is a temperature on $\Omega(p_0; c_0)$ and a subtemperature on an open superset of $\overline{\Omega}(p_0; c_0)$.

THEOREM 6.45. *Let w be a subtemperature on an open superset E of $\overline{\Omega}(p_0; c_0)$, and a temperature on $\Omega(p_0; c_0)$. Then $w(p_0) = \mathcal{M}(w; p_0; c)$ whenever $0 < c \leq c_0$.*

PROOF. We choose a contracting sequence of open circular cylinders $\{D_k\}$ such that $\overline{D}_k \subseteq E$ for all k, $\bigcap_{k=1}^{\infty} D_k = \{p_0\}$, and for each k the set $\partial D_k \cap \Omega(p_0; c_0)$ is a subset of some hyperplane $\mathbb{R}^n \times \{t_k\}$. For each k, we denote by w_k the function $\pi_{D_k} w$ of Theorem 3.21. Then w_k is a temperature on D_k, a subtemperature that majorizes w on E, and equal to w on $\partial_n D_k \cup (E \setminus \overline{D}_k)$. Since the sequence $\{D_k\}$ is contracting, the sequence $\{w_k\}$ is decreasing. Furthermore, in view of Theorem 2.14, each w_k is a temperature on $D_k \cup \Omega(p_0; c_0)$. Therefore $w_k(p_0) = \mathcal{M}(w_k; p_0; c)$ whenever $0 < c < c_0$, by Theorem 1.6. The function $\lim_{k\to\infty} w_k$ is equal to w on $E \setminus \{p_0\}$, and is a subtemperature on E by Theorem 3.60. Therefore $w = \lim_{k\to\infty} w_k$ everywhere on E, by Theorem 3.59. It now follows from Lebesgue's monotone convergence theorem that

$$w(p_0) = \lim_{k\to\infty} w_k(p_0) = \lim_{k\to\infty} \mathcal{M}(w_k; p_0; c) = \mathcal{M}(w; p_0; c)$$

whenever $0 < c < c_0$. It follows from Theorem 6.43 that the function $\mathcal{M}(w; p_0; \cdot)$ is continuous on $]0, c_0]$, so that $w(p_0) = \mathcal{M}(w; p_0; c_0)$ also. □

6.7. Further Characterizations of Subtemperatures

In this section, we give two characterizations of subtemperatures in terms of the means $\mathcal{V}_m$ for $m \geq 1$. The first is analogous to Theorem 3.51, which involved the means $\mathcal{V}$.

THEOREM 6.46. *Let m be an integer with $m \geq 1$, and let w be an upper finite and upper semicontinuous function on an open set E, that is finite on a dense subset of E. Suppose that, given any point $p \in E$ and $\epsilon > 0$, we can find a positive number $c < \epsilon$ such that the inequality $w(p) \leq \mathcal{V}_m(w; p; c)$ holds. Then w is a subtemperature on E.*

Conversely, if w is a subtemperature on E and $p \in E$, then the inequality $w(p) \leq \mathcal{V}_m(w; p; c)$ holds for all $c > 0$ such that $\overline{\Omega}_m(p; c) \subseteq E$.

PROOF. For the first part, since w is finite on a dense subset of E, we have only to prove that w is a hypotemperature on E. The proof of this is similar to the proof of the first part of Theorem 3.48, using Theorem 1.25 instead of Theorem 1.6.

We now suppose, conversely, that w is a subtemperature on E, and that $\overline{\Omega}_m(p;c)$ is contained in E. We choose a bounded open set D such that $\overline{\Omega}_m(p;c) \subseteq D$ and $\overline{D} \subseteq E$. By Theorem 6.42, there is a decreasing sequence $\{w_k\}$ of smooth subtemperatures that converges pointwise to w on D. By Theorem 1.25, since $\overline{\Omega}_m(p;c) \subseteq D$ the inequality $w_k(p) \leq \mathcal{V}_m(w_k;p;c)$ holds for all k. Therefore, by Lebesgue's monotone convergence theorem,

$$w(p) = \lim_{k \to \infty} w_k(p) \leq \lim_{k \to \infty} \mathcal{V}_m(w_k;p;c) = \mathcal{V}_m(w;p;c).$$

□

COROLLARY 6.47. *Let m be an integer with $m \geq 1$, let w be a subtemperature on an open set E, and let $p_0 \in E$. Then*

$$\lim_{c \to 0+} \mathcal{V}_m(w;p_0;c) = w(p_0).$$

PROOF. Let $A > w(p_0)$. Since w is upper semicontinuous, we can find a neighbourhood N of p_0 such that $w(p) < A$ for all $p \in N$. So, whenever $\overline{\Omega}_m(p_0;c)$ is contained in N, we have

$$w(p_0) \leq \mathcal{V}_m(w;p_0;c) \leq \mathcal{V}_m(A;p_0;c) = A,$$

using Theorem 6.46. This implies the result. □

THEOREM 6.48. *Let m be an integer with $m \geq 1$, and let w be an extended real-valued function on an open set E. Then w is a subtemperature on E if and only if it satisfies the following four conditions:*
(a) $-\infty \leq w(p) < +\infty$ for all $p \in E$;
(b) w is upper semicontinuous on E;
(c) given any point $p \in E$, we can find a point $q \in E$ such that $p \in \Lambda(q;E)$ and $w(q) > -\infty$;
(d) given any point $p \in E$, the inequality $w(p) \leq \mathcal{V}_m(w;p;c)$ holds whenever $\overline{\Omega}_m(p;c) \subseteq E$.

PROOF. Theorem 6.46 shows that any subtemperature on E satisfies condition (d).

For the converse, conditions (a) and (b) imply that w is locally upper bounded on E. Therefore condition (d) and Lemma 3.54 show that w is locally integrable on $\Lambda(q;E)$ whenever $w(q) > -\infty$. Now condition (c) shows that w is locally integrable on E, and hence finite on a dense subset of E. Now Theorem 6.46 shows that w is a subtemperature on E. □

6.8. Supertemperatures on an Infinite Strip or Half-Space

In this section, we look at some consequences of the Riesz decomposition for supertemperatures on an infinite strip or half-space $\mathbb{R}^n \times \,]0,a[$. The first gives a characterization of heat potentials in terms of the hyperplane means M_b, which were defined in Section 5.1.

6.8. SUPERTEMPERATURES ON AN INFINITE STRIP OR HALF-SPACE

THEOREM 6.49. *If $G\mu$ is a heat potential on the strip or half-space $\mathbb{R}^n \times {]}0, a[$, and $0 < b < a$, then the hyperplane mean $M_b(G\mu; \cdot)$ is real-valued and increasing on ${]}0, b[$, with*
$$\lim_{t \to 0+} M_b(G\mu; t) = 0.$$

Conversely, if u is a nonnegative supertemperature on $\mathbb{R}^n \times {]}0, a[$, and

(6.11) $$\liminf_{t \to 0+} M_b(u; t) = 0$$

for some $b < a$, then u is a heat potential.

PROOF. If $G\mu$ is a heat potential on $\mathbb{R}^n \times {]}0, a[$, and $0 < t < b < a$, then by Tonelli's theorem and Theorem 4.10, we have

$$M_b(G\mu; t) = \int_{\mathbb{R}^n} W(x, b-t) \left(\int\int_{\mathbb{R}^n \times {]}0,t[} W(x-y, t-s) \, d\mu(y, s) \right) dx$$
$$= \int\int_{\mathbb{R}^n \times {]}0,t[} \left(\int_{\mathbb{R}^n} W(x, b-t) W(x-y, t-s) \, dx \right) d\mu(y, s)$$
$$= \int\int_{\mathbb{R}^n \times {]}0,t[} W(y, b-s) \, d\mu(y, s)$$
$$= G\mu_{S(t)}(0, b),$$

where $\mu_{S(t)}$ is the restriction of μ to the strip $S(t) = \mathbb{R}^n \times {]}0, t[$. The function $G\mu_{S(\cdot)}$ is increasing on ${]}0, a]$, and $G\mu_{S(a)} = G\mu$ is finite almost everywhere by Theorem 3.56. Therefore the mean $M_b(G\mu; \cdot)$ is increasing on ${]}0, b[$, and $G\mu_{S(t)}$ is finite a.e., and hence is a heat potential, for each t. By Corollary 6.22, $G\mu_{S(t)}$ is a temperature on $\mathbb{R}^n \times {]}t, a[$, and so $G\mu_{S(t)}(0, b) < +\infty$ whenever $t < b < a$. Hence $M_b(G\mu; \cdot)$ is real-valued.

Next, the function $G\mu_{S(\cdot)}$ is lower bounded by 0, and $G\mu_{S(t)}$ decreases as t decreases to 0. Since $G\mu_{S(t)}$ is a temperature on $\mathbb{R}^n \times {]}t, a[$, it follows from the Harnack monotone convergence theorem that the function $v = \lim_{t \to 0+} G\mu_{S(t)}$ is a temperature on $\mathbb{R}^n \times {]}0, a[$. Furthermore, $G\mu$ is a heat potential, and $0 \leq v \leq G\mu$ on $\mathbb{R}^n \times {]}0, a[$, so Theorem 6.19 implies that $v = 0$. Hence

$$\lim_{t \to 0+} M_b(G\mu; t) = \lim_{t \to 0+} G\mu_{S(t)}(0, b) = v(0, b) = 0.$$

For the converse, we suppose that u is a nonnegative supertemperature on $\mathbb{R}^n \times {]}0, a[$. The Riesz decomposition theorem shows that $G\mu_u$ is a heat potential and $u = G\mu_u + h$ on $\mathbb{R}^n \times {]}0, a[$, where μ_u is the Riesz measure associated with u, and h is the greatest thermic minorant of u on $\mathbb{R}^n \times {]}0, a[$. Let b be such that (6.11) holds. Since $h \geq 0$, we have $M_b(h; t) = h(0, b)$ by Theorems 4.18 and 5.5. Therefore $M_b(u; t) = M_b(G\mu_u; t) + h(0, b)$ whenever $0 < t < b$. Hence, by (6.11) and the first part of the theorem,

$$0 = \liminf_{t \to 0+} M_b(u; t) = \lim_{t \to 0+} M_b(G\mu_u; t) + h(0, b) = h(0, b).$$

Now Theorem 4.17 shows that $h = 0$, which completes the proof. □

COROLLARY 6.50. *If $G\mu$ is a heat potential on the strip or half-space $\mathbb{R}^n \times {]}0, a[$, then $\liminf_{t \to 0+} G\mu(x, t) = 0$ for almost every $x \in \mathbb{R}^n$.*

PROOF. By Fatou's lemma and Theorem 6.49, we have

$$0 \leq \int_{\mathbb{R}^n} \liminf_{t \to 0+} \left(W(x, b-t) G\mu(x,t)\right) dx \leq \lim_{t \to 0+} M_b(G\mu; t) = 0$$

for each $b < a$. The result follows. □

It is natural to conjecture that, in the above corollary, $\liminf_{t \to 0+} G\mu(x,t) = 0$ might be replaced by $\lim_{t \to 0+} G\mu(x,t) = 0$, but this is not the case. In Example 6.52 below, we show that it is possible to have $\limsup_{t \to 0+} G\mu(x,t) = +\infty$ for all x in an interval. Example 6.51 contains a preliminary result, which is also of interest.

EXAMPLE 6.51. There exist a heat potential v on the half-plane $\mathbb{R} \times {]}0, +\infty{[}$, and a positive, decreasing function g on $[0, 1[$, such that $v(x, g(x)) = +\infty$ whenever $0 \leq x < 1$.

Given any point $x \in [0, 1[$, we write the binary expansion of x as

$$x = \sum_{k=1}^{\infty} a_k(x) 2^{-k},$$

where $a_k(x) \in \{0, 1\}$ for all k, and $\{k : a_k(x) = 0\}$ is an infinite set. Given any positive number α, we define a decreasing function f_α on $[0, 1[$ by putting

$$f_\alpha(x) = -\alpha \sum_{k=1}^{\infty} a_k(x) 4^{-k},$$

and a heat potential u_α on $\mathbb{R}^2$ by putting

(6.12) $$u_\alpha(x, t) = \alpha \int_0^1 W(x - y, t - f_\alpha(y)) \, dy.$$

(The number α plays no essential role in this example, but it is useful to have it for Example 6.52 below.)

We show that $u_\alpha(x, f_\alpha(x)) = +\infty$ whenever $0 \leq x < 1$. For a fixed x, we let i be any integer such that $i > 2$ and $a_i(x) = 0$. Then, for all $y \in [0, 1[$ such that $a_k(y) = a_k(x)$ whenever $1 \leq k < i$, and $a_i(y) = 1$, we have

$$0 < y - x = 2^{-i} + \sum_{k=i+1}^{\infty} (a_k(y) - a_k(x)) 2^{-k} \leq 2^{-i} + \sum_{k=i+1}^{\infty} 2^{-k} = 2^{1-i},$$

$$f_\alpha(x) - f_\alpha(y) = \alpha 4^{-i} - \alpha \sum_{k=i+1}^{\infty} (a_k(x) - a_k(y)) 4^{-k} \geq \alpha 4^{-i} - \alpha \sum_{k=i+1}^{\infty} 4^{-k} = \alpha \frac{2}{3} 4^{-i},$$

and

$$f_\alpha(x) - f_\alpha(y) \leq \alpha 4^{-i} + \alpha \sum_{k=i+1}^{\infty} 4^{-k} = \alpha \frac{4}{3} 4^{-i}.$$

Therefore

(6.13) $$W(x - y, f_\alpha(x) - f_\alpha(y)) \geq C_\alpha 2^i$$

where C_α is positive and depends only on α.

The set of points $y \in [0, 1[$ such that $a_k(y) = a_k(x)$ whenever $1 \leq k < i$, and $a_i(y) = 1$, is an interval I_i of length 2^{-i}. Furthermore, if $i \neq j$ then $I_i \cap I_j = \emptyset$. For suppose that $j > i$, $y \in I_i$ and $z \in I_j$. Then $a_k(y) = a_k(z)$ whenever $1 \leq k < i$,

$a_i(y) = 1$, $a_i(z) = a_i(x) = 0$, and we can choose an integer $l > i+1$ such that $a_l(z) = 0$. It follows that

$$y - z = 2^{-i} + \sum_{k=i+1}^{\infty} \left(a_k(y) - a_k(z)\right)2^{-k} \geq 2^{-i} - \sum_{k=i+1}^{\infty} 2^{-k} + 2^{-l} = 2^{-l} > 0.$$

Now the inequality (6.13) implies that

$$u_\alpha(x, f_\alpha(x)) \geq \alpha \sum_{\{i>2:a_i(x)=0\}} \int_{I_i} W(x-y, f_\alpha(x) - f_\alpha(y)) \, dy$$

$$\geq \alpha \sum_{\{i>2:a_i(x)=0\}} \int_{I_i} C_\alpha 2^i \, dy$$

(6.14)
$$= +\infty.$$

We now put $g = 1 + f_1$ and $v(x,t) = \int_0^1 W(x-y, t-g(y)) \, dy = u_1(x, t-1)$. Then $g > 0$ and $v(x, g(x)) = u_1(x, f_1(x)) = +\infty$ whenever $0 \leq x < 1$.

EXAMPLE 6.52. Let the functions f_α and heat potentials u_α be as in Example 6.51. For each positive integer j, we take $\alpha = 2^{-j}$ and put $\phi_j = f_\alpha = \alpha f_1$, $h_j = u_\alpha$, and

$$v_j(x,t) = h_j(x, t - 2^{-j}) = 2^{-j} \int_0^1 W(x - y, t - (1 + f_1(y))2^{-j}) \, dy$$

for all $(x,t) \in \mathbb{R} \times \,]0, +\infty[$. Then, whenever $0 \leq x < 1$, we have

$$v_j(x, (1 + f_1(x))2^{-j}) = h_j(x, f_1(x)2^{-j}) = h_j(x, \phi_j(x)) = u_\alpha(x, f_\alpha(x)) = +\infty,$$

by (6.14). We now put

$$w(x,t) = \sum_{j=1}^{\infty} v_j(x,t),$$

so that $w(x, (1 + f_1(x))2^{-k}) = +\infty$ for every positive integer k and all $x \in [0, 1[$. It follows that

$$\limsup_{t \to 0+} w(x,t) = +\infty$$

whenever $0 \leq x < 1$. Note that, if $t \geq 1$, the fact that $f_1 < 0$ implies that $W(y, t - (1 + f_1(y))2^{-j}) \leq (2\pi)^{-\frac{1}{2}}$, so that

$$w(0,t) = \sum_{j=1}^{\infty} h_j(0, t - 2^{-j}) = \sum_{j=1}^{\infty} 2^{-j} \int_0^1 W(y, t - (1 + f_1(y))2^{-j}) \, dy \leq (2\pi)^{-\frac{1}{2}}.$$

Hence Theorem 6.16 shows that w is a heat potential on $\mathbb{R} \times \,]0, +\infty[$. Thus, in Corollary 6.50, we cannot replace the $\liminf_{t \to 0+}$ by $\lim_{t \to 0+}$.

Theorem 6.49 allows us to remove the finiteness conditions from some results in Chapter 5.

THEOREM 6.53. *Suppose that $0 < a \leq +\infty$, and that $w \in \bigcap_{b<a} \Sigma_b$. Then, for each $b < a$, the function $M_b(w; \cdot)$ is decreasing, real-valued, and left continuous on $\,]0, b[$.*

PROOF. We first suppose that w belongs to the class $\bigcap_{b<a} \Phi_b$ of Section 5.3. Then w has a nonnegative thermic majorant u on $\mathbb{R}^n \times {]}0, a[$, by Theorem 5.29. The function $u - w$ is a nonnegative supertemperature, and therefore can be written in the form $u - w = G\mu + h$, where μ is its associated Riesz measure and h is a nonnegative temperature, by Theorem 6.34. Thus $-w = G\mu + (h - u)$ is the sum of a heat potential and a temperature which can be expressed as the difference of two nonnegative temperatures. Hence, by Theorems 4.18 and 5.5, whenever $0 < t < b < a$ we have

$$M_b(w; t) = -M_b(G\mu; t) - h(0, b) + u(0, b).$$

Theorem 6.49 now implies that $M_b(w; \cdot)$ is real-valued on ${]}0, b[$.

We now suppose that $w \in \bigcap_{b<a} \Sigma_b$. We fix $b < a$, take any positive number $s < a - b$, and define a subtemperature w_s on the substrip $\mathbb{R}^n \times {]}0, a - s[$ by putting $w_s(x, t) = w(x, t + s)$. Then $0 < b < a - s$, and whenever $0 < t < c < a - s$ we have

$$\begin{aligned} M_c(w_s^+; t) &= \int_{\mathbb{R}^n} W(x, c - t) w_s^+(x, t) \, dx \\ &= \int_{\mathbb{R}^n} W(x, (c + s) - (t + s)) w^+(x, t + s) \, dx \\ &= M_{c+s}(w^+; t + s). \end{aligned}$$

Since $0 < c + s < a$, our hypothesis and Theorem 5.17 show that $M_{c+s}(w^+; \cdot)$ is bounded on ${]}s, c + s[$. Hence $M_c(w_s^+; \cdot)$ is bounded on ${]}0, c[$ whenever $c < a - s$, so that $w_s \in \bigcap_{c<a-s} \Phi_c$. Therefore, by the first part of this proof, $M_c(w_s; \cdot)$ is real-valued on ${]}0, c[$, whenever $c < a - s$. Since $b - s < a - s$, it follows that $M_{b-s}(w_s; \cdot)$ is finite on ${]}0, b - s[$, so that $M_b(w; t + s) = M_{b-s}(w_s; t)$ is finite whenever $0 < t < b - s$, and hence $M_b(w; t)$ is finite whenever $s < t < b$. Since s can be arbitrarily close to 0, it follows that $M_b(w; \cdot)$ is finite on ${]}0, b[$. The other properties now follow from Theorem 5.17. □

Theorem 6.53 allows us to prove the following variant of Theorem 5.9, in which the necessary and sufficient condition for a subtemperature w to belong to $\bigcap_{b<a} \Sigma_b$ involves w itself rather than w^+.

THEOREM 6.54. *Suppose that $0 < a \leq +\infty$, and that w is a subtemperature on $\mathbb{R}^n \times {]}0, a[$. Then $w \in \bigcap_{b<a} \Sigma_b$ if and only if the inequality*

$$(6.15) \qquad w(x, t) \leq \int_{\mathbb{R}^n} W(x - y, t - s) w(y, s) \, dy$$

holds whenever $x \in \mathbb{R}^n$, $0 < s < t < a$, and the integrals are finite.

PROOF. If $w \in \bigcap_{b<a} \Sigma_b$, then Theorem 6.53 shows that $M_b(w; \cdot)$ is real-valued whenever $b < a$, so that (6.15) holds and the integrals are finite, by Theorem 5.8.

Conversely, if (6.15) holds and the integrals are finite, then

$$w^+(x, t) \leq \int_{\mathbb{R}^n} W(x - y, t - s) w^+(y, s) \, dy$$

holds whenever $x \in \mathbb{R}^n$, $0 < s < t < a$, and these integrals are finite. Therefore $w \in \bigcap_{b<a} \Sigma_b$, by Theorem 5.9. □

COROLLARY 6.55. *Suppose that $0 < a \leq +\infty$, and that u is a temperature on $\mathbb{R}^n \times\,]0, a[$. Then u has the semigroup property on $\mathbb{R}^n \times\,]0, a[$ if and only if the inequality*

$$u(x,t) \leq \int_{\mathbb{R}^n} W(x-y, t-s) u(y,s) \, dy$$

holds whenever $x \in \mathbb{R}^n$, $0 < s < t < a$, and the integrals are finite.

PROOF. Since u has the semigroup property if and only if $u \in \bigcap_{b<a} \Sigma_b$, by Theorem 5.11, the result follows from Theorem 6.54. □

6.9. Notes and Comments

The results in sections 1-3 come from Watson [**72**] and Doob [**14**].

Section 4 and the proof of Theorem 6.34 are adapted from chapter 4 of Armitage & Gardiner. A less elementary distributional approach to the Riesz decomposition theorem was given in Watson [**72**]. A different approach is indicated in Doob [**14**]. The dual of Theorem 6.30 was proved by Smyrnélis [**62**]. Corollary 6.40 was proved by Watson [**80, 81**].

Generalizations of the means $\mathcal{M}$ to means $\mathcal{M}_D$ over level surfaces of Green functions $G_D(p_0;\cdot)$ for any open set D which is regular for the Dirichlet problem for the adjoint equation, were introduced in Watson [**83, 84**]. The generalization of Corollary 6.40 was proved by Brzezina [**11**].

Theorem 6.41 first appeared in Watson [**85**], and was generalized to the means $\mathcal{M}_D$, described above, in Watson [**88**].

Theorem 6.42 appeared in a stronger form in Doob [**14**], and given a different proof in Garofalo & Lanconelli [**26**]. The elegant proof given in the text comes from Lanconelli & Pascucci [**47**].

The convexity property of the means $\mathcal{M}$, given in Theorem 6.43, is due to Watson [**80, 81, 84**], as is its generalization to the means $\mathcal{M}_D$, described above, in [**83**]. Its consequence, Theorem 6.45, appeared in Watson [**80**], and is generalized in Watson [**83**].

The **open question**, about whether any of the L^r-mean values of nonnegative subtemperatures for $r > 1$, given by $\mathcal{M}_r(u; x_0, t_0; c) = \mathcal{M}(u^r; x_0, t_0; c)^{1/r}$, have the convexity property of the means $\mathcal{M}$, is worth repeating here. The case where u is the characteristic function of $\mathbb{R}^n \times\,]-\infty, 0]$ and $t_0 > 0$, shows that the property fails if r is sufficiently large. For the case corresponding to $r = +\infty$, where the integral mean is replaced by a supremum, Watson [**83**] has proved a result using *coheat* spheres instead of heat spheres.

Theorem 6.48 is new in its generality. A partial result was given in Watson [**87**].

The theorems in section 6.8 come from Watson [**75**], but the examples are due to Kaufman & Wu [**39**]. Further examples of interesting graphs can be found in Dont [**13**] and Taylor & Watson [**67**].

CHAPTER 7

Polar Sets and Thermal Capacity

Sets on which a supertemperature can take the value $+\infty$ are called polar. Because any supertemperature is locally integrable, any polar set has Lebesgue measure zero. Indeed, polar sets play a role reminiscient of that played by null sets in measure theory, in that they are negligible in certain contexts. For example, in the boundary maximum principle for hypotemperatures, we can relax the standard limit superior condition on a polar subset of the boundary (cf. Theorem 7.9). As another example, a lower bounded supertemperature defined on $E\backslash Z$, where E is open and Z is polar and relatively closed, can be extended to one on the whole of E (cf. Theorem 7.14). A Borel polar set cannot support a non-null measure whose heat potential is bounded (cf. Theorem 7.52).

A measure gives a way of estimating the size of a set, and the thermal capacity does a similar job in a way the is especially appropriate to the current situation. The sets with thermal capacity zero are precisely the polar sets, and so are sometimes negligible, as outlined above. But more is true, because the cothermal capacity (the corresponding notion for the adjoint equation) of any set coincides with its thermal capacity, and so the copolar sets coincide with the polar sets (cf. Theorem 7.46). Moreover, for subsets of a hyperplane of the form $\mathbb{R}^n \times \{a\}$, the thermal capacity coincides with the n-dimensional Lebesgue measure on the hyperplane (cf. Theorem 7.55).

7.1. Polar Sets

DEFINITION 7.1. Let Z be any subset of $\mathbb{R}^{n+1}$. If there is an open set E containing Z, and a supertemperature w on E such that

$$Z \subseteq \{p \in E : w(p) = +\infty\},$$

then Z is called a *polar set*.

Obviously any subset of a polar set is itself polar. Since any supertemperature is locally integrable (by Theorem 3.56), any polar set has Lebesgue measure zero. Furthermore, Corollary 3.22 shows that, if B is an open ball in $\mathbb{R}^n$, then any set of the form $B \times \{a\}$ is not polar. Moreover, Corollary 6.40 shows that any polar subset of a heat sphere has surface area measure zero.

The polar sets are the negligible sets of heat potential theory. A statement about points of a set $S \subseteq \mathbb{R}^{n+1}$, which is true except for the points of some polar subset of S, is said to be true *quasi-everywhere* on S, or *q.e.* on S.

We already have an example of a polar set in $\mathbb{R}^2$, in Example 6.51. Here is a much simpler one, in $\mathbb{R}^{n+1}$ for any n.

EXAMPLE 7.2. Any singleton is a polar set. To show this, we take any point $(x_0, t_0) \in \mathbb{R}^{n+1}$, put $p_k = (x_0, t_0 - 2^{-k})$ for every positive integer k, and put $A = \{p_k : k \in \mathbb{N}\}$. We let μ be the measure supported by A such that $\mu(\{p_k\}) = k^{-2}$ for all k. Then $\mu(\mathbb{R}^{n+1}) = \sum_{k=1}^{\infty} k^{-2} < +\infty$, so that $G\mu$ is a heat potential, by Theorem 6.18. Moreover,

$$G\mu(x_0, t_0) = \int_A (4\pi(t_0 - s))^{-\frac{n}{2}} d\mu(y, s) = (4\pi)^{-\frac{n}{2}} \sum_{k=1}^{\infty} \left(2^{-k}\right)^{-\frac{n}{2}} = +\infty,$$

so that $\{(x_0, t_0)\}$ is polar.

In the definition of a polar set Z, the supertemperature w can be chosen to be the heat potential of a finite measure on any open superset of Z, and finite at any pre-assigned point outside Z, as we now show.

THEOREM 7.3. *Let Z be a polar subset of an open set E, and let $p_0 \in E\backslash Z$. Then there is a heat potential $G_E\mu$ of a finite measure μ, such that $G_E\mu(p) = +\infty$ for all $p \in Z$ and $G_E\mu(p_0) < +\infty$.*

PROOF. Because Z is polar, there exist an open set D containing Z, and a supertemperature w on D, such that $w(p) = +\infty$ for all $p \in Z$. Replacing D with $(D \cap E)\backslash\{p_0\}$, if necessary, we can assume that $D \subseteq E$ and $p_0 \notin D$. We choose a sequence $\{B_k\}$ of open balls, such that $\overline{B}_k \subseteq D$ for all k and $\bigcup_{k=1}^{\infty} B_k = D$. For each k, we define a finite measure ν_k by putting

$$\nu_k(A) = \frac{\mu_w(A \cap B_k)}{\mu_w(B_k) + 1},$$

for any Borel subset of $\mathbb{R}^{n+1}$, where μ_w is the Riesz measure associated with w. By Corollary 6.37, there is a temperature h_k on B_k such that

$$w(p) = \int_{B_k} G(p; q) \, d\mu_w(q) + h_k(p)$$

for all $p \in B_k$. Therefore the heat potential w_k, defined on $\mathbb{R}^{n+1}$ by

$$w_k(p) = \int_{\mathbb{R}^{n+1}} G(p; q) \, d\nu_k(q),$$

takes the value $+\infty$ at every point of $Z \cap B_k$. Furthermore, since $p_0 \notin D$ and $\overline{B}_k \subseteq D$, the point p_0 is outside the support of ν_k. Therefore Corollary 6.22 implies that $w_k(p_0) < +\infty$.

Now we put

$$\mu = \sum_{k=1}^{\infty} \frac{2^{-k}\nu_k}{1 + w_k(p_0)}.$$

Then $\mu(\mathbb{R}^{n+1}) \leq 1$, so that $G\mu$ is a heat potential by Theorem 6.18. Furthermore, $G\mu(p) = +\infty$ for all $p \in Z$, and $G\mu(p_0) < +\infty$. Since $\mu(\mathbb{R}^{n+1}\backslash E) = 0$, it follows from Theorem 6.24 that the restriction μ_E of μ to E satisfies $G\mu = G_E\mu_E + u$ on E, for some temperature u on E. It follows that $G_E\mu_E$ has all the properties stated in the theorem. □

THEOREM 7.4. *The union of any sequence of polar sets is itself a polar set.*

PROOF. Let $\{Z_k\}$ be a sequence of polar sets. By Theorem 7.3, for each k we can find a heat potential $G\mu_k$ such that $G\mu_k = +\infty$ on Z_k and $\mu_k(\mathbb{R}^{n+1}) \leq 2^{-k}$. The measure $\mu = \sum_{k=1}^{\infty} \mu_k$ satisfies $\mu(\mathbb{R}^{n+1}) \leq 1$, so that $G\mu$ is a heat potential, by Theorem 6.18. Furthermore, $G\mu(p) = \sum_{k=1}^{\infty} G\mu_k(p)$ for all $p \in \mathbb{R}^{n+1}$, so that $G\mu(p) = +\infty$ for all $p \in \bigcup_{k=1}^{\infty} Z_k$. Hence $\bigcup_{k=1}^{\infty} Z_k$ is a polar set. $\square$

EXAMPLE 7.5. Any countable set is polar. For, by Example 7.2, any singleton is polar, and so Theorem 7.4 gives the result.

THEOREM 7.6. *If E is a connected open set, and Z is a relatively closed polar subset of E, then $E \backslash Z$ is connected.*

PROOF. The set $E \backslash Z$ is open, and so its components are also open. Let D be a component of $E \backslash Z$. By Theorem 7.3, we can choose a heat potential u on $\mathbb{R}^{n+1}$ such that $u(p) = +\infty$ for all $p \in Z$. We define a function v on E by putting

$$v(p) = \begin{cases} u(p) & \text{if } p \in D, \\ +\infty & \text{if } p \in E \backslash D. \end{cases}$$

Then v is lower finite and lower semicontinuous on E. Moreover, the inequality $v(p) \geq \mathcal{V}(v; p; c)$ holds whenever $\overline{\Omega}(p; c) \subseteq D$, and whenever $\overline{\Omega}(p; c) \subseteq E \backslash D$, so that given any point $p \in E$ and $\epsilon > 0$, we can find a positive number $c < \epsilon$ such that the inequality holds. If $q \in D$, then $v(p) < +\infty$ on a dense subset of $\Lambda(q; D)$, and so for any point p in the larger set $\Lambda(q; E)$ we can find a point $r \in \Lambda(q; E)$ such that $p \in \Lambda(r; \Lambda(q; E)) = \Lambda(r; E)$ and $v(r) < +\infty$. Hence, by Theorem 3.56, v is a supertemperature on $\Lambda(q; E)$, which implies that $\Lambda(q; E) \subseteq D$.

Suppose that $D \neq E \backslash Z$. Then we can find a component C of $E \cap \partial D$ that forms part of the boundary between D and another component of $E \backslash Z$. Since $\Lambda(q; E) \subseteq D$ for all $q \in D$, the set C is contained in some hyperplane $\mathbb{R}^n \times \{a\}$, and is a relatively open subset because of how C was chosen. Therefore C is not polar, which contradicts the fact that $C \subseteq Z$. Hence $D = E \backslash Z$, which proves the theorem. $\square$

EXAMPLE 7.7. It follows from Theorem 7.6 that a nonempty, relatively open subset of *any* hyperplane in $\mathbb{R}^{n+1}$ (not necessarily of the form $\mathbb{R}^n \times \{a\}$) cannot be polar.

As an example of the negligibility of polar sets, we shall give a refinement of the boundary maximum principle for hypotemperatures (Theorem 3.13), using the following terminology.

DEFINITION 7.8. A sequence $\{p_k\}$ of points in an open set E is called a *Λ-sequence* if it satisfies the condition $p_{k+1} \in \Lambda(p_k; E)$ for all k.

THEOREM 7.9. *Let w be a hypotemperature on an open set E, and let Z be a polar subset of ∂E. Suppose that*

$$\limsup_{k \to \infty} w(p_k) \leq A$$

for every Λ-sequence $\{p_k\}$ in E that tends either to a point of $\partial E \backslash Z$ or to the point at infinity, and that

$$\limsup_{k \to \infty} w(p_k) < +\infty$$

for every Λ-sequence $\{p_k\}$ in E that tends to a point of Z. Then $w(p) \leq A$ for all $p \in E$.

PROOF. In view of Theorem 7.3, we can find a heat potential v on $\mathbb{R}^{n+1}$ such that $v(q) = +\infty$ for all $q \in Z$. Because v is lower semicontinuous, we have $\lim_{p \to q} v(p) = +\infty$ for all $q \in Z$. Therefore, for any $\epsilon > 0$, the function $w - \epsilon v$ is a hypotemperature on E such that $\limsup_{k \to \infty}(w - \epsilon v)(p_k) \leq A$ for every Λ-sequence $\{p_k\}$ in E that tends either to a point of ∂E or to the point at infinity. Hence $w(p) - \epsilon v(p) \leq A$ for all $p \in E$, by Theorem 3.13. Making $\epsilon \to 0$, we see that $w(p) \leq A$ for all $p \in E$ such that $v(p) < +\infty$, and hence almost everywhere on E by Theorem 3.56. Now Theorem 3.51 shows that whenever $\overline{\Omega}(p; c) \subseteq E$ we have

$$w(p) \leq \mathcal{V}(w; p; c) \leq \mathcal{V}(A; p; c) \leq A,$$

which completes the proof. $\square$

EXAMPLE 7.10. The result of Theorem 7.9 cannot be obtained if we allow

$$\liminf_{k \to \infty} w(p_k) = +\infty$$

for Λ-sequences $\{p_k\}$ in E that tend to even a single point of Z. For example, if $w = G(\cdot; q_0)$, $E = \mathbb{R}^{n+1} \backslash \{q_0\}$, $Z = \{q_0\}$ and $A = 0$, then w is a temperature on E that satisfies the hypotheses of Theorem 7.9 except at the point q_0, but $w \geq A$ on E.

7.2. Families of Supertemperatures

By Theorem 3.63, if $\mathcal{F}$ is an upward-directed family of supertemperatures on E, then $\sup \mathcal{F}$ is usually also a supertemperature on E. However, we shall also need to consider more general families of supertemperatures, and whether their infima are supertemperatures. This is a more delicate matter, as the infimum of a family of lower semicontinuous functions is not usually lower semicontinuous. It transpires that the lack of lower semicontinuity is the only obstacle, and this can effectively be overcome by replacing the infimum by what is called its lower semicontinuous smoothing. We now define this concept, and proceed towards the result we require (Theorem 7.13).

DEFINITION 7.11. If u is an extended real-valued function on an open set E, then the *lower semicontinuous smoothing* $\widehat{u}$ of u is defined by

$$\widehat{u}(p) = u(p) \wedge \liminf_{q \to p} u(q),$$

for all $p \in E$.

As the name suggests, $\widehat{u}$ is lower semicontinuous on E, and clearly $\widehat{u} \leq u$ on E. Furthermore, if w is a lower semicontinuous minorant of u on E, then $w = \widehat{w} \leq \widehat{u}$ on E, so that $\widehat{u}$ is the greatest such minorant.

The following result is called *Choquet's topological lemma*. It plays a similar rôle to that of Lemma 3.62, but in the present more complex situation.

LEMMA 7.12. Let $\{u_\alpha : \alpha \in I\}$ be a family of extended real-valued functions on an open set E, and for each subset J of I let $u_J(p) = \inf\{u_\alpha(p) : \alpha \in J\}$ for all $p \in E$. Then there is a countable set $K \subseteq I$ such that $\widehat{u}_K = \widehat{u}_I$ on E.

PROOF. The assertion is that, if g is lower semicontinuous and $g \leq u_K$ on E, then $g \leq u_I$ on E. Thus the result is concerned only with the order properties of functions, and so we can assume that $u_\alpha(E) \subseteq [-\frac{\pi}{2}, \frac{\pi}{2}]$ for all $\alpha \in I$. This can be achieved by replacing each function u_α with $\tan^{-1} u_\alpha$, if necessary, because $\tan^{-1}$ is an order-preserving map of the extended real number system.

Let $\{B_k\}$ be a countable base for the Euclidean topology of E. For each k, we choose a point $p_k \in B_k$ such that
$$u_I(p_k) < \inf\{u_I(q) : q \in B_k\} + \frac{1}{k},$$
and then an index $\alpha_k \in I$ such that
$$u_{\alpha_k}(p_k) < u_I(p_k) + \frac{1}{k}.$$
Then we have
$$(7.1) \quad \inf\{u_{\alpha_k}(q) : q \in B_k\} \leq u_{\alpha_k}(p_k) < u_I(p_k) + \frac{1}{k} < \inf\{u_I(q) : q \in B_k\} + \frac{2}{k}.$$
We put $K = \{\alpha_k : k \in \mathbb{N}\}$, so that K is a countable subset of I, and show that $\widehat{u}_K = \widehat{u}_I$ on E. Let w be a lower semicontinuous minorant of u_K on E, let $p \in E$, and let $\epsilon > 0$. Then there is $l \in \mathbb{N}$ such that $p \in B_l$, $w(q) > w(p) - \frac{\epsilon}{2}$ for all $q \in B_l$, and $\frac{2}{l} < \frac{\epsilon}{2}$. Therefore
$$w(p) - \inf\{w(q) : q \in B_l\} \leq \frac{\epsilon}{2}.$$
Furthermore, because $w \leq u_K \leq u_{\alpha_l}$, we have
$$\inf\{w(q) : q \in B_l\} - \inf\{u_{\alpha_l}(q) : q \in B_l\} \leq 0.$$
Putting $k = l$ in (7.1), we also have
$$\inf\{u_{\alpha_l}(q) : q \in B_l\} - \inf\{u_I(q) : q \in B_l\} < \frac{2}{l} < \frac{\epsilon}{2}.$$
Adding these last three inequalities, we obtain
$$w(p) - \inf\{u_I(q) : q \in B_l\} < \epsilon.$$
The fact that $p \in B_l$ now implies that
$$w(p) < \inf\{u_I(q) : q \in B_l\} + \epsilon \leq u_I(p) + \epsilon.$$
It follows that $w \leq u_I$ whenever w is a lower semicontinuous minorant of u_K, so that $\widehat{u}_K \leq \widehat{u}_I$. Since $K \subseteq I$ we also have $\widehat{u}_I \leq \widehat{u}_K$, and so equality holds. □

THEOREM 7.13. *Let $\mathcal{F} = \{u_\alpha : \alpha \in I\}$ be a family of supertemperatures on an open set E, and suppose that the function $u = \inf \mathcal{F}$ is locally lower bounded. Then the lower semicontinuous smoothing $\widehat{u}$ is a supertemperature on E, is equal to u almost everywhere on E, and satisfies*
$$\widehat{u}(p) = \liminf_{q \to p} u(q)$$
for all $p \in E$.

PROOF. Since the family $\mathcal{F}$ may be uncountable, the function u may not be measurable. We therefore seek an appropriate countable family that still has $\widehat{u}$ as the lower semicontinuous smoothing of its infimum. To this end, we first let $\mathcal{G} = \{v_\alpha : \alpha \in J\}$ denote the family of all pointwise minima $u_{\alpha_1} \wedge ... \wedge u_{\alpha_l}$ ($l \geq 1$) that can be formed using finitely many elements of $\mathcal{F}$. Corollary 3.18 shows that

$\mathcal{G}$ is a family of supertemperatures on E, and clearly $u = \inf \mathcal{G}$. By Lemma 7.12, there is a sequence $\{w_k\}$ of functions in $\mathcal{G}$ whose infimum has lower semicontinuous smoothing $\widehat{u}$. We put $w = \inf\{w_k : k \in \mathbb{N}\}$, and note that $u \leq w$ and $\widehat{u} = \widehat{w}$.

For each $\alpha \in J$ and $p \in E$, it follows from Lemma 3.16 that
$$v_\alpha(p) = \liminf_{q \to p} v_\alpha(q) \geq \liminf_{q \to p} u(q),$$
so that $u(p) \geq \liminf_{q \to p} u(q)$, and hence $\widehat{u}(p) = \liminf_{q \to p} u(q)$.

We now choose an integer $m \geq 5$. By Theorem 6.46, whenever $\overline{\Omega}_m(p;c) \subseteq E$ we have $\mathcal{V}_m(w_k; p; c) \leq w_k(p)$ for all k, so that
$$(7.2) \qquad \mathcal{V}_m(\widehat{w}; p; c) \leq \mathcal{V}_m(w; p; c) \leq w(p).$$

Because $w \geq u$, w is locally lower bounded. Since w is also majorized by each supertemperature w_k, Theorem 3.56 implies that w is locally integrable on E. Therefore Theorem 1.28 implies that the function $p \mapsto \mathcal{V}_m(w; p; c)$ is continuous on $\{p : \overline{\Omega}_m(p;c) \subseteq E\}$, and so it follows from (7.2) that
$$(7.3) \qquad \mathcal{V}_m(\widehat{w}; p; c) \leq \mathcal{V}_m(w; p; c) \leq \widehat{w}(p),$$
because $\widehat{w}$ is the greatest lower semicontinuous minorant of w on any open subset of E. It now follows from Theorem 6.46 that $\widehat{w}$ is a supertemperature on E.

Now Corollary 6.47 and (7.3) show that
$$\widehat{w}(p) = \lim_{c \to 0+} \mathcal{V}_m(\widehat{w}; p; c) \leq \liminf_{c \to 0+} \mathcal{V}_m(w; p; c) \leq \limsup_{c \to 0+} \mathcal{V}_m(w; p; c) \leq \widehat{w}(p).$$

Thus $\lim_{c \to 0+} \mathcal{V}_m(w; p; c) = \widehat{w}(p)$ for all $p \in E$. In view of Theorem 1.28(b) and the local integrability of w, given any compact subset K of E we have
$$\lim_{c \to 0+} \int_K |\mathcal{V}_m(w; q; c) - w(q)|\, dq = 0,$$
and so we can find a null sequence $\{c_i\}$ such that $\lim_{i \to \infty} \mathcal{V}_m(w; q; c_i) = w(q)$ for almost every point $q \in K$. Therefore $\widehat{w}(p) = w(p)$ for almost every $p \in E$. Thus $\widehat{w}(p) = \widehat{u}(p) \leq u(p) \leq w(p) = \widehat{w}(p)$, and hence $\widehat{u}(p) = u(p)$, for almost all $p \in E$. $\square$

Using Theorem 7.13, we now prove an extension theorem for supertemperatures across a relatively closed polar subset of E.

THEOREM 7.14. *Let E be an open set, let Z be a relatively closed polar subset of E, and let u be a supertemperature on $E \backslash Z$ that is locally lower bounded on E as a function defined almost everywhere on E. Then the function $\bar{u}$, defined for all $p \in E$ by*
$$\bar{u}(p) = \liminf_{q \to p,\, q \in E \backslash Z} u(q),$$
is the unique extension of u to a supertemperature on E.

PROOF. In view of Theorem 7.3, we can find a heat potential v on E such that $v(p) = +\infty$ for all $p \in Z$. Since v is lower semicontinuous on E, for all $p \in Z$ we have $+\infty = \lim_{q \to p} v(q)$. Given any positive integer k, we put
$$u_k(p) = \begin{cases} u(p) + \frac{1}{k} v(p) & \text{if } p \in E \backslash Z, \\ +\infty & \text{if } p \in Z. \end{cases}$$

Because u is locally lower bounded on E, we have $u_k(p) = +\infty = \lim_{q\to p} u_k(q)$ for all $p \in Z$. Therefore each function u_k is lower semicontinuous on E, and hence a supertemperature on E. We now put

$$w(p) = \inf\{u_k(p) : k \in \mathbb{N}\}$$

for all $p \in E$. For each k we have $u \leq u_k$ on $E\backslash Z$, so that $u \leq w$ there. Hence w is locally lower bounded on E. Theorem 7.13 now shows that the lower semicontinuous smoothing $\widehat{w}$ is a supertemperature on E. Since u is a lower semicontinuous as well as a minorant of w on $E\backslash Z$, we have $u \leq \widehat{w}$ there. Moreover, $w(p) = u(p)$ for all $p \in E\backslash Z$ such that $v(p) < +\infty$, and hence almost everywhere on E, so that $u = \widehat{w} = w$ a.e. The functions u and $\widehat{w}$ are both supertemperatures on $E\backslash Z$, so that $u = \widehat{w}$ everywhere on $E\backslash Z$, by Theorem 3.59. Thus $\widehat{w}$ is an extension of u to a supertemperature on E. It is unique, because any other such extension would be equal to $\widehat{w}$ a.e. on E, and hence everywhere on E. Finally, because $w(p) = +\infty$ for all $p \in Z$, we have

$$\widehat{w}(p) = \liminf_{q\to p} w(q) = \liminf_{q\to p,\, q\in E\backslash Z} w(q) \geq \liminf_{q\to p,\, q\in E\backslash Z} \widehat{w}(q) \geq \widehat{w}(p)$$

for all $p \in E$, so that

$$\widehat{w}(p) = \liminf_{q\to p,\, q\in E\backslash Z} \widehat{w}(q) = \liminf_{q\to p,\, q\in E\backslash Z} u(q).$$

Taking $\bar{u} = \widehat{w}$, we obtain the result. □

COROLLARY 7.15. *Let E be an open set, let Z be a relatively closed polar subset of E, and let u be a temperature on $E\backslash Z$ that is locally bounded on E as a function defined almost everywhere on E. Then u has a unique extension to a temperature on E.*

PROOF. By Theorem 7.14, the functions u and $v = -u$ can be extended to supertemperatures $\bar{u}$ and $\bar{v}$ on E. Since $\bar{u} = u = -v = -\bar{v}$ on $E\backslash Z$, we have $\bar{u} = -\bar{v}$ almost everywhere on E. Therefore, for all $p \in E$ we have

$$\bar{u}(p) = \lim_{c\to 0+} \mathcal{V}(\bar{u};p;c) = \lim_{c\to 0+} \mathcal{V}(-\bar{v};p;c) = -\bar{v}(p),$$

by Theorem 3.59. Hence $\bar{u} = -\bar{v}$ everywhere on E. Since $\bar{u}$ is a supertemperature and $-\bar{v}$ is a subtemperature, $\bar{u}$ is a temperature on E. □

As another application of Theorem 7.14, we show that the Green function with pole at p_0 is a minimal temperature for $E\backslash\{p_0\}$ (in the sense of Section 4.4).

THEOREM 7.16. *Let E be an open set and let $p_0 \in E$. Then the restriction of $G_E(\cdot;p_0)$ to $E\backslash\{p_0\}$ is a minimal temperature for $E\backslash\{p_0\}$.*

PROOF. Let u be a nonnegative temperature such that $u \leq G_E(\cdot;p_0)$ on $E\backslash\{p_0\}$. Theorem 7.14 shows that u can be extended to a supertemperature $\bar{u}$ on E. By Theorem 6.34, if $\mu_{\bar{u}}$ denotes the Riesz measure associated with $\bar{u}$, then $G_E\mu_{\bar{u}}$ is a heat potential and $\bar{u} = G_E\mu_{\bar{u}} + h$ on E, where h is the greatest thermic minorant of $\bar{u}$ on E. Because $\bar{u}$ is a temperature on $E\backslash\{p_0\}$, the measure $\mu_{\bar{u}}$ is supported by $\{p_0\}$, in view of Theorem 6.25. Therefore $\bar{u} = cG_E(\cdot;p_0) + h$ for some nonnegative number c. Furthermore, $0 \leq h \leq \bar{u} = u \leq G_E(\cdot;p_0)$ on $E\backslash\{p_0\}$, so that $h(p_0) = 0$ in view of Theorem 6.7 and the continuity of h. Hence $h \leq G_E(\cdot;p_0)$ on E. It follows that $h = 0$ on E, so that $u = cG_E(\cdot;p_0)$ on $E\backslash\{p_0\}$, as required. □

7.3. The Natural Order Decomposition

The natural order decomposition theorem for nonnegative supertemperatures is an essential tool for proving basic properties of thermal capacity. In its proof we use the following characterization of supertemperatures, which shows, under certain conditions, that the lower semicontinuity requirement can be weakened in line with the result of Lemma 3.16.

THEOREM 7.17. *Let u be an extended real-valued function on an open set E. Then u is a supertemperature on E if and only if it satisfies the following four conditions:*

(a) u is locally lower bounded on E;
(b) for each point $(x_0, t_0) \in E$, the inequality

$$\liminf_{(x,t)\to(x_0,t_0-)} u(x,t) \geq u(x_0,t_0)$$

holds;
(c) u is finite on a dense subset of E;
(d) the inequality $u(p) \geq \mathcal{L}(u;p;c)$ holds whenever $\overline{\Delta}(p;c) \subseteq E$.

PROOF. If u is a supertemperature on E, then because it is lower finite and lower semicontinuous, it is locally lower bounded, by Lemma 3.4. Thus u satisfies conditions (a), (b) and (c), and Theorem 3.17 shows that it also satisfies (d).

To prove the converse, we have to prove that, if u satisfies the four conditions, then it is lower semicontinuous at each point of E. Let $p_0 = (x_0, t_0) \in E$. Given any subset S of E, we adopt the notation

$$S^+ = \{(x,t) \in S : t \geq t_0\}, \qquad S^- = \{(x,t) \in S : t < t_0\}.$$

Since u satisfies condition (b), given any number $A < u(x_0, t_0)$, we can find an open ball V, with centre (x_0, t_0), such that $\overline{V} \subseteq E$ and $u(y,s) > A$ for all $(y,s) \in V^-$. Let U be a neighbourhood of (x_0, t_0) such that $\overline{U} \subseteq V$. Then the distance between U and $\mathbb{R}^{n+1}\backslash V$ is positive, and so we can find $c_0 > 0$ such that $\overline{\Delta}(x,t;c_0) \subseteq V$ for all $(x,t) \in U$. Since u satisfies condition (a), we can find a number $M > 0$ such that $u(y,s) > -M$ for all $(y,s) \in V$.

For any point $p_1 = (x_1, t_1)$, we let μ_{p_1} denote the caloric measure at p_1 for the heat cylinder $\Delta(p_1; c_0)$. By Lemma 2.10, we have

$$\mu_{p_1}(\{(y,s) \in \partial_n \Delta(p_1;c_0) : s \geq t_1\}) = 0,$$

and so for any given $\epsilon > 0$ we can find a number $\delta_0 > 0$ such that

$$M\mu_{p_1}(\{(y,s) \in \partial_n \Delta(p_1;c_0) : s \geq t_1 - r\}) < \epsilon$$

whenever $0 < r < \delta_0$. Furthermore, because

$$\mu_{p_1}(\{(y,s) \in \partial_n \Delta(p_1;c_0) : s < t_1\}) = \int_{\partial_n \Delta(p_1;c_0)} d\mu_{p_1} = \mathcal{L}(1;p_1;c_0) = 1,$$

we can find a number δ_1, satisfying $0 < \delta_1 < \delta_0$, such that

$$\mu_{p_1}(\{(y,s) \in \partial_n \Delta(p_1;c_0) : s < t_1 - r\}) > 1 - \epsilon$$

whenever $0 < r < \delta_1$. It now follows from the translation invariance of caloric measure that, if the point $p = (x,t)$ is such that $t_0 < t < t_0 + \delta_1$, then

$$M \int_{\partial_n \Delta(p;c_0)^+} d\mu_p = M\mu_p(\{(y,s) \in \partial_n \Delta(p;c_0) : s \geq t_0\}) < \epsilon$$

and

$$\int_{\partial_n \Delta(p;c_0)^-} d\mu_p = \mu_p(\{(y,s) \in \partial_n \Delta(p;c_0) : s < t_0\}) > 1 - \epsilon.$$

Therefore, for every point $p = (x,t) \in U^+$ such that $t < t_0 + \delta_1$, condition (d) implies that

$$\begin{aligned}
u(p) &\geq \int_{\partial_n \Delta(p;c_0)} u \, d\mu_p \\
&\geq -M \int_{\partial_n \Delta(p;c_0)^+} d\mu_p + A \int_{\partial_n \Delta(p;c_0)^-} d\mu_p \\
&> -\epsilon + (A - |A|\epsilon) \\
&= A - (1 + |A|)\epsilon.
\end{aligned}$$

Hence

$$\liminf_{p \to p_0, \, p \in U^+} u(p) > A - (1 + |A|)\epsilon$$

for every $\epsilon > 0$ and number $A < u(p_0)$. Thus

$$\liminf_{p \to p_0, \, p \in U^+} u(p) \geq u(p_0),$$

which implies that u is lower semicontinuous at p_0, in view of condition (b). Hence u is a supertemperature on E. □

COROLLARY 7.18. *Let u be a locally bounded function on an open set E. Then u is a temperature on E if and only if, at every point $(x_0, t_0) \in E$, u satisfies*

(a) $u(x_0, t_0) = \lim_{(x,t) \to (x_0, t_0-)} u(x,t)$

and

(b) $u(x_0, t_0) = \mathcal{L}(w; x_0, t_0; c)$ *whenever* $\overline{\Delta}(x_0, t_0; c) \subseteq E$.

PROOF. If u is a temperature on E, then condition (a) is obviously satisfied, and Theorem 2.14 shows that condition (b) is too.

Conversely, if conditions (a) and (b) are satisfied, then Theorem 7.17 shows that u is both a supertemperature and a subtemperature on E, so that u is a temperature on E. □

REMARK 7.19. Results similar to Theorem 7.17 and Corollary 7.18, can be obtained using any of the other mean values $\mathcal{M}$, $\mathcal{V}$ or $\mathcal{V}_m$.

The proof of the Natural Order Decomposition Theorem depends also on the following result about piecing together two supertemperatures.

LEMMA 7.20. *Let u be a supertemperature on an open set E, and let v be a supertemperature on an open subset V of E. If*

(7.4) $$\liminf_{p \to q, \, p \in V} v(p) \geq u(q) \quad \text{for all} \quad q \in E \cap \partial V,$$

and w is defined on E by

$$w(p) = \begin{cases} (u \wedge v)(p) & \text{if } p \in V, \\ u(p) & \text{if } p \in E \backslash V, \end{cases}$$

then w is a supertemperature on E.

PROOF. It is clear that w is a supertemperature on $E \backslash \partial V$, that $w(p) > -\infty$ for all $p \in E$, and that $w < +\infty$ on a dense subset of E. Condition (7.4) ensures that, for each point $q \in E \cap \partial V$,

$$\liminf_{p \to q} w(p) = \left(\liminf_{p \to q, \, p \in V} v(p) \right) \wedge \left(\liminf_{p \to q} u(p) \right) \geq u(q) = w(q),$$

so that w is lower semicontinuous on E. It remains only to check that the inequality $w(p) \geq \mathcal{L}(w; p; c)$ holds whenever $p \in E \cap \partial V$ and $\overline{\Delta}(p; c) \subseteq E$. In this case we have, by Theorem 3.17,

$$w(p) = u(p) \geq \mathcal{L}(u; p, c) \geq \mathcal{L}(w; p, c).$$

Hence w is a supertemperature on E. □

We now come to the Natural Order Decomposition Theorem.

THEOREM 7.21. *If u, u_1 and u_2 are nonnegative supertemperatures such that $u \leq u_1 + u_2$ on an open set E, then there are nonnegative supertemperatures v_1 and v_2 which satisfy*

$$v_1 \leq u_1, \quad v_2 \leq u_2, \quad u = v_1 + v_2$$

on E.

PROOF. Let $\mathcal{F}$ denote the class of nonnegative supertemperatures v such that $u \leq v + u_2$ on E, and let w_1 denote the lower semicontinuous smoothing of $\inf \mathcal{F}$. Then $u \leq \inf \mathcal{F} + u_2$ on E, so that $u \leq w_1 + u_2$ almost everywhere on E, by Theorem 7.13. Since w_1 is a supertemperature on E, it follows from Theorem 3.59 that $u \leq w_1 + u_2$ everywhere on E. Furthermore, because $u_1 \in \mathcal{F}$, we have $w_1 \leq u_1$ on E.

Now we let $\mathcal{G}$ denote the class of nonnegative supertemperatures v such that $u \leq w_1 + v$ on E, and w_2 denote the lower semicontinuous smoothing of $\inf \mathcal{G}$. Then, by a similar argument, w_2 is a supertemperature such that $u \leq w_1 + w_2$ on E. Furthermore, because $u_2 \in \mathcal{G}$, we have $w_2 \leq u_2$ on E.

We now show that $u - w_2$, if defined appropriately on the set where $w_2 = +\infty$, is a supertemperature on E. Clearly $u \in \mathcal{G}$, so that $w_2 \leq u$ on E. We put $F = \{p \in E : w_2(p) < +\infty\}$, and define

$$(u - w_2)(p) = \begin{cases} u(p) - w_2(p) & \text{if } p \in F, \\ \liminf_{q \to p, \, q \in F} \left(u(q) - w_2(q) \right) & \text{if } p \in E \backslash F. \end{cases}$$

Then $u - w_2 \geq 0$ on E, and if $p \in E \backslash F$ then $u - w_2$ is lower semicontinuous at p.

For any $p \in E$, we let $\Delta = \Delta(p; c)$ be a heat cylinder with closure in E, and let π_Δ be the operator in Theorem 3.21. The inequality $u \leq w_1 + w_2$ implies that $\pi_\Delta u \leq \pi_\Delta w_1 + \pi_\Delta w_2 \leq w_1 + \pi_\Delta w_2$ on E, so that

$$u = \pi_\Delta u + u - \pi_\Delta u \leq w_1 + (\pi_\Delta w_2 + u - \pi_\Delta u)$$

on $\overline{\Delta}\backslash\partial_n\Delta$. The function $w_3 = \pi_\Delta w_2 + u - \pi_\Delta u = u - (\pi_\Delta u - \pi_\Delta w_2)$ is the difference of u and a nonnegative temperature on $\overline{\Delta}\backslash\partial_n\Delta$, and the inequality $u \geq \pi_\Delta u$ implies that $w_3 \geq \pi_\Delta w_2$ there. We now put

$$w_4(r) = \begin{cases} (w_2 \wedge w_3)(r) & \text{if } r \in \overline{\Delta}\backslash\partial_n\Delta, \\ w_2(r) & \text{if } r \in E\backslash(\overline{\Delta}\backslash\partial_n\Delta). \end{cases}$$

Since $\pi_\Delta w_2$ is lower semicontinuous on E, and is equal to w_2 on $\partial_n\Delta$, we have

$$\liminf_{r\to q,\, r\in\Delta} w_3(r) \geq \liminf_{r\to q,\, r\in\Delta} \pi_\Delta w_2(r) \geq w_2(q)$$

for all $q \in \partial_n\Delta$. It now follows from Lemma 7.20 that w_4 is a supertemperature on $E\backslash\overline{\Xi}$, where $\Xi = \partial\Delta\backslash\partial_n\Delta$. If $q \in \Xi$, then

$$\liminf_{r\to q} w_4(r) = \left(\liminf_{r\to q} w_2(r)\right) \wedge \left(\liminf_{r\to q,\, r\in\Delta} w_3(r)\right) \geq w_2(q) \wedge w_3(q) = w_4(q),$$

so that w_4 is lower semicontinuous at q. Moreover, if $q \in \overline{\Xi}\backslash\Xi$ then

$$\liminf_{r\to q} w_4(r) \geq \left(\liminf_{r\to q} w_2(r)\right) \wedge \left(\liminf_{r\to q,\, r\in\Delta} \pi_\Delta w_2(r)\right) \geq w_2(q) = w_4(q),$$

so that w_4 is lower semicontinuous at q. Hence w_4 is lower semicontinuous on E. Furthermore, whenever $q \in \Xi$ and $\overline{\Delta}(q;d) \subseteq \overline{\Delta}\backslash\partial_n\Delta$, we have

$$w_4(q) = w_2(q) \wedge w_3(q) \geq \mathcal{L}(w_2;q;d) \wedge \mathcal{L}(w_3;q;d) \geq \mathcal{L}(w_4;q;d).$$

Moreover, whenever $q \in \overline{\Xi}\backslash\Xi$ and $\overline{\Delta}(q;d) \subseteq E$, we have

$$w_4(q) = w_2(q) \geq \mathcal{L}(w_2;q;d) \geq \mathcal{L}(w_4;q;d).$$

It follows that w_4 is a supertemperature on E.

Moreover, because $u \leq w_1 + w_2$ on E and $u \leq w_1 + w_3$ on $\overline{\Delta}\backslash\partial_n\Delta$, we have $u \leq w_1 + w_4$ on E. Therefore $w_4 \in \mathcal{G}$, so that $w_4 \geq w_2$ on E, and hence

$$\pi_\Delta w_2 + u - \pi_\Delta u = w_3 \geq w_2$$

on $\overline{\Delta}\backslash\partial_n\Delta$. Thus

$$u - w_2 \geq \pi_\Delta u - \pi_\Delta w_2$$

on $\overline{\Delta}\backslash\partial_n\Delta$. It follows that $(u - w_2)(p) \geq \mathcal{L}(u - w_2; p; c)$, and that

$$\liminf_{r\to p,\, r\in\Delta} (u - w_2)(r) \geq \lim_{r\to p,\, r\in\Delta} (\pi_\Delta u - \pi_\Delta w_2)(r)$$
$$= (\pi_\Delta u - \pi_\Delta w_2)(p)$$
$$= \mathcal{L}(u;p;c) - \mathcal{L}(w_2;p;c).$$

Because u and w_2 are supertemperatures on E, Corollary 3.20 shows that

$$\mathcal{L}(u;p;c) - \mathcal{L}(w_2;p;c) \to u(p) - w_2(p)$$

as $c \to 0+$, if $p \in F$. Hence the inequality

$$\liminf_{r\to p,\, r\in\Delta} (u - w_2)(r) \geq (u - w_2)(p)$$

holds whenever $p \in F$, and it holds by definition whenever $p \in E\backslash F$. It now follows from Theorem 7.17 that $u - w_2$ is a supertemperature on E.

The relations $(u - w_2) \leq w_1$ and $u = (u - w_2) + w_2$ are valid on F, and therefore almost everywhere on E. Hence, because $u - w_2$, w_1, u and w_2 are all supertemperature on E, these relations hold everywhere on E. Taking $v_1 = u - w_2$ and $v_2 = w_2$, we obtain the result of the theorem. $\square$

7.4. Reductions and Smoothed Reductions

DEFINITION 7.22. Let u be a nonnegative supertemperature on an open set E. If $L \subseteq E$, then the *reduction of u over L* (relative to E), denoted by R_u^L, is the infimum of the family of nonnegative supertemperatures on E that majorize u on L.

Note that $u = R_u^L$ on L.

By Theorem 7.13, the lower semicontinuous smoothing $\widehat{R}_u^L$ of the reduction of u over L is a supertemperature on E, is equal to R_u^L almost everywhere on E, and satisfies the equality

$$\widehat{R}_u^L(p) = \liminf_{q \to p} R_u^L(q)$$

at every point $p \in E$.

DEFINITION 7.23. We call $\widehat{R}_u^L$ the *smoothed reduction of u over L* (relative to E).

Note that $u \geq R_u^L \geq \widehat{R}_u^L \geq 0$ on E, and that $R_u^L = \widehat{R}_u^L$ on the interior of L. Smoothed reductions allow the following characterization of polar sets.

THEOREM 7.24. *For any subset L of the open set E, the following statements are equivalent:*
(a) L is polar.
(b) There is a positive supertemperature u on E such that $\widehat{R}_u^L = 0$.
(c) $\widehat{R}_u^L = 0$ for every nonnegative supertemperature u on E.

PROOF. We suppose first that L is a polar subset of E, and take any point $p_0 \in E \backslash L$. In view of Theorem 7.3, we can find a heat potential v on E such that $v(p) = +\infty$ for all $p \in L$ and $v(p_0) < +\infty$. For any nonnegative supertemperature u on E, we have $\epsilon v \geq u$ on L for each $\epsilon > 0$, so that $R_u^L(p_0) \leq \epsilon v(p_0)$ and hence $R_u^L(p_0) = 0$. Since p_0 is arbitrary, we have $R_u^L(p) = 0$ for all $p \in E \backslash L$. The set L has no interior points, so that $\liminf_{q \to p} R_u^L(q) = 0$ for all $p \in E$, and hence $\widehat{R}_u^L = 0$. Thus (a) implies (c).

Clearly (c) implies (b), so it remains only to prove that (b) implies (a). Let u be a positive supertemperature on E such that $\widehat{R}_u^L = 0$. By Theorem 7.13, $R_u^L = 0$ almost everywhere on E. We put $S = \{p \in E : R_u^L(p) = 0\}$, and let $p_0 \in S$. Then for each positive integer k, we can find a supertemperature v_k on E such that $v_k \geq u$ on L and $v_k(p_0) \leq 2^{-k}$. We put $v = \sum_{k=1}^{\infty} v_k$ on E, so that $v(p) = +\infty$ for all $p \in L$, and $v(p_0) \leq 1$. By Corollary 3.57 and Theorem 3.60, v is a supertemperature on $\Lambda(p_0; E)$, and so $L \cap \Lambda(p_0; E)$ is polar. The set S is dense in E, and so $\bigcup_{p \in S} \Lambda(p; E) = E$. By Lindelöf's theorem, we can find a sequence $\{q_j\}$ of points of S, such that $\bigcup_{j=1}^{\infty} \Lambda(q_j; E) = E$. Since p_0 is an arbitrary point of S, it follows that $L \cap \Lambda(q_j; E)$ is polar for all j. Therefore, by Theorem 7.4, the set $L = \bigcup_{j=1}^{\infty} \left(L \cap \Lambda(q_j; E) \right)$ is polar. □

LEMMA 7.25. *Let E be an open set, and let $G_E \mu$ be the heat potential of a measure μ whose support F is a compact subset of E.*
(a) If D is an open superset of F such that $\overline{D} \subseteq E$, and v is a nonnegative supertemperature on E such that $v \geq G_E \mu$ on ∂D, then $v \geq G_E \mu$ on $E \backslash \overline{D}$ also. In

particular, if D is bounded, then $G_E\mu$ is bounded on $E\backslash D$.

(b) If L is a neighbourhood of F such that $\overline{L} \subseteq E$, then $R^L_{G_E\mu} = G_E\mu = \widehat{R}^L_{G_E\mu}$ on E.

PROOF. (a) Since F is compact, we have $\mu(F) < +\infty$. Therefore $G\mu$ is a heat potential, by Theorem 6.18. Let h denote the greatest thermic minorant of $G\mu$ on E. Theorem 6.31 shows that the Riesz measure associated with $G\mu$ is μ itself, and so the Riesz Decomposition Theorem shows that $G\mu = G_E\mu + h$ on E. Therefore $G\mu - h = G_E\mu \leq v$ on ∂D. We now put

$$w = \begin{cases} h \vee (G\mu - v) & \text{on } E\backslash \overline{D}, \\ h & \text{on } \overline{D}. \end{cases}$$

The function $G\mu$ is a temperature on $\mathbb{R}^{n+1}\backslash F$, by Corollary 6.22, so that $G\mu - v$ is a subtemperature on $E\backslash F$. Because $F \subseteq D$, D is open and $\overline{D} \subseteq E$, we have $\partial D \subseteq E\backslash F$, and therefore

$$\limsup_{p \to q,\, p \in E\backslash \overline{D}} (G\mu - v)(p) \leq (G\mu - v)(q) \leq h(q)$$

for all $q \in \partial D$. Hence Lemma 7.20 shows that w is a subtemperature on E. Moreover, $w = h \leq G\mu$ on $\overline{D}$ and $w \leq h \vee G\mu = G\mu$ on $E\backslash \overline{D}$, so that $w \leq G\mu$ on E. Thus w is a subtemperature which minorizes $G\mu$ on E, which implies that $w \leq h$. Therefore $G\mu - v \leq h$ on $E\backslash \overline{D}$, and hence $G_E\mu = G\mu - h \leq v$ there.

If D is bounded, then $G_E\mu$ is bounded on ∂D because it is a temperature on $E\backslash F$. So we can take v to be any upper bound for $G_E\mu$ on ∂D.

(b) Let v be a nonnegative supertemperature on E such that $v \geq G_E\mu$ on L. Let D be an open set such that $F \subseteq D$ and $\overline{D} \subseteq L$. Then $v \geq G_E\mu$ on $\overline{D}$, so that it follows from part (a) that $v \geq G_E\mu$ on $E\backslash \overline{D}$ also, and hence on the whole of E. Thus $R^L_{G_E\mu} = G_E\mu$ on E, and the lower semicontinuity of $G_E\mu$ now implies that $\widehat{R}^L_{G_E\mu} = G_E\mu$ on E. $\square$

EXAMPLE 7.26. Let E be an open set, let $p_0 \in E$, and let L be a neighbourhood of p_0 such that $\overline{L} \subseteq E$. Then

$$R^L_{G_E(\cdot;p_0)} = G_E(\cdot;p_0) = \widehat{R}^L_{G_E(\cdot;p_0)}$$

on E. To see this, take μ to be the unit mass at p_0 in Lemma 7.25(b).

Our next result gives some of the basic properties of reductions and smoothed reductions.

THEOREM 7.27. Let u and v be nonnegative supertemperatures on an open set E, and let L, M, Z, be subsets of E.

(a) If $u \leq v$ on L, then $R^L_u \leq R^L_v$ and $\widehat{R}^L_u \leq \widehat{R}^L_v$ on E.
(b) If $L \subseteq M$, then $R^L_u \leq R^M_u$ and $\widehat{R}^L_u \leq \widehat{R}^M_u$ on E.
(c) If α is a positive number, then $R^L_{\alpha u} = \alpha R^L_u$ and $\widehat{R}^L_{\alpha u} = \alpha \widehat{R}^L_u$ on E.
(d) On $E\backslash \overline{L}$, the functions R^L_u and $\widehat{R}^L_u$ are equal and are temperatures.
(e) If Z is polar, then $\widehat{R}^{L \cup Z}_u = \widehat{R}^L_u$ on E.

PROOF. The proofs of (a) and (b) are elementary.

To prove (c), we observe that v is a nonnegative supertemperature on E that majorizes αu on L, if and only if v is α times a nonnegative supertemperature on

E that majorizes u on L. Therefore, if v and w are nonnegative supertemperatures on E, we have
$$R^L_{\alpha u} = \inf\{v : v \geq \alpha u \text{ on } L\} = \inf\{\alpha w : w \geq u \text{ on } L\} = \alpha R^L_u.$$
It now follows from Theorem 7.13 that the result for smoothed reductions holds almost everywhere on E, and therefore Theorem 3.59 shows that it holds everywhere on E because both sides of the equation are supertemperatures.

For (d), we let $\mathcal{F}$ denote the family of all restrictions to $E \backslash \overline{L}$ of nonnegative supertemperatures on E that majorize u on L. Then $\mathcal{F}$ is a saturated family (in the sense of Section 3.3), and so its infimum (which is the restriction to $E \backslash \overline{L}$ of R^L_u) is a temperature on $E \backslash \overline{L}$, by Theorem 3.26. Hence R^L_u is continuous on $E \backslash \overline{L}$, which implies that $\widehat{R}^L_u = R^L_u$ there.

If Z is a polar set, we can find a heat potential w on E such that $w(p) = +\infty$ for all $p \in Z$, by Theorem 7.3. If v is a nonnegative supertemperature on E that majorizes u on L, then for each $\epsilon > 0$ the function $v + \epsilon w$ is a nonnegative supertemperature on E that majorizes u on $L \cup Z$, and so $v + \epsilon w \geq R^{L \cup Z}_u$. Making $\epsilon \to 0+$, we deduce that $v \geq R^{L \cup Z}_u$ on the set $F = \{p \in E : w(p) < +\infty\}$. It follows that $R^L_u \geq R^{L \cup Z}_u$ on F, and (b) shows that $R^L_u \leq R^{L \cup Z}_u$ on E. Therefore $R^L_u = R^{L \cup Z}_u$ almost everywhere on E, so that $\widehat{R}^L_u = \widehat{R}^{L \cup Z}_u$ almost everywhere (by Theorem 7.13), and hence everywhere (by Theorem 3.59), on E. □

Our next two theorems contain some consequences of Theorem 7.27(d) that involve heat potentials.

THEOREM 7.28. *If u is a nonnegative supertemperature on an open set E, and $\overline{L}$ is a compact subset of E, then $\widehat{R}^L_u$ is a heat potential.*

PROOF. We show that the greatest thermic minorant of $\widehat{R}^L_u$ is zero. Let U and V be bounded open sets such that $\overline{L} \subseteq U, \overline{U} \subseteq V$, and $\overline{V} \subseteq E$. We put $v = \widehat{R}^U_u$, and note that v is a temperature on $E \backslash \overline{U} \supseteq \partial V$, by Theorem 7.27(d). The collection of open sets $\{\Lambda^*(p; E) : p \in E\}$ covers the compact set ∂V, and so we can find a finite set $\{p_1, ..., p_l\}$ of points of E such that $\partial V \subseteq \bigcup_{i=1}^l \Lambda^*(p_i; E)$. By Theorem 6.7, $G_E(\cdot; p_i) > 0$ on $\Lambda^*(p_i; E)$ for each i, and so the supertemperature
$$g = \sum_{i=1}^l G_E(\cdot; p_i) > 0$$
on an open superset of ∂V. We can therefore choose a positive number α such that $\alpha g \geq v$ on ∂V, since $v \in C(\partial V)$. We now define
$$w(p) = \begin{cases} v(p) & \text{if } p \in \overline{V}, \\ (\alpha g(p)) \wedge v(p) & \text{if } p \in E \backslash \overline{V}. \end{cases}$$
Since
$$\liminf_{p \to q} \alpha g(p) \geq \alpha g(q) \geq v(q)$$
for all $q \in \partial V$, it follows from Lemma 7.20 that w is a supertemperature on E. Moreover $w \geq 0$ on E, and $w = v = u$ on $U \supseteq L$, so that $w \geq R^L_u \geq \widehat{R}^L_u$ on E. If h is a thermic minorant of $\widehat{R}^L_u$ on E, then $h \leq w \leq \alpha g$ on $E \backslash \overline{V}$, and so $h \leq \alpha g$ on E by the minimum principle. By Theorem 3.66, the greatest thermic minorant of

αg on E is the sum of the greatest thermic minorants of $\alpha G_E(\cdot;p_i)$ ($i \in \{1,...,l\}$) on E, and hence is zero. Thus $h \leq 0$, and so $\widehat{R}_u^L$ is a heat potential by Corollary 6.39. □

THEOREM 7.29. *Let L be a relatively closed subset of the open set E. If L is not polar, then there is a bounded heat potential $G_E\mu$ such that μ is not null and is supported by a compact subset of L.*

PROOF. We can write L as the union of a sequence of compact subsets of E. If every set in that sequence was polar, then L would also be polar, by Theorem 7.4. Thus we can find a compact subset K of L such that K is not polar. We put $1(p) = 1$ for all $p \in E$. Then $\widehat{R}_1^K$ is a heat potential by Theorem 7.28, and is not identically 0 by Theorem 7.24. Moreover, $\widehat{R}_1^K$ is bounded on E, and is a temperature on $E \setminus K$ by Theorem 7.27(d). Therefore we can write $\widehat{R}_1^K = G_E\mu$, where $\mu(E \setminus K) = 0$ by Theorem 6.25. □

The following corollary reflects on the sharpness of Corollary 7.15.

COROLLARY 7.30. *If L is a compact subset of the open set E, and is not polar, then there is a bounded temperature on $E \setminus L$ that cannot be extended to a supertemperature on E.*

PROOF. By the theorem, we can find a bounded heat potential $G_E\mu$ such that μ is not null and is supported by a compact subset K of L. If $u = -G_E\mu$, then u is a bounded subtemperature on E, and a temperature on $E \setminus K \supseteq E \setminus L$ by Theorem 7.27(d). Suppose that there is a supertemperature v on E such that $v = u$ on $E \setminus L$. Then $v - u$ is a supertemperature on E, and $v - u = 0$ on $E \setminus L$. Since L is compact, the minimum principle shows first that $v - u \geq 0$ on E, then that $v = u$ on E. Therefore $G_E\mu$ is a temperature on E, and Corollary 6.20 shows that μ is null, a contradiction. □

We now present some properties of reductions over open subsets of E.

THEOREM 7.31. *Let u and v be nonnegative supertemperatures on an open set E, and let C, D be open subsets of E.*

(a) The equality $R_u^D = \widehat{R}_u^D$ holds on E, and so R_u^D is a supertemperature on E.

(b) The equality $R_{u+v}^D = R_u^D + R_v^D$ holds on E.

(c) If $C \subseteq D$, then the equalities
$$R_{R_u^C}^D = R_{R_u^D}^C = R_u^C$$
hold on E.

(d) If L is any subset of E, and $u \in C(D)$ for some open superset D of L, then
$$R_u^L = \inf\{R_u^C : C \text{ is an open superset of } L\}.$$

PROOF. Since D is an open set, we have $u = R_u^D = \widehat{R}_u^D$ on D. Since $\widehat{R}_u^D$ is itself a nonnegative supertemperature on E, it follows that $\widehat{R}_u^D \geq R_u^D$ on E. The reverse inequality is always true, and so (a) holds.

Since $u = R_u^D$ and $v = R_v^D$ on D, the nonnegative supertemperature $R_u^D + R_v^D$ majorizes $u + v$ on D, and so $R_u^D + R_v^D \geq R_{u+v}^D$ on E. Therefore, by the natural

order decomposition (Theorem 7.21), there are nonnegative supertemperatures u^* and v^* on E such that
$$u^* \leq R_u^D, \quad v^* \leq R_v^D, \quad R_{u+v}^D = u^* + v^*.$$
On D, we have $u = R_u^D \geq u^*$, $v = R_v^D \geq v^*$, and $u + v = R_{u+v}^D = u^* + v^*$, so that $u^* = u$ and $v^* = v$. Hence $u^* \geq R_u^D$ and $v^* \geq R_v^D$ on E, and so equality holds. This proves (b).

For any nonnegative supertemperatures v and w such that $v \leq w$ on E, we have $R_v^C \leq R_v^D \leq v$ and $R_v^C \leq R_w^C$, by Theorem 7.27(a),(b). Therefore, because R_u^C is a supertemperature on E by (a), we have
$$R_{R_u^C}^C \leq R_{R_u^C}^D \leq R_u^C \quad \text{and} \quad R_{R_u^C}^C \leq R_{R_u^D}^C \leq R_u^C,$$
so that the result of (c) will follow if we prove the special case $D = C$. Since $R_u^C = u$ on C, a supertemperature w majorizes u on C if and only if it majorizes R_u^C on C, from which the case $D = C$ follows.

For (d), let $v \geq u$ on L and let $\epsilon > 0$. Since $u \in C(D)$, the function $v - u$ is lower semicontinuous on D, and so the set $V = \{p \in D : v(p) - u(p) > -\epsilon\}$ is an open superset of L. Because $v + \epsilon \geq u$ on V, we have $v + \epsilon \geq R_u^V$ on E, so that
$$v + \epsilon \geq \inf\{R_u^C : C \text{ is an open superset of } L\}.$$
Therefore
$$R_u^L + \epsilon \geq \inf\{R_u^C : C \text{ is an open superset of } L\}$$
for every $\epsilon > 0$, so that
$$R_u^L \geq \inf\{R_u^C : C \text{ is an open superset of } L\}.$$
The reverse inequality follows from Theorem 7.27(b). □

Our next result is the important *Strong Subadditivity Property* of reductions and smoothed reductions. We shall eventually be able to show, in Theorem 9.32, that the condition $v \in C(E)$ is unnecessary.

THEOREM 7.32. *If v is a nonnegative supertemperature on E, and a member of $C(E)$, then for any subsets L, M of E we have*

(7.5) $$R_v^{L \cup M} + R_v^{L \cap M} \leq R_v^L + R_v^M,$$

and

(7.6) $$\widehat{R}_v^{L \cup M} + \widehat{R}_v^{L \cap M} \leq \widehat{R}_v^L + \widehat{R}_v^M$$

on E. Moreover, if $L_+ = \{p \in L : v(p) > 0\}$ then $R_v^L = R_v^{L_+}$ and $\widehat{R}_v^L = \widehat{R}_v^{L_+}$ on E.

PROOF. We first consider the case of open subsets of E, with a view to using Theorem 7.31(d). Let A and B be open subsets of E, and put $w = R_v^A \wedge R_v^B$. On A, we have $R_v^A + R_v^B = v + R_v^B$ and $R_v^A = v \geq R_v^B$. Similarly, on B we have $R_v^A + R_v^B = R_v^A + v$ and $R_v^B = v \geq R_v^A$. Hence, on $A \cup B$ we have
$$R_v^A + R_v^B = v + R_v^A \wedge R_v^B = v + w.$$
It therefore follows from Theorem 7.31(b),(c) that

(7.7) $$R_v^{A \cup B} + R_w^{A \cup B} = R_{v+w}^{A \cup B} = R_{R_v^A + R_v^B}^{A \cup B} = R_{R_v^A}^{A \cup B} + R_{R_v^B}^{A \cup B} = R_v^A + R_v^B$$

on E. Furthermore, on $A \cap B$ we have $R_w^{A \cup B} = w = R_v^A \wedge R_v^B = v$. Therefore, because $R_w^{A \cup B}$ is a nonnegative supertemperature on E by Theorem 7.31(a), we have $R_w^{A \cup B} \geq R_v^{A \cap B}$ on E. Now (7.7) implies that

$$R_v^{A \cup B} + R_v^{A \cap B} \leq R_v^A + R_v^B,$$

for open sets A and B.

By Theorem 7.31(d), we have

$$R_v^L = \inf\{R_v^C : C \text{ is an open superset of } L\}.$$

Moreover, if $A \supseteq L$ and $B \supseteq M$, then Theorem 7.27(b) and the inequality just proved give us

$$R_v^{L \cup M} + R_v^{L \cap M} \leq R_v^{A \cup B} + R_v^{A \cap B} \leq R_v^A + R_v^B.$$

Therefore

$$R_v^{L \cup M} + R_v^{L \cap M} \leq R_v^A + \inf\{R_v^C : C \text{ is an open superset of } M\} = R_v^A + R_v^M,$$

and hence

$$R_v^{L \cup M} + R_v^{L \cap M} \leq \inf\{R_v^C : C \text{ is an open superset of } L\} + R_v^M = R_v^L + R_v^M,$$

so that (7.5) holds. The inequality (7.6) therefore holds almost everywhere on E, and hence everywhere on E because all the functions therein are supertemperatures.

For the last part, it follows from Theorem 11(b) and the strong subadditivity property just proved that

$$R_v^{L+} \leq R_v^L \leq R_v^{L+} + R_v^{L \setminus L+} = R_v^{L+} + R_0^{L \setminus L+} = R_v^{L+}.$$

$\square$

7.5. The Thermal Capacity of Compact Sets

Throughout the remainder of this chapter, E denotes a fixed open set and all reductions are relative to E. The function 1 is defined by $1(p) = 1$ for all $p \in E$.

If K is a compact subset of E, then $\widehat{R}_1^K$ is a heat potential on E by Theorem 7.28, and is a temperature on $E \setminus \partial K$ by Theorem 7.27(d). Therefore the support of its associated Riesz measure is contained in ∂K, by Theorem 6.25.

DEFINITION 7.33. We call $\widehat{R}_1^K$ the *thermal capacitary potential of* K, and its associated Riesz measure ω_K the *thermal capacitary distribution of* K. The *thermal capacity* $\mathcal{C}(K)$ of K is defined by

$$\mathcal{C}(K) = \omega_K(E).$$

Theorem 7.24 shows that $\mathcal{C}(K) = 0$ if and only if K is polar.

We also introduce the corresponding concepts relative to the adjoint equation. In Section 7.7, we see some nontrivial interaction between the two theories.

For a compact subset K of E, we denote by $\widehat{R}_1^{K*}$ the smoothed reduction of 1 over K relative to the adjoint equation. It is a coheat potential on E and a cotemperature on $E \setminus \partial K$. Its associated Riesz measure has its support in ∂K.

DEFINITION 7.34. We call $\widehat{R}_1^{K*}$ the *cothermal capacitary potential of K*, and its associated Riesz measure ω_K^* the *cothermal capacitary distribution of K*. The *cothermal capacity $\mathcal{C}^*(K)$ of K* is defined by

$$\mathcal{C}^*(K) = \omega_K^*(E).$$

EXAMPLE 7.35. Relative to $E = \mathbb{R}^{n+1}$, the cothermal capacity of the closed heat ball $K = \overline{\Omega}(0;c)$, and the heat sphere $\partial\Omega(0;c)$, is $(4\pi c)^{n/2}$. To see this, we let v be a nonnegative cosupertemperature on $\mathbb{R}^{n+1}$ such that $v \geq 1$ on $\overline{\Omega}(0;c)$ (equivalently $v \geq 1$ on $\partial\Omega(0;c)$). Since $G(0;\cdot)$ is a cotemperature on $\mathbb{R}^{n+1}\backslash\overline{\Omega}(0;c)$, and $G(0;q) \to 0$ as q tends to the point at infinity, the minimum principle shows that $v \geq (4\pi c)^{n/2} G(0;\cdot)$ on $\mathbb{R}^{n+1}\backslash\overline{\Omega}(0;c)$. Hence

$$v \geq (4\pi c)^{n/2}\bigl(G(0;\cdot) \wedge (4\pi c)^{-n/2}\bigr)$$

on $\mathbb{R}^{n+1}$. The function $(4\pi c)^{n/2}\bigl(G(0;\cdot) \wedge (4\pi c)^{-n/2}\bigr)$ is a cosupertemperature on $\mathbb{R}^{n+1}$, and is therefore $\widehat{R}_1^{K*}$. By Example 6.15, if

$$d\mu(x,t) = (4\pi c)^{-n/2} Q(x,-t)\, d\sigma(x,t)$$

on $\partial\Omega(0;c)$, where σ denotes the surface area measure, then the coheat potential of μ is

$$G^*\mu = G(0;\cdot) \wedge (4\pi c)^{-n/2}$$

on $\mathbb{R}^{n+1}$. Hence

$$\widehat{R}_1^{K*} = (4\pi c)^{n/2} G^*\mu,$$

and so

$$\mathcal{C}^*(K) = (4\pi c)^{n/2}\mu(K) = (4\pi c)^{n/2}.$$

The next three results are used to prove some important properties of thermal capacity.

THEOREM 7.36. *If μ and ν are nonnegative measures on E, then*

$$\int_E G_E\mu\, d\nu = \int_E G_E^*\nu\, d\mu.$$

PROOF. Using Theorem 6.10, and interchanging the order of the integrals, we obtain

$$\int_E G_E\mu\, d\nu = \int_E \left(\int_E G_E(p;q)\, d\mu(q)\right) d\nu(p)$$
$$= \int_E \left(\int_E G_E^*(q;p)\, d\nu(p)\right) d\mu(q)$$
$$= \int_E G_E^*\nu\, d\mu.$$

□

LEMMA 7.37. *Let K be a compact subset of E.*
(a) If μ and ν are nonnegative measures on E such that $G_E\mu \leq G_E\nu$ on E, and μ has its support in K, then $\mu(E) \leq \nu(E)$.
(b) If $\{G_E\mu_j\}$ is a monotone sequence of heat potentials which converges on

$E\setminus K$ to a heat potential $G_E\mu$, and each measure μ_j has its support in K, then $\mu_j(E) \to \mu(E)$ as $j \to \infty$.

PROOF. Let M be a compact subset of E whose interior contains K. Then the cothermal capacitary potential of M satisfies $G_E^*\omega_M^*(p) = 1$ for all $p \in K$.

(a) Because μ has its support in K, we have
$$\mu(E) = \int G_E^*\omega_M^* \, d\mu = \int G_E\mu \, d\omega_M^*,$$
by Theorem 7.36. Furthermore, because $G_E^*\omega_M^* \leq 1$ on E, we have
$$\nu(E) \geq \int G_E^*\omega_M^* \, d\nu = \int G_E\nu \, d\omega_M^*,$$
again by Theorem 7.36. Now the inequality $G_E\mu \leq G_E\nu$ implies that $\mu(E) \leq \nu(E)$.

(b) Because each measure μ_j has its support in K, each heat potential $G_E\mu_j$ is a temperature on $E\setminus K$, by Theorem 6.25. The sequence $\{G_E\mu_j\}$ converges monotonically to $G_E\mu$ on $E\setminus K$, and so $G_E\mu$ is also a temperature on $E\setminus K$, by the Harnack monotone convergence theorem. Therefore μ has its support in K, by Theorem 6.25. Hence, by Theorem 7.36,
$$\mu_j(E) = \int G_E^*\omega_M^* \, d\mu_j = \int G_E\mu_j \, d\omega_M^*$$
and
$$\mu(E) = \int G_E^*\omega_M^* \, d\mu = \int G_E\mu \, d\omega_M^*.$$
Since the support of ω_M^* is contained in $\partial M \subseteq E\setminus K$, the Lebesgue monotone convergence theorem now shows that $\mu_j(E) \to \mu(E)$ as $j \to \infty$. □

LEMMA 7.38. *Let u be a nonnegative supertemperature on an open set E.*

(a) If $\{L_i\}$ is an expanding sequence of subsets of E whose union D is open, then
$$\lim_{i\to\infty} \widehat{R}_u^{L_i} = \widehat{R}_u^D$$
on E.

(b) If $\{K_i\}$ is a contracting sequence of compact subsets of E with intersection K, and there is an open superset D of K such that $u \in C(D)$, then
$$\lim_{i\to\infty} R_u^{K_i} = R_u^K$$
on E.

PROOF. If $\{L_i\}$ is expanding with union D, then Theorem 7.27(b) shows that the sequence $\{\widehat{R}_u^{L_i}\}$ is increasing and majorized by $\widehat{R}_u^D$. Therefore the function $v = \lim_{i\to\infty} \widehat{R}_u^{L_i}$ is also majorized by $\widehat{R}_u^D$, and is thus finite on a dense subset of E. Theorem 3.60 now shows that v is a supertemperature on E. Moreover, for each i we have $\widehat{R}_u^{L_i} = R_u^{L_i} = u$ almost everywhere on L_i, by Theorem 7.13. It follows that $v = u$ almost everywhere on D, and hence everywhere on D because both are supertemperatures. Thus $v \geq R_u^D = \widehat{R}_u^D \geq v$ on E, which proves (a).

If $\{K_i\}$ is a contracting sequence of compact subsets of E with intersection K, then given any open superset C of K we can find a number i_0 such that $K_i \subseteq C$ whenever $i > i_0$. Therefore Theorem 7.27(b) shows that $R_u^K \leq R_u^{K_i} \leq R_u^C$ on E whenever $i > i_0$, and that the sequence $\{R_u^{K_i}\}$ is decreasing. It follows that

$R_u^K \leq \lim_{i\to\infty} R_u^{K_i} \leq R_u^C$ on E. Since C is arbitrary, Theorem 7.31(d) now shows that
$$R_u^K \leq \lim_{i\to\infty} R_u^{K_i} \leq \inf\{R_u^C : C \text{ is an open superset of } K\} = R_u^K,$$
which proves (b). □

The following theorem gives the basic properties of the thermal capacity of compact sets.

THEOREM 7.39. *The nonnegative, finite-valued set function $\mathcal{C}$, on the class of compact subsets of E, has the following properties.*

(a) $\mathcal{C}(\emptyset) = 0$, and if $K \subseteq L$ then $\mathcal{C}(K) \leq \mathcal{C}(L)$.

(b) If $\{K_j\}$ is a contracting sequence with intersection K, then
$$\mathcal{C}(K) = \lim_{j\to\infty} \mathcal{C}(K_j).$$

(c) $\mathcal{C}(K \cup L) + \mathcal{C}(K \cap L) \leq \mathcal{C}(K) + \mathcal{C}(L)$.

PROOF. (a) Since $R_1^\emptyset = 0$ on E, we have $G_E \omega_\emptyset = \widehat{R}_1^\emptyset = 0$, and so $\omega_\emptyset$ is null.

If $K \subseteq L$, then $\widehat{R}_1^K \leq \widehat{R}_1^L$; that is, $G_E \omega_K \leq G_E \omega_L$. Therefore, by Lemma 7.37(a), $\omega_K(E) \leq \omega_L(E)$; that is, $\mathcal{C}(K) \leq \mathcal{C}(L)$.

(b) Since $\{K_j\}$ is a contracting sequence with intersection K, the sequence $\{\widehat{R}_1^{K_j}\} = \{G_E \omega_{K_j}\}$ is decreasing and Lemma 7.38(b) shows that $\lim_{j\to\infty} R_1^{K_j} = R_1^K$ on E. Furthermore, the equalities $R_1^{K_j} = G_E \omega_{K_j}$ and $R_1^K = G_E \omega_K$ hold on $E \setminus K_1$, by Theorem 7.27(d), and hence $\lim_{j\to\infty} G_E \omega_{K_j} = G_E \omega_K$ on $E \setminus K_1$. The supports of the measures ω_{K_j} are all contained in K_1, and so we can apply Lemma 7.37(b) to obtain
$$\mathcal{C}(K) = \omega_K(E) = \lim_{j\to\infty} \omega_{K_j}(E) = \lim_{j\to\infty} \mathcal{C}(K_j).$$

(c) By Theorem 7.32, $\widehat{R}_1^{K \cup L} + \widehat{R}_1^{K \cap L} \leq \widehat{R}_1^K + \widehat{R}_1^L$; that is,
$$G_E \omega_{K \cup L} + G_E \omega_{K \cap L} \leq G_E \omega_K + G_E \omega_L.$$

The support of the measure $\omega_{K \cup L} + \omega_{K \cap L}$ is contained in $K \cup L$, and so it follows from Lemma 7.37(a) that $(\omega_{K \cup L} + \omega_{K \cap L})(E) \leq (\omega_K + \omega_L)(E)$; that is,
$$\mathcal{C}(K \cup L) + \mathcal{C}(K \cap L) \leq \mathcal{C}(K) + \mathcal{C}(L).$$
□

7.6. The Thermal Capacity of More General Sets

We now consider a definition of thermal capacity for sets that are perhaps not compact. The definition involves the notions of inner and outer thermal capacity.

DEFINITION 7.40. *If S is an arbitrary subset of E, the inner thermal capacity of S is defined by*
$$\mathcal{C}_-(S) = \sup\{\mathcal{C}(K) : K \text{ is a compact subset of } S\},$$
and the outer thermal capacity of S by
$$\mathcal{C}_+(S) = \inf\{\mathcal{C}_-(D) : D \text{ is an open superset of } S\}.$$

These two set functions take nonnegative, extended real values. If $S \subseteq T \subseteq E$, then $\mathcal{C}_-(S) \leq \mathcal{C}_-(T)$ and $\mathcal{C}_+(S) \leq \mathcal{C}_+(T)$. Furthermore, if K is a compact subset of S and D is an open superset of S, then $K \subseteq D$ so that $\mathcal{C}(K) \leq \mathcal{C}_-(D)$. Taking the supremum over all choices of K, we get $\mathcal{C}_-(S) \leq \mathcal{C}_-(D)$. Now taking the infimum over all choices of D, we obtain $\mathcal{C}_-(S) \leq \mathcal{C}_+(S)$.

DEFINITION 7.41. If $S \subseteq E$ and $\mathcal{C}_-(S) = \mathcal{C}_+(S)$, then S is called *(thermal) capacitable*.

Note that, if S is open and D is an open superset of S, then $\mathcal{C}_-(S) \leq \mathcal{C}_-(D)$, so that $\mathcal{C}_+(S) = \mathcal{C}_-(S)$ and S is capacitable.

LEMMA 7.42. *If K is a compact subset of E, then K is capacitable and*
$$\mathcal{C}_-(K) = \mathcal{C}_+(K) = \mathcal{C}(K).$$

PROOF. Let $\{K_i\}$ be a contracting sequence of compact subsets of E, such that $K \subseteq K_i^\circ$ for all i and $\bigcap_{i=1}^\infty K_i = K$. Then, by Theorem 7.39(b),
$$\mathcal{C}_-(K) \leq \mathcal{C}_+(K) \leq \mathcal{C}_-(K_i^\circ) \leq \mathcal{C}_-(K_i) = \mathcal{C}(K_i) \to \mathcal{C}(K) = \mathcal{C}_-(K)$$
as $i \to \infty$. Therefore $\mathcal{C}_+(K) = \mathcal{C}_-(K) = \mathcal{C}(K)$. □

DEFINITION 7.43. If a subset S of E is capacitable, we write $\mathcal{C}(S)$ for the common value of $\mathcal{C}_+(S)$ and $\mathcal{C}_-(S)$, and call it the *thermal capacity* of S.

Lemma 7.42 shows that this definition of thermal capacity is consistent with the one given earlier for compact sets.

The corresponding notions related to the adjoint equation are called the *inner cothermal capacity*, the *outer cothermal capacity*, the *(cothermal) capacitable* and the *cothermal capacity*. They are denoted with a $\mathcal{C}^*$ rather than a $\mathcal{C}$, and with subscripts if appropriate.

We now consider the properties of thermal capacity for the class of open subsets of E.

THEOREM 7.44. *(a) If $\{U_i\}$ is an expanding sequence of open sets, then*
$$\lim_{i \to \infty} \mathcal{C}(U_i) = \mathcal{C}\left(\bigcup_{i=1}^\infty U_i\right).$$

(b) If U and V are open sets, then
$$\mathcal{C}(U \cup V) + \mathcal{C}(U \cap V) \leq \mathcal{C}(U) + \mathcal{C}(V).$$

(c) If $\{V_i\}$ is an arbitrary sequence of open sets, then
$$\mathcal{C}\left(\bigcup_{i=1}^\infty V_i\right) \leq \sum_{i=1}^\infty \mathcal{C}(V_i).$$

(d) If U is a bounded open set such that $\overline{U} \subseteq E$, then R_1^U is a heat potential whose associated Riesz measure ω_U satisfies $\omega_U(E) = \mathcal{C}(U)$.

PROOF. (a) We first note that the sequence $\{\mathcal{C}(U_i)\}$ is increasing, and that $\lim_{i\to\infty} \mathcal{C}(U_i) \leq \mathcal{C}(\cup_{i=1}^{\infty} U_i)$. To prove the reverse inequality, we take an arbitrary compact subset K of $\cup_{i=1}^{\infty} U_i$, and note that $K \subseteq U_m$ for some integer m, so that $\mathcal{C}(K) \leq \mathcal{C}(U_m) \leq \lim_{i\to\infty} \mathcal{C}(U_i)$. Taking the supremum over all choices of K, we obtain $\mathcal{C}(\bigcup_{i=1}^{\infty} U_i) \leq \lim_{i\to\infty} \mathcal{C}(U_i)$, as required.

(b) We take any compact subset K of $U \cap V$, any compact subset L of $U \cup V$, choose disjoint open sets C and D such that $L \backslash V \subseteq C \subseteq U$ and $L \backslash U \subseteq D \subseteq V$, and put $L_1 = L \backslash D$ and $L_2 = L \backslash C$. Then $L_1 \subseteq U$, $L_2 \subseteq V$, and $L_1 \cup L_2 = L$. It therefore follows from Theorem 7.39(c) that

$$\begin{aligned}\mathcal{C}(L) + \mathcal{C}(K) &\leq \mathcal{C}(K \cup L) + \mathcal{C}(K \cup (L_1 \cap L_2)) \\ &= \mathcal{C}((K \cup L_1) \cup (K \cup L_2)) + \mathcal{C}((K \cup L_1) \cap (K \cup L_2)) \\ &\leq \mathcal{C}(K \cup L_1) + \mathcal{C}(K \cup L_2) \\ &\leq \mathcal{C}(U) + \mathcal{C}(V).\end{aligned}$$

Taking the suprema over all choices of K and L, we obtain the result.

(c) It follows from (b) that $\mathcal{C}(V_1 \cup V_2) \leq \mathcal{C}(V_1) + \mathcal{C}(V_2)$, and hence, by induction, that

$$\mathcal{C}\left(\bigcup_{i=1}^{m} V_i\right) \leq \sum_{i=1}^{m} \mathcal{C}(V_i) \leq \sum_{i=1}^{\infty} \mathcal{C}(V_i)$$

for every integer m. Putting $D_m = \bigcup_{i=1}^{m} V_i$ for all m, and using (a), we obtain

$$\mathcal{C}\left(\bigcup_{i=1}^{\infty} V_i\right) = \lim_{i\to\infty} \mathcal{C}(D_m) \leq \sum_{i=1}^{\infty} \mathcal{C}(V_i).$$

(d) By Theorem 7.31(a), we have $R_1^U = \widehat{R}_1^U$ on E, so that Theorem 7.28 shows that R_1^U is a heat potential. Now let $\{U_i\}$ be an expanding sequence of bounded open sets such that $\overline{U}_i \subseteq U_{i+1}$ for all i and $\bigcup_{i=1}^{\infty} U_i = U$. Put $K_i = \overline{U}_i$ for all i. Then $\{K_i\}$ is an expanding sequence of sets whose union U is open, so that $\lim_{i\to\infty} \widehat{R}_1^{K_i} = \widehat{R}_1^U$ on E, by Lemma 7.38(a); that is, $\lim_{i\to\infty} G_E \omega_{K_i} = G_E \omega_U$. Therefore, since all these measures have their supports in the compact set $\overline{U}$, Lemma 7.37(b) shows that $\lim_{i\to\infty} \omega_{K_i}(E) = \omega_U(E)$; that is, $\lim_{i\to\infty} \mathcal{C}(K_i) = \omega_U(E)$. If K is any compact subset of U, then there is an integer m such that $K \subseteq K_m$, and hence $\mathcal{C}(K) \leq \mathcal{C}(K_m) \leq \mathcal{C}(U)$. Taking the supremum over all choices of K, we obtain $\mathcal{C}(U) = \lim_{m\to\infty} \mathcal{C}(K_m) = \omega_U(E)$. $\square$

We now prove an extension of Theorem 7.44 to arbitrary subsets of E. Such sets may not be capacitable, and so the theorem is about outer thermal capacity rather than thermal capacity itself.

THEOREM 7.45. *(a) If $\{S_i\}$ is an expanding sequence of sets, then*

$$\lim_{i\to\infty} \mathcal{C}_+(S_i) = \mathcal{C}_+\left(\bigcup_{i=1}^{\infty} S_i\right).$$

(b) For any sets S and T, we have

$$\mathcal{C}_+(S \cup T) + \mathcal{C}_+(S \cap T) \leq \mathcal{C}_+(S) + \mathcal{C}_+(T).$$

(c) If $\{T_i\}$ is an arbitrary sequence of sets, then
$$\mathcal{C}_+\left(\bigcup_{i=1}^{\infty} T_i\right) \le \sum_{i=1}^{\infty} \mathcal{C}_+(T_i).$$

(d) If S is a bounded set such that $\overline{S} \subseteq E$, then $\widehat{R}_1^S$ is a heat potential whose associated Riesz measure ω_S satisfies $\omega_S(E) = \mathcal{C}_+(S)$.

PROOF. (a) The conclusion is trivial if $\mathcal{C}_+(S_i) = +\infty$ for some i, so we suppose otherwise. Furthermore, it is obvious that
$$\lim_{i \to \infty} \mathcal{C}_+(S_i) \le \mathcal{C}_+\left(\bigcup_{i=1}^{\infty} S_i\right),$$

and so we have only to prove the reverse inequality. We begin this by taking any $\epsilon > 0$, and for each positive integer i choosing an open set U_i such that $S_i \subseteq U_i$ and $\mathcal{C}(U_i) < \mathcal{C}_+(S_i) + 2^{-i}\epsilon$.

The proof depends on the inequality
$$\mathcal{C}\left(\bigcup_{i=1}^{m} U_i\right) < \mathcal{C}_+(S_m) + (1 - 2^{-m})\epsilon,$$

for every positive integer m, which we prove by induction. It is obviously true when $m = 1$. Suppose that it holds when $m = k$. Since
$$S_k \subseteq U_k \cap S_{k+1} \subseteq \left(\bigcup_{i=1}^{k} U_i\right) \cap U_{k+1},$$

it follows from Theorem 7.44(b) that
$$\mathcal{C}\left(\bigcup_{i=1}^{k+1} U_i\right) + \mathcal{C}_+(S_k) \le \mathcal{C}\left(\bigcup_{i=1}^{k+1} U_i\right) + \mathcal{C}\left(\left(\bigcup_{i=1}^{k} U_i\right) \cap U_{k+1}\right)$$
$$\le \mathcal{C}\left(\bigcup_{i=1}^{k} U_i\right) + \mathcal{C}(U_{k+1})$$
$$< \mathcal{C}_+(S_k) + \mathcal{C}_+(S_{k+1}) + (1 - 2^{-k-1})\epsilon.$$

Since $\mathcal{C}_+(S_k) < +\infty$, we can cancel it to obtain the required inequality when $m = k + 1$. By induction, the inequality holds for all m.

We now make $m \to \infty$, and use Theorem 7.44(a), to obtain
$$\mathcal{C}_+\left(\bigcup_{i=1}^{\infty} S_i\right) \le \mathcal{C}\left(\bigcup_{i=1}^{\infty} U_i\right) = \lim_{m \to \infty} \mathcal{C}\left(\bigcup_{i=1}^{m} U_i\right) \le \lim_{m \to \infty} \mathcal{C}_+(S_m) + \epsilon.$$

Since ϵ is arbitrary, we deduce the required reverse inequality, and so the proof of (a) is complete.

(b) Let U and V be open sets such that $S \subseteq U$ and $T \subseteq V$. Then
$$\mathcal{C}_+(S \cup T) + \mathcal{C}_+(S \cap T) \le \mathcal{C}(U \cup V) + \mathcal{C}(U \cap V) \le \mathcal{C}(U) + \mathcal{C}(V),$$

by Theorem 7.44(b). Taking the infima over all choices of U and V, we obtain the result.

(c) It follows from (b) that $\mathcal{C}_+(T_1 \cup T_2) \leq \mathcal{C}_+(T_1) + \mathcal{C}_+(T_2)$, and hence, by induction, that
$$\mathcal{C}_+\left(\bigcup_{i=1}^m T_i\right) \leq \sum_{i=1}^m \mathcal{C}_+(T_i) \leq \sum_{i=1}^\infty \mathcal{C}_+(T_i)$$
for every integer m. Using (a), we deduce that
$$\mathcal{C}_+\left(\bigcup_{i=1}^\infty T_i\right) = \lim_{i \to \infty} \mathcal{C}_+\left(\bigcup_{i=1}^m T_i\right) \leq \sum_{i=1}^\infty \mathcal{C}_+(T_i).$$

(d) Theorem 7.28 shows that $\widehat{R}_1^S$ is a heat potential. Let $0 < \epsilon < 1$, and let U be a bounded open set such that $S \subseteq U$, $\overline{U} \subseteq E$, and $\mathcal{C}(U) < \mathcal{C}_+(S) + \epsilon$. Let $\mathcal{F}$ denote the family of nonnegative supertemperatures v on E that satisfy the condition $v \geq 1$ on S, so that $R_1^S = \inf \mathcal{F}$. By Lemma 7.12, there is a sequence $\{u_k\}$ of functions in $\mathcal{F}$ such that, if $u = \inf\{u_k : k \in \mathbb{N}\}$ then $\widehat{u} = \widehat{R}_1^S$. We can take $u_1 = 1$. We now put
$$v_k = u_1 \wedge u_2 \wedge \ldots \wedge u_k, \qquad w_k = R_{v_k}^U, \qquad U_k = \{p \in U : w_k(p) > 1 - \epsilon\},$$
and note that $v_k, w_k \in \mathcal{F}$ and $U_k \supseteq S$, for all k. Moreover, the sequences $\{v_k\}$ and $\{w_k\}$ are decreasing, with $\lim_{k \to \infty} v_k = u$ on E. Each function w_k is a heat potential, by Theorems 7.31(a) and 7.28, and is a temperature on $E \backslash \overline{U}$ by Theorem 7.27(d). Since $v_k \geq 1$ on S and $S \subseteq U$, we have $R_1^S \leq R_{v_k}^U = w_k \leq v_k$ on E, and hence
$$\widehat{R}_1^S \leq \lim_{k \to \infty} w_k \leq \lim_{k \to \infty} v_k = u$$
on E. By the Harnack monotone convergence theorem, the function $w = \lim_{k \to \infty} w_k$ is a temperature on $E \backslash \overline{U}$, and therefore $w = \widehat{w}$ there. Hence
$$\widehat{R}_1^S \leq w = \widehat{w} \leq \widehat{u} = \widehat{R}_1^S,$$
and so $w = \widehat{R}_1^S$, on $E \backslash \overline{U}$.

Furthermore, since the sequence $\{w_k\}$ is decreasing, the sequence of sets $\{U_k\}$ is contracting, and so the sequence of reductions $\{R_1^{U_k}\}$ is decreasing. Since each w_k is lower semicontinuous on E, each set U_k is open. Hence, using Theorems 7.31(a), 7.28 and 6.25, the sequence $\{R_1^{U_k}\} = \{G_E \omega_{U_k}\}$ is a decreasing sequence of heat potentials of measures supported in the compact set $\overline{U}$. Therefore Lemma 7.37(a) shows that the sequence $\{\omega_{U_k}(E)\}$ is decreasing. Since $w_k = R_{v_k}^U$ is a heat potential $G_E \nu_k$ and a temperature on $E \backslash \overline{U}$, the measure ν_k has its support in $\overline{U}$. Moreover, since w is a temperature on $E \backslash \overline{U}$, and $w = \widehat{R}_1^S = G_E \omega_S$ there, the measure ω_S also has its support in $\overline{U}$. It therefore follows from Lemma 7.37(b) that $\nu_k(E) \to \omega_S(E)$ as $k \to \infty$.

Because $R_1^U \geq \widehat{R}_1^S$, Lemma 7.37(a) implies that $\omega_U(E) \geq \omega_S(E)$. Furthermore, the definition of U_k implies that $w_k \geq (1 - \epsilon) R_1^{U_k}$ on E; that is,
$$G_E \nu_k \geq (1 - \epsilon) G_E \omega_{U_k}.$$
So it follows from Lemma 7.37(a) that $\nu_k(E) \geq (1 - \epsilon) \omega_{U_k}(E)$, and hence that
$$\omega_S(E) = \lim_{k \to \infty} \nu_k(E) \geq (1 - \epsilon) \lim_{k \to \infty} \omega_{U_k}(E) = (1 - \epsilon) \lim_{k \to \infty} \mathcal{C}(U_k),$$
by Theorem 7.44(d). It now follows, using Theorem 7.44(d) again, that
$$\mathcal{C}_+(S) + \epsilon > \mathcal{C}(U) = \omega_U(E) \geq \omega_S(E) \geq (1 - \epsilon) \lim_{k \to \infty} \mathcal{C}(U_k) \geq (1 - \epsilon) \mathcal{C}_+(S).$$

Making $\epsilon \to 0+$, we obtain $\mathcal{C}_+(S) = \omega_S(E)$, as required. □

7.7. Thermal and Cothermal Capacities

In this section, we use Theorem 7.36 to show that the thermal and cothermal capacities coincide, and that the polar sets are the same as the copolar sets.

THEOREM 7.46. *If S is an arbitrary subset of E, then:*
(a) $\mathcal{C}_-(S) = \mathcal{C}_-^*(S)$ and $\mathcal{C}_+(S) = \mathcal{C}_+^*(S)$,
(b) S is polar if and only if $\mathcal{C}(S) = 0$,
and
(c) S is polar if and only if S is copolar.

PROOF. (a) Let K be a compact subset of E, and let U be any bounded open superset of K such that $\overline{U} \subseteq E$. By Theorem 7.44(d), R_1^U is a heat potential whose associated Riesz measure ω_U satisfies $\omega_U(E) = \mathcal{C}(U)$. Therefore, because $\widehat{R}_1^{K*} = G_E^* \omega_K^* \leq 1$ on E, we have

$$\mathcal{C}(U) = \omega_U(E) \geq \int_E G_E^* \omega_K^* \, d\omega_U = \int_K G_E \omega_U \, d\omega_K^* = \omega_K^*(K) = \mathcal{C}^*(K)$$

by Theorem 7.36, since ω_K^* is supported in K and $G_E \omega_U = R_1^U = 1$ on $U \supseteq K$. Taking the infimum over all choices of U and using Lemma 7.42, we obtain the inequality $\mathcal{C}(K) \geq \mathcal{C}^*(K)$. The dual of this result is the reverse inequality, and so equality holds. It now follows that $\mathcal{C}_-(S) = \mathcal{C}_-^*(S)$ for any S, and then that $\mathcal{C}_+(S) = \mathcal{C}_+^*(S)$.

(b) Let $\{U_i\}$ be a sequence of bounded open sets, such that $\overline{U}_i \subseteq E$ for all i and $\bigcup_{i=1}^\infty U_i = E$. By Theorem 7.24, each set $S \cap U_i$ is polar if and only if $\widehat{R}_1^{S \cap U_i} = 0$, which occurs if and only if the associated Riesz measure $\omega_{S \cap U_i}$ satisfies $\omega_{S \cap U_i}(E) = 0$. By Theorem 7.45(d), we have $\omega_{S \cap U_i}(E) = \mathcal{C}_+(S \cap U_i)$, and so $S \cap U_i$ is polar if and only if $\mathcal{C}_+(S \cap U_i) = 0$.

It follows that, if S is polar then

$$\mathcal{C}_+(S) = \mathcal{C}_+\left(\bigcup_{i=1}^\infty (S \cap U_i)\right) \leq \sum_{i=1}^\infty \mathcal{C}_+(S \cap U_i) = 0,$$

by Theorem 7.45(c). Conversely, if $\mathcal{C}_+(S) = 0$ then $\mathcal{C}_+(S \cap U_i) = 0$ for all i, so that $S \cap U_i$ is polar for all i, and so S is polar by Theorem 7.4.

(c) This follows from (b), (a), and the dual of (b). □

REMARK 7.47. It follows from Theorem 7.46(a) and Example 7.35 that, relative to $E = \mathbb{R}^{n+1}$, the *thermal* capacity of the closed heat ball $\overline{\Omega}(0;c)$, and the heat sphere $\partial \Omega(0;c)$, is $(4\pi c)^{n/2}$.

7.8. Capacitable Sets

We already know that the compact sets, the open sets, and the polar sets are capacitable. In this section, we prove that the collection of capacitable sets is very large, and includes all the Borel sets. This involves the concept of an analytic set, which we define after introducing some notation.

We use the standard notation Y^X for the set of all functions from X into Y. Thus $\mathbb{N}^{\mathbb{N}}$ denotes the collection of all sequences of positive integers. This should not be confused with $\bigcup_{k=1}^{\infty} \mathbb{N}^k$, which is the collection of all *finite* sequences of positive integers. We also denote by $\mathcal{K}$ the collection of all compact subsets of $\mathbb{R}^{n+1}$.

DEFINITION 7.48. A subset A of $\mathbb{R}^{n+1}$ is called *analytic* if there exists a mapping $\phi : \bigcup_{k=1}^{\infty} \mathbb{N}^k \to \mathcal{K}$ such that

(7.8) $$A = \bigcup_{\{m_i\} \in \mathbb{N}^{\mathbb{N}}} \big(\phi(m_1) \cap \phi(m_1, m_2) \cap \phi(m_1, m_2, m_3) \cap ...\big).$$

We denote by $\mathcal{A}$ the collection of all analytic subsets of $\mathbb{R}^{n+1}$.

LEMMA 7.49. *(a) If $\{A_j\}$ is a sequence of analytic sets, then $\bigcup_{j=1}^{\infty} A_j$ and $\bigcap_{j=1}^{\infty} A_j$ are also analytic.*
(b) Every Borel set is analytic.
(c) If A is an analytic subset of E, then the compact sets $\phi(m_1, ..., m_k)$ in (7.8) can be chosen to be subsets of E.

PROOF. (a) For each j, there is a mapping $\phi_j : \bigcup_{k=1}^{\infty} \mathbb{N}^k \to \mathcal{K}$ such that

$$A_j = \bigcup_{\{m_i\} \in \mathbb{N}^{\mathbb{N}}} \big(\phi_j(m_1) \cap \phi_j(m_1, m_2) \cap \phi_j(m_1, m_2, m_3) \cap ...\big).$$

Let the mapping $k \mapsto (\alpha(k), \beta(k))$ be a bijection from $\mathbb{N}$ to $\mathbb{N}^2$. If we define the map $\phi : \bigcup_{k=1}^{\infty} \mathbb{N}^k \to \mathcal{K}$ by putting

$$\phi(m_1, ..., m_k) = \phi_{\alpha(m_1)}(\beta(m_1), m_2, m_3, ..., m_k),$$

then we can write A_j as

$$A_j = \bigcup_{\{\{m_i\} : \alpha(m_1) = j\}} \big(\phi(m_1) \cap \phi(m_1, m_2) \cap \phi(m_1, m_2, m_3) \cap ...\big).$$

We can now write $\bigcup_{j=1}^{\infty} A_j$ as

$$\bigcup_{j=1}^{\infty} A_j = \bigcup_{\{m_i\} \in \mathbb{N}^{\mathbb{N}}} \big(\phi(m_1) \cap \phi(m_1, m_2) \cap \phi(m_1, m_2, m_3) \cap ...\big),$$

which shows that the union is analytic.

We now show that the intersection is also analytic. A point p belongs to $\bigcap_{j=1}^{\infty} A_j$ if and only if, for each $j \in \mathbb{N}$ there is an element $\{m_i^{(j)}\}$ of $\mathbb{N}^{\mathbb{N}}$ such that

$$p \in \phi_j(m_1^{(j)}) \cap \phi_j(m_1^{(j)}, m_2^{(j)}) \cap \phi_j(m_1^{(j)}, m_2^{(j)}, m_3^{(j)}) \cap \ldots.$$

That is, if and only if there is a function $f : \mathbb{N}^2 \to \mathbb{N}$ such that

$$p \in \bigcap_{j=1}^{\infty} \big(\phi_j(f(j,1)) \cap \phi_j(f(j,1), f(j,2)) \cap \phi_j(f(j,1), f(j,2), f(j,3)) \cap \ldots\big).$$

Given any element $\{m_k\}$ of $\mathbb{N}^{\mathbb{N}}$, we let $E(j, i)$ denote the (j, i) entry of the infinite matrix

$$\begin{matrix} m_1 & m_2 & m_4 & m_7 & \ldots \\ m_3 & m_5 & m_8 & \ldots & \ldots \\ m_6 & m_9 & \ldots & \ldots & \ldots \\ m_{10} & \ldots & \ldots & \ldots & \ldots \\ \ldots & \ldots & \ldots & \ldots & \ldots \end{matrix}$$

We then define $\phi(m_1, ..., m_k)$ by looking at the position of m_k in the above matrix. If $m_k = E(j, i)$, we define
$$\phi(m_1, ..., m_k) = \phi_j(f(j, 1), \ldots, f(j, i)).$$
Then
$$\bigcap_{j=1}^{\infty} \bigl(\phi_j(f(j,1)) \cap \phi_j(f(j,1), f(j,2)) \cap \phi_j(f(j,1), f(j,2), f(j,3)) \cap \ldots\bigr)$$
$$= \bigcap_{k=1}^{\infty} \phi(m_1, ..., m_k).$$
Thus
$$\bigcap_{j=1}^{\infty} A_j = \bigcup_{\{m_k\} \in \mathbb{N}^{\mathbb{N}}} \bigl(\phi(m_1) \cap \phi(m_1, m_2) \cap \phi(m_1, m_2, m_3) \cap \ldots\bigr),$$
as required.

(b) Any compact set K is analytic, because it can be written in the form (7.8) by taking $\phi(m_1, ..., m_k) = K$ for any choice of $(m_1, ..., m_k)$. Since any open or closed set can be written as the union of a sequence of compact sets, it follows from part (a) that such sets are also analytic. We now consider $\mathcal{C}$, the collection of all analytic sets A such that $\mathbb{R}^{n+1} \setminus A$ is also analytic. If $\{A_k\}$ is a sequence of sets in $\mathcal{C}$, then part (a) shows that $\bigcup_{k=1}^{\infty} A_k$ is analytic, and that
$$\mathbb{R}^{n+1} \setminus \bigcup_{k=1}^{\infty} A_k = \bigcap_{k=1}^{\infty} \left(\mathbb{R}^{n+1} \setminus A_k\right)$$
is too. Hence $\bigcup_{k=1}^{\infty} A_k \in \mathcal{C}$, and so $\mathcal{C}$ is a σ-algebra that contains the open sets. Therefore $\mathcal{C}$, and hence $\mathcal{A}$, contains the Borel sets.

(c) Let A be an analytic subset of E. Since A is analytic, there is a mapping $\psi : \bigcup_{k=1}^{\infty} \mathbb{N}_k \to \mathcal{K}$ such that
$$A = \bigcup_{\{m_i\} \in \mathbb{N}^{\mathbb{N}}} \bigl(\psi(m_1) \cap \psi(m_1, m_2) \cap \psi(m_1, m_2, m_3) \cap \ldots\bigr).$$
Let $\{K_j\}$ be a sequence of compact sets with union E. Given any $\{m_i\} \in \mathbb{N}^{\mathbb{N}}$, we put $\phi(m_1) = K_{m_1}$ and $\phi(m_1, \ldots, m_k) = K_{m_1} \cap \psi(m_2, \ldots, m_k)$ whenever $k \geq 2$. Then each $\phi(m_1, \ldots, m_k)$ is a compact subset of E, and
$$\phi(m_1) \cap \phi(m_1, m_2) \cap \phi(m_1, m_2, m_3) \cap \ldots$$
$$= K_{m_1} \cap \psi(m_2) \cap \psi(m_2, m_3) \cap \psi(m_2, m_3, m_4) \cap \ldots,$$
so that
$$\bigcup_{\{m_i\} \in \mathbb{N}^{\mathbb{N}}} \bigl(\phi(m_1) \cap \phi(m_1, m_2) \cap \phi(m_1, m_2, m_3) \cap \ldots\bigr)$$
$$= \bigcup_{j=1}^{\infty} K_j \cap \bigcup_{\{m_i\} \in \mathbb{N}^{\mathbb{N}}} \bigl(\psi(m_2) \cap \psi(m_2, m_3) \cap \psi(m_2, m_3, m_4) \cap \ldots\bigr)$$
$$= E \cap A$$
$$= A.$$

□

LEMMA 7.50. *Let A be given by (7.8), and let $\{n_i\} \in \mathbb{N}^{\mathbb{N}}$. For all $k \in \mathbb{N}$, we define*

$$S_k = \bigcup_{\{\{m_i\} \in \mathbb{N}^{\mathbb{N}} : m_i \leq n_i \text{ if } i \leq k\}} \big(\phi(m_1) \cap \phi(m_1, m_2) \cap \phi(m_1, m_2, m_3) \cap \dots\big)$$

and

$$T_k = \bigcup_{\{\{m_i\} \in \mathbb{N}^k : m_i \leq n_i \text{ if } i \leq k\}} \big(\phi(m_1) \cap \dots \cap \phi(m_1, \dots, m_k)\big).$$

Then

(a) $S_k \subseteq A \cap T_k$ for all k, and the sequence $\{S_k\}$ is contracting.

(b) $\{T_k\}$ is a contracting sequence of compact sets whose intersection is a subset of A.

PROOF. It is clear that (a) holds. Furthermore, because each set T_k is a union of finitely many compact sets, each T_k is compact. It is also clear that $\{T_k\}$ is contracting, and so it remains only to prove that its intersection T is contained in A.

Let $p \in T$. Then for each $k \in \mathbb{N}$, there is an element $(m_1^{(k)}, \dots, m_k^{(k)})$ of $\mathbb{N}^k$ such that $m_i^{(k)} \leq n_i$ whenever $i \leq k$ and such that

$$p \in \phi(m_1^{(k)}) \cap \dots \cap \phi(m_1^{(k)}, \dots, m_k^{(k)}).$$

Since $1 \leq m_1^{(k)} \leq n_1$ for all $k \in \mathbb{N}$, there is an integer $m_1' \leq n_1$ such that $m_1^{(k)} = m_1'$ for infinitely many values of k. Since $1 \leq m_2^{(k)} \leq n_2$ for all those values of k for which $m_1^{(k)} = m_1'$, we similarly deduce that there is an integer $m_2' \leq n_2$ such that $(m_1^{(k)}, m_2^{(k)}) = (m_1', m_2')$ for infinitely many values of k. Continuing in this manner indefinitely, we obtain a sequence $\{m_i'\}$ such that

$$p \in \phi(m_1') \cap \phi(m_1', m_2') \cap \phi(m_1', m_2', m_3') \cap \dots \subseteq A.$$

Hence $T \subseteq A$, and the proof is complete. □

THEOREM 7.51. *Every analytic subset of E is (thermal) capacitable.*

PROOF. Let A be an analytic subset of E. Then, by Lemma 7.49(c), we can write A in the form (7.8) with the compact sets $\phi(m_1, \dots, m_k)$ contained in E. We choose any number $\alpha < \mathcal{C}_+(A)$, and define a sequence $\{n_i\} \in \mathbb{N}^{\mathbb{N}}$ inductively, as follows.

At stage 1 of the process, we define the sequence of sets $\{I_{1,j}\}$ by putting

$$I_{1,j} = \{\{m_i\} \in \mathbb{N}^{\mathbb{N}} : m_1 \leq j\}$$

for all j. The sequence is expanding with union $\mathbb{N}^{\mathbb{N}}$. Hence the sequence of sets $\{A_{1,j}\}$, where

$$A_{1,j} = \bigcup_{\{m_i\} \in I_{1,j}} \big(\phi(m_1) \cap \phi(m_1, m_2) \cap \phi(m_1, m_2, m_3) \cap \dots\big)$$

for all j, is also expanding, and its union is A. It therefore follows from Theorem 7.45(a) that we can choose an integer n_1 such that $\mathcal{C}_+(A_{1,n_1}) > \alpha$.

At stage $k+1$ of the process, we have integers $n_1, \dots, n_k$ and sets

$$I_{k,n_k} = \{\{m_i\} \in \mathbb{N}^{\mathbb{N}} : m_1 \leq n_1, \dots, m_k \leq n_k\}$$

such that the set
$$A_{k,n_k} = \bigcup_{\{m_i\} \in I_{k,n_k}} \big(\phi(m_1) \cap \phi(m_1, m_2) \cap \phi(m_1, m_2, m_3) \cap \ldots\big)$$
satisfies $\mathcal{C}_+(A_{k,n_k}) > \alpha$. The sequence of sets $\{I_{k+1,j}\}$, defined by
$$I_{k+1,j} = \{\{m_i\} \in \mathbb{N}^\mathbb{N} : m_1 \leq n_1, \ldots, m_k \leq n_k, m_{k+1} \leq j\}$$
for all j, is expanding with union I_{k,n_k}. Therefore the sequence of sets $\{A_{k+1,j}\}$, defined by
$$A_{k+1,j} = \bigcup_{\{m_i\} \in I_{k+1,j}} \big(\phi(m_1) \cap \phi(m_1, m_2) \cap \phi(m_1, m_2, m_3) \cap \ldots\big)$$
for all j, is also expanding, and its union is A_{k,n_k}. It follows that we can find an integer n_{k+1} such that $\mathcal{C}_+(A_{k+1,n_{k+1}}) > \alpha$.

The sets A_{k,n_k} are the sets S_k of Lemma 7.49, so that $\mathcal{C}_+(S_k) > \alpha$ for all k. Having chosen the sequence $\{n_k\}$ in this way, we use it to define the sets T_k as in Lemma 7.49. By that result, $\{T_k\}$ is a contracting sequence of compact sets whose intersection T is contained in A, and $S_k \subseteq T_k$ for all k. Therefore, by Theorem 7.39(b),
$$\mathcal{C}_-(A) \geq \mathcal{C}(T) = \lim_{k \to \infty} \mathcal{C}(T_k) \geq \lim_{k \to \infty} \mathcal{C}_+(S_k) \geq \alpha.$$
Thus $\mathcal{C}_-(A) \geq \alpha$ for any number $\alpha < \mathcal{C}_+(A)$, so that $\mathcal{C}_+(A) \leq \mathcal{C}_-(A)$. The reverse inequality is always true, and so A is capacitable. $\square$

7.9. Polar Sets and Heat Potentials

The first theorem in this section characterizes the polar Borel sets as those that cannot support a nontrivial, bounded heat potential. Part of this result generalizes Theorem 7.29 from relatively closed sets to Borel sets.

THEOREM 7.52. *If S is a Borel subset of E, then the following statements are equivalent.*

(a) S is polar.

(b) If μ is a nonnegative measure on E such that the heat potential $G_E\mu$ is bounded, then $\mu(S) = 0$.

PROOF. Suppose first that S is polar, and let μ be a nonnegative measure on E such that $G_E\mu$ is bounded. By Theorem 7.46, S is also copolar. Therefore, by the dual of Theorem 7.3, there is a coheat potential $G_E^*\nu^*$ such that $G_E^*\nu^*(p) = +\infty$ for all $p \in S$, and $\nu^*(E) < +\infty$. By Theorem 7.36,
$$\int_E G_E^*\nu^* \, d\mu = \int_E G_E\mu \, d\nu^* < +\infty,$$
because $G_E\mu$ is bounded and $\nu^*(E) < +\infty$. Since $G_E^*\nu^*(p) = +\infty$ for all $p \in S$, this implies that $\mu(S) = 0$.

Now suppose, conversely, that S is not polar. By Lemma 7.49(b) and Theorem 7.51, the Borel sets are capacitable, so that $\mathcal{C}_-(S) = \mathcal{C}(S) > 0$ by Theorem 7.46. Therefore there exists a compact subset K of S such that $\mathcal{C}(K) > 0$. The thermal capacitary potential $\widehat{R}_1^K = G_E\omega_K$ of K is bounded on E. Moreover, the thermal capacitary distribution ω_K has its support in $K \subseteq S$ and $\omega_K(E) = \mathcal{C}(K) > 0$. Thus ω_K is a nonnegative measure on E such that $G_E\omega_K$ is bounded and $\omega_K(S) > 0$. $\square$

If Z is a polar subset of E and $p_0 \in E\backslash Z$, then we know from Theorem 7.3 that there is a heat potential $G_E\mu$ such that $G_E\mu(p) = +\infty$ for all $p \in Z$ and $G_E\mu(p_0) < +\infty$. Our next result shows that, if Z is a *compact* polar subset of E, then μ can be chosen such that $G_E\mu(p) < +\infty$ for *all* $p \in E\backslash Z$.

THEOREM 7.53. *If Z is a compact polar subset of E, then there is a nonnegative measure μ on E such that $G_E\mu(p) = +\infty$ for all $p \in Z$ and $G_E\mu(p) < +\infty$ for all $p \in E\backslash Z$.*

PROOF. We take a contracting sequence of bounded open sets $\{V_j\}$ such that $\overline{V}_{j+1} \subseteq V_j$ and $\overline{V}_j \subseteq E$ for all j, and $\bigcap_{j=1}^{\infty} V_j = Z$. We put $K_j = \overline{V}_j$ and $v_j = \widehat{R}_1^{K_j}$ for all j. Then $\{v_j\}$ is a decreasing sequence of heat potentials, by Theorem 7.28, such that $v_j(p) = 1$ for all $p \in Z$ and v_j is a temperature on $E\backslash K_j$, by Theorem 7.27(d). We put $v = \lim_{j\to\infty} v_j$ on E.

Given any integer k, the functions v_j for $j \geq k$ are all temperatures on $E\backslash K_k$, and so v is a temperature on $E\backslash K_k$ by the Harnack monotone convergence theorem. Since k is arbitrary, v is a temperature on $E\backslash Z$. Because Z is a closed polar set and $0 \leq v \leq 1$ on E, the restriction of v to $E\backslash Z$ has a unique extension to a temperature $\bar{v}$ on E, by Corollary 7.15. Moreover, $0 \leq \bar{v} \leq v_1$ on E, and v_1 is a heat potential, so it follows from Theorem 6.19 that $\bar{v} = 0$ on E. Hence $v = 0$ on $E\backslash Z$.

We now take an expanding sequence of bounded open sets $\{D_i\}$ such that $V_1 \subseteq D_1$, $\overline{D}_i \subseteq E$ for all i, and $\bigcup_{i=1}^{\infty} D_i = E$. Given any integer l, the heat potentials v_j for $j > l$ are temperatures on $E\backslash \overline{V}_j$, and hence are continuous on the compact set $\overline{D}_l\backslash V_l$. The sequence $\{v_j\}$ decreases to the constant 0, and so Dini's theorem implies that the convergence is uniform on $\overline{D}_l\backslash V_l$. Hence there is an integer $j_l \geq l$ such that $v_{j_l} < 2^{-l}$ on $\overline{D}_l\backslash V_l$. We now put $w = \sum_{l=1}^{\infty} v_{j_l}$, and note that if $p \in Z$ then $w(p) = +\infty$ because $v_{j_l}(p) = 1$ for all l. If $q \in E\backslash Z$, then there is an integer m such that $q \in \overline{D}_l\backslash V_l$ for all $l \geq m$. Hence

$$w(q) = \sum_{l=1}^{m-1} v_{j_l}(q) + \sum_{l=m}^{\infty} v_{j_l}(q) \leq (m-1) + \sum_{l=m}^{\infty} 2^{-l} < +\infty.$$

Therefore w is a nonnegative supertemperature on E, by Theorem 3.60, and is finite on $E\backslash Z$. By the Riesz decomposition theorem, $G_E\mu_w$ is a heat potential and $w = G_E\mu_w + h$ on E, where h is the greatest thermic minorant of w on E. Thus $G_E\mu_w(p) = +\infty$ if and only if $p \in Z$. □

7.10. Thermal Capacity and Lebesgue Measure

In this section, we replace the arbitrary open set E by a strip $E_s = \mathbb{R}^n \times]a,b[$, where $-\infty \leq a < b \leq +\infty$. All reductions, and hence the thermal capacity, are relative to E_s. By Example 6.2, the Green function for E_s is G.

The main result is that, for analytic subsets of a hyperplane of the form $\mathbb{R}^n \times \{c\}$, the thermal capacity is the same as the n-dimensional Lebesgue measure.

We denote by $M^+(K)$ the class of nonnegative measures on E_s which have their support in the compact set K. We know that $\widehat{R}_1^K = G\omega_K \leq 1$ on E_s, and that $\omega_K \in M^+(K)$. The next lemma shows that $G\omega_K$ is the largest such heat potential.

7.10. THERMAL CAPACITY AND LEBESGUE MEASURE

LEMMA 7.54. *For any compact subset K of E_s, we have*
$$G\omega_K = \sup\{G\mu : G\mu \leq 1 \text{ on } E_s, \ \mu \in M^+(K)\}.$$

PROOF. Let K be a compact subset of E_s, and let $\mu \in M^+(K)$ and satisfy $G\mu \leq 1$ on E_s.

We claim that $G\mu(x,t) \to 0$ as (x,t) tends to the point at infinity in such a way that t remains upper bounded. Clearly $G\mu(x,t) = 0$ if $t < s$ for every point $(y,s) \in K$. Let $D = \{(y,s) \in E_s : |y| < \rho, \ c < s < d\}$ be a circular cylinder that contains K. There is a constant C such that
$$G\mu(x,t) = \int_D W(x-y, t-s)\, d\mu(y,s) \leq C \int_D |x-y|^{-n}\, d\mu(y,s).$$
If $|x| > \rho + r$, then $|x - y| > r$ whenever $(y,s) \in D$, and so it follows that
$$G\mu(x,t) \leq Cr^{-n}\mu(D) \to 0 \text{ as } r \to +\infty.$$

We now take any nonnegative supertemperature v on E_s such that $v \geq 1$ on K. Then $v - G\mu \geq 0$ on K, and we need to show that this inequality is also true on $E_s\backslash K$. For all points $q \in \partial K$, we have
$$\liminf_{p \to q} (v(p) - G\mu(p)) \geq \liminf_{p \to q} v(p) - 1 \geq v(q) - 1 \geq 0.$$
Furthermore, as $p = (x,t)$ tends to the point at infinity in such a way that t remains upper bounded, we have
$$\liminf (v(p) - G\mu(p)) \geq \liminf v(p) - \lim G\mu(p) \geq 0.$$
The function $v - G\mu$ is a supertemperature on $E_s \backslash K$, because $G\mu$ is a temperature there (by Corollary 6.22). It therefore follows from the minimum principle that $v - G\mu \geq 0$ on $E_s \backslash K$. Hence $v \geq G\mu$ on E_s, and it follows that $R_1^K \geq G\mu$. Since $G\mu$ is lower semicontinuous, we obtain $G\omega_K = \widehat{R}_1^K \geq G\mu$, as required. □

THEOREM 7.55. *Let A be an analytic subset of $\mathbb{R}^n$, let $c \in\,]a,b[$, and let m_n denote n-dimensional Lebesgue measure on $\mathbb{R}^n$. Then $\mathcal{C}(A \times \{c\}) = m_n(A)$, and if A is compact then the thermal capacitary distribution $\omega_{A \times \{c\}}$ is the product of the restriction of m_n to A with the unit mass at c.*

PROOF. Let K be a compact subset of $\mathbb{R}^n$, and let $G\omega_{K \times \{c\}}$ denote the thermal capacitary potential of $K \times \{c\}$. Then $\omega_{K \times \{c\}}$ is supported by $K \times \{c\}$, so that we can write $\omega_{K \times \{c\}} = \nu_K \times \delta_c$, where ν_K is supported by K and δ_c is the unit mass at c. Hence
$$G\omega_{K \times \{c\}}(x,t) = \int_{K \times \{c\}} W(x-y, t-s)\, d\omega_{K \times \{c\}}(y,s) = \int_K W(x-y, t-c)\, d\nu_K(y)$$
if $c < t < b$, and $G\omega_{K \times \{c\}}(x,t) = 0$ if $a < t \leq c$. In particular, $G\omega_{K \times \{c\}}$ is the Gauss-Weierstrass integral of ν_K on $\mathbb{R}^n \times\,]c,b[$. Therefore, since $0 \leq G\omega_{K \times \{c\}} \leq 1$ on $\mathbb{R}^n \times\,]c,b[$, it follows from Corollary 5.35 that there is a function f on $\mathbb{R}^n \times \{c\}$ such that $|f| \leq 1$ and
$$G\omega_{K \times \{c\}}(x,t) = \int_{\mathbb{R}^n} W(x-y, t-c) f(y,c)\, dy$$
for all $(x,t) \in \mathbb{R}^n \times\,]c,b[$. Gauss-Weierstrass integrals have unique representing measures, by Theorem 4.11, and hence $f(y,c)\, dy = d\nu_K(y)$. In particular, f is

supported by K, so that

(7.9) $$G\omega_{K\times\{c\}}(x,t) \leq \int_K W(x-y,t-c)\,dy$$

on $\mathbb{R}^n \times {]c,b[}$. By Lemma 7.54,

$$G\omega_{K\times\{c\}} = \sup\{G\mu : G\mu \leq 1 \text{ on } E_s,\ \mu \in M^+(K\times\{c\})\},$$

so that equality holds in (7.9). Therefore the product of the restriction of m_n to K with δ_c is the thermal capacitary distribution of $K \times \{c\}$, and $\mathcal{C}(K\times\{c\}) = m_n(K)$. Now the inner regularity of Lebesgue measure implies that, for any analytic subset A of $\mathbb{R}^n$, we have

$$m_n(A) = \sup\{m_n(K) : K \text{ is a compact subset of } A\} = \mathcal{C}_-(A \times \{c\}),$$

and the result follows because $A \times \{c\}$ is thermal capacitable (Theorem 7.51). □

It is important to realize that the result of Theorem 7.55 does *not* extend to subsets of more than one hyperplane of the form $\mathbb{R}^n \times \{c\}$, and that although Lebesgue measure is additive, thermal capacity is only strongly subbadditive. The following example illustrates these points.

EXAMPLE 7.56. Let A and B be compact subsets of $\mathbb{R}^n$ with positive Lebesgue measure, let $-\infty < a < b < +\infty$, and let $K = (A \times \{a\}) \cup (B \times \{b\})$. Then K is compact, and its thermal capacitary distribution ω_K (relative to $\mathbb{R}^{n+1}$) is supported by K, so that we can write $\omega_K = (\mu_A \times \delta_a) + (\mu_B \times \delta_b)$, where δ_c denotes the unit mass at $c \in \{a,b\}$. Therefore $G\omega_K(x,t) = 0$ if $t \leq a$,

$$G\omega_K(x,t) = \int_A W(x-y,t-a)\,d\mu_A(y)$$

if $a < t \leq b$, and

$$G\omega_K(x,t) = \int_A W(x-y,t-a)\,d\mu_A(y) + \int_B W(x-y,t-b)\,d\mu_B(y)$$

if $t > b$.

We consider first $G\omega_K$ on the strip $\mathbb{R}^n \times {]a,b]}$. Since $0 \leq G\omega_K \leq 1$, it follows from Corollary 5.35 that there is a function f on $\mathbb{R}^n \times \{a\}$ such that $|f| \leq 1$ and

$$G\omega_K(x,t) = \int_{\mathbb{R}^n} W(x-y,t-a)f(y,a)\,dy$$

for all $(x,t) \in \mathbb{R}^n \times {]a,b[}$. The uniqueness of the Gauss-Weierstrass representation (Theorem 4.11) shows that $f(y,a)\,dy = d\mu_A(y)$, so that f is supported by A and

$$G\omega_K(x,t) \leq \int_A W(x-y,t-a)\,dy$$

on $\mathbb{R}^n \times {]a,b[}$. By Theorem 2.2, this integral represents a continuous function on $\mathbb{R}^n \times {]a,+\infty[}$, and it therefore follows from Lemma 3.16 that this inequality holds on $\mathbb{R}^n \times {]a,b]}$. Moreover Lemma 7.37(a), with $E = \mathbb{R}^n \times {]-\infty,b[}$, shows that $\mu_A(A) = \omega_K(A \times \{a\}) \leq m_n(A)$.

We now put

$$g(x,b) = \int_A W(x-y,b-a)\,d\mu_A(y)$$

for all $x \in \mathbb{R}^n$, so that

$$G\omega_K(x,t) = \int_{\mathbb{R}^n} W(x-y,t-b)g(y,b)\,dy + \int_B W(x-y,t-b)\,d\mu_B(y)$$

if $t > b$, in view of Theorem 4.10. Since $0 \leq G\omega_K \leq 1$, Corollary 5.35 shows that there is a function h on $\mathbb{R}^n \times \{b\}$ such that $|h| \leq 1$ and

$$G\omega_K(x,t) = \int_{\mathbb{R}^n} W(x-y,t-b)h(y,b)\,dy$$

whenever $t > b$. By Theorem 4.11 the representing measure is unique, so that $d\mu_B(y) = (h(y,b) - g(y,b))\,dy$. Therefore $h - g$ is supported by $B \times \{b\}$, and in particular $h(\cdot,b) = g(\cdot,b)$ on $\mathbb{R}^n \backslash B$. Moreover, since $h - g \leq 1 - g$ on $B \times \{b\}$, we have $d\mu_B(y) \leq (1 - g(y,b))\chi_B(y)\,dy$. Since $m_n(A) > 0$ we have $g > 0$, so that since $m_n(B) > 0$ it follows that $\mu_B(B) < m_n(B)$. Hence

$$\mathcal{C}((A \times \{a\}) \cup (B \times \{b\})) = \mu_A(A) + \mu_B(B)$$
$$< m_n(A) + m_n(B) = \mathcal{C}(A \times \{a\}) + \mathcal{C}(B \times \{b\}).$$

Using Theorem 7.55 and the equivalence of polarity and zero thermal capacity, we deduce the following result, which we then apply to improve upon Theorem 5.1.

THEOREM 7.57. *If $Z \subseteq \mathbb{R}^n$, $m_n(Z) = 0$ and $c \in \mathbb{R}$, then there is a positive temperature u on $\mathbb{R}^n \times]c, +\infty[$ such that*

$$\lim_{(x,t) \to (y,c+)} u(x,t) = +\infty$$

for all $y \in Z$.

PROOF. We choose a, b such that $-\infty \leq a < c < b \leq +\infty$, and consider the thermal capacity relative to $E_s = \mathbb{R}^n \times]a,b[$. By Theorem 7.55, $\mathcal{C}(Z \times \{c\}) = 0$. Therefore $Z \times \{c\}$ is polar, by Theorem 7.46, and so there is a heat potential $G\mu$ of a finite measure such that $G\mu(x,c) = +\infty$ for all $x \in Z$, by Theorem 7.3. Let ν denote the restriction of μ to the half-space $\mathbb{R}^n \times]-\infty, c]$. Then ν is finite, and so $G\nu$ is a heat potential by Theorem 6.18. Therefore $G\nu$ is a temperature on $\mathbb{R}^n \times]c, +\infty[$, by Corollary 6.22. Furthermore, for any $y \in Z$ we have

$$G\nu(y,c) = \int_{\mathbb{R}^n \times]-\infty,c[} W(y-z,c-s)\,d\nu(y,s)$$
$$= \int_{\mathbb{R}^n \times]-\infty,c[} W(y-z,c-s)\,d\mu(y,s)$$
$$= G\mu(y,c)$$
$$= +\infty,$$

so that $G\nu = +\infty$ on $Z \times \{c\}$. The lower semicontinuity of $G\nu$ now implies that $G\nu(x,t) \to +\infty$ as $(x,t) \to (y,c)$ for each $y \in Z$. Hence the restriction of $G\nu$ to $\mathbb{R}^n \times]-\infty, c[$ is the required temperature. $\square$

THEOREM 7.58. *Suppose that $0 \leq s < b$, and that w is a subtemperature on $\mathbb{R}^n \times]s,b[$. If the hyperplane mean $M_b(w^+;\cdot)$ is a locally integrable function on the half-closed interval $[s,b[$, if*

(7.10) $$\limsup_{(x,t) \to (\xi,s+)} w(x,t) < +\infty$$

for all $\xi \in \mathbb{R}^n$, and if
$$\limsup_{(x,t)\to(\xi,s+)} w(x,t) \leq A \tag{7.11}$$
for almost all $\xi \in \mathbb{R}^n$, then $w \leq A$ on $\mathbb{R}^n \times]s,b[$.

PROOF. Let Z denote the set of all ξ such that (7.11) does not hold. Then $m_n(Z) = 0$, and so there is a positive temperature u on $\mathbb{R}^n \times]s,b[$ such that
$$\lim_{(x,t)\to(\xi,s+)} u(x,t) = +\infty \tag{7.12}$$
for all $\xi \in Z$, by Theorem 7.57. Given any positive number ϵ, we put $w_\epsilon = w - \epsilon u$ on $\mathbb{R}^n \times]s,b[$. Then w_ϵ is a subtemperature on $\mathbb{R}^n \times]s,b[$, and since $w_\epsilon \leq w$ the hyperplane mean $M_b(w_\epsilon^+;\cdot)$ is locally integrable on $[s,b[$. Furthermore, for all $\xi \in Z$ we have
$$\limsup_{(x,t)\to(\xi,s+)} w_\epsilon(x,t) = \limsup_{(x,t)\to(\xi,s+)} w(x,t) - \epsilon \lim_{(x,t)\to(\xi,s+)} u(x,t) = -\infty,$$
in view of (7.10) and (7.12). Moreover, for all $\xi \in \mathbb{R}^n \backslash Z$ we have
$$\limsup_{(x,t)\to(\xi,s+)} w_\epsilon(x,t) \leq \limsup_{(x,t)\to(\xi,s+)} w(x,t) \leq A,$$
in view of (7.11). Thus
$$\limsup_{(x,t)\to(\xi,s+)} w_\epsilon(x,t) \leq A$$
for all $\xi \in \mathbb{R}^n$, and so it follows from Theorem 5.1 that $w_\epsilon \leq A$ on $\mathbb{R}^n \times]s,b[$. Making $\epsilon \to 0+$, we deduce that $w \leq A$ on $\mathbb{R}^n \times]s,b[$, as required. □

7.11. Notes and Comments

The main references for the results in this chapter are Watson [**72, 73**] and Doob [**14**]. Some results about the thermal capacity of compact sets were also proved by Landis [**49**] and Lanconelli [**46**] for $E = \mathbb{R}^{n+1}$, taking as the definition the characterization given in Lemma 7.54. However, the treatment here is heavily influenced by Armitage & Gardiner [**3**], both in the methods used and in the order many of the results are proved. This is especially true of sections, 7.5, 7.6 and 7.8.

Theorem 7.6 is new, as are Examples 7.35 and 7.56.

Corollary 7.15 can be referred to as a removable sets theorem for temperatures. Several authors have considered such results, including Král [**43**], Kaiser & Müller [**37**], Umanskiĭ [**68**], Watson [**86**] and Hui [**36**].

A slightly different version of Theorem 7.17, using the means $\mathcal{M}$ or $\mathcal{V}$ instead of $\mathcal{L}$, was given by Watson [**71**]. In [**72**], Watson showed that there are bounded supertemperatures u such that
$$\limsup_{(x,t)\to(x_0,t_0-)} u(x,t) > u(x_0,t_0),$$
which shows that supertemperatures do not generally satisfy the corresponding one-sided continuity condition.

In the case where $E = \mathbb{R}^{n+1}$ and S is analytic, Theorem 7.46(a) was proved in a different form by Gariepy & Ziemer [**25**]. Related work can be found in Maeda [**51**]. Theorem 7.46(b),(c) were proved by Watson [**73**] for subsets of a quasi-regular (called 'admissible' in [**73**]) open set E. Since $\mathbb{R}^{n+1}$ is regular, and a subset of E is polar if and only if it is polar as a subset of $\mathbb{R}^{n+1}$, the equivalence of polar and copolar follows for any E. That equivalence was later proved by Doob [**14**].

Theorem 7.55 was first proved by Lanconelli [**46**], for the case where $E_s = \mathbb{R}^{n+1}$ and A is compact. The general case was proved independently by Watson [**73**].

Kaufman & Wu [**39**] began the comparison of polarity with classical capacities by showing that, if $n = 1$ and $S \subseteq \{x_0\} \times \mathbb{R}$, then S is polar if and only if it has zero Riesz $\frac{1}{2}$-capacity. On the other hand, if $S \subseteq \mathbb{R}^n \times \{t_0\}$ for any n, then Theorem 7.55 shows that S is polar if and only if it has n-dimensional Lebesgue measure zero. Such a distinct difference between the coordinates means that any spherically symmetric measure or capacity will be of little use here, since for $n = 1$ the critical dimension is 1 in the x-coordinate and $\frac{1}{2}$ in the t-coordinate. To overcome this, Taylor & Watson [**67**] defined measures, of Hausdorff type, using a restricted class of covering sets. These measures effectively double the classical dimension in the t-direction whilst leaving it unchanged in the x-direction. They found a class of sets in $\mathbb{R}^{n+1}$, not all subsets of $\{x_0\} \times \mathbb{R}$ for any x_0, for which the thermal capacity is zero if and only if the Riesz $\frac{n}{2}$-capacity of the projection onto the t-axis is zero. They found another class of sets, not all subsets of $\mathbb{R}^n \times \{t_0\}$ for any t_0, for which the thermal capacity is zero if and only if the n-dimensional Lebesgue measure of the projection onto the hyperplane $\mathbb{R}^n \times \{0\}$ is zero. They obtained comparison theorems in both directions, and left some **open questions**. Mysovskikh [**54**] also obtained comparison theorems for polarity, using both anisotropic and classical Hausdorff measures. Wu [**93**] considered the product of compact subsets X and T of $\mathbb{R}$, and gave criteria for the thermal capacity of $X \times T$ to be positive in terms of the classical Hausdorff measures and Riesz capacities of X and T. She also left an **open question**.

CHAPTER 8

The Dirichlet Problem on Arbitrary Open Sets

We first recall, from Section 3.3, the Dirichlet Problem on a convex domain of revolution R. Show that, for an arbitrary function $f \in C(\partial_n R)$, there is a function $u_f \in C(R \cup \partial_n R)$ which is a temperature on R and equal to f on $\partial_n R$. The maximum principle guarantees uniqueness.

In trying to generalize this problem to an arbitrary open set E in $\mathbb{R}^{n+1}$, we immediately encounter the question of which part of the boundary we can expect to recover the given values on. That is, which part of ∂E corresponds to $\partial_n R$. Of course, we could prescribe f on the whole of ∂E, but that would give us a problem that could not be solved even for so simple a domain as a circular cylinder, as Theorem 2.3 shows.

We therefore need to decide which part of ∂E it is reasonable to prescribe f on. A clue is provided by the boundary maximum principle in Theorem 3.13. If the maximum principle is to guarantee uniqueness, then we must include all points $p \in \partial E$ for which there is a sequence $\{p_k\}$ of points in E such that $p_{k+1} \in \Lambda(p_k; E)$ for all k, which converges to p. This does not give a simple, local criterion for the inclusion of p, but suggests that we should include all boundary points (x, t) for which the upper half-ball $\{(y, s) : |y - x|^2 + (s - t)^2 < r^2, s > t\}$ meets E for every $r > 0$. This does give a simple, local criterion.

However, we also need to consider how the boundary values can be expected to be taken. We illustrate this by the following simple example, where E consists of two circular cylinders one on top of the other. Let B be a ball in $\mathbb{R}^n$, and let $E = B \times (]a, b[\cup]b, c[)$. We put $E_1 = B \times]a, b[$, and $E_2 = B \times]b, c[$. If $f \in C(\partial E)$, then the restriction of f to $\partial_n E_1$ is continuous and real-valued, and hence Theorem 2.3 gives us a function $u_f^{(1)} \in C(E_1 \cup \partial_n E_1)$ which is a temperature on $\overline{E_1} \backslash \partial_n E_1$ and equal to f on $\partial_n E_1$. Thus we cannot prescribe the boundary values of $u_f^{(1)}$ on $\partial E_1 \backslash \partial_n E_1$. Similarly, there is a function $u_f^{(2)} \in C(E_2 \cup \partial_n E_2)$ which is a temperature on $\overline{E_2} \backslash \partial_n E_2$ and equal to f on $\partial_n E_2$. The temperature u_f on E that corresponds to f is given by $u_f = u_f^{(i)}$ on E_i for all $i \in \{1, 2\}$. However, for each point $x \in B$, we have

$$u_f(y, s) \to f(x, b) \quad \text{as} \quad (y, s) \to (x, b+),$$

but in general

$$u_f(y, s) \not\to f(x, b) \quad \text{as} \quad (y, s) \to (x, b-).$$

Thus, we can expect the boundary values to be attained on approach from above, but not on approach from below. By contrast, at all points of $\partial B \times (]a, b[\cup]b, c[)$, the boundary values are attained on any approach through E.

We shall give a classification of boundary points which separates out those points where the boundary values cannot be expected to be recovered, those where the boundary values can be expected to be recovered only through approach from above, and those where the boundary values can be expected to be recovered through any approach. This will be done using simple, local criteria.

We shall use the PWB method to solve the problem, as we did in the case of convex domains of revolution in Chapter 3. However, as well as adding the extra complication of arbitrary open sets, we also generalize the problem by allowing arbitrary boundary functions. For this we must use hypotemperatures rather than just subtemperatures, to ensure that the upper and lower classes are not empty. Of course, if the boundary functions are not continuous, then we cannot expect the boundary values to be taken in a continuous manner. We therefore split the problem into two. Given a boundary function f, we use the PWB method to associate with f a temperature u_f. We then investigate the relation between the boundary behaviour of u_f and the function f, especially at points of continuity of f.

8.1. Classification of Boundary Points

Before classifying the boundary points of our arbitrary open set E, we need to show that the temperatures obtained by the PWB method are unaffected if ∂E is changed by a polar set.

The PWB method on E is akin to that on a convex domain of revolution. Given a function f on some particular part of ∂E, we define an upper class $\mathfrak{U}_f^E$ of lower bounded hypertemperatures u on E such that $\liminf_{p \to q} u(p) \geq f(q)$ in some appropriate sense, for all revelant points q of ∂E. We also define U_f^E to be $\inf \mathfrak{U}_f^E$. Let Z denote a relatively closed polar subset of E, so that the set $D = E \backslash Z$ is also open. Given any point $p_0 \in D$, we choose a heat potential v on E, such that $v(p) = +\infty$ for all $p \in Z$ and $v(p_0) < +\infty$ (see Theorem 7.3). If $u \in \mathfrak{U}_f^E$, and g is any extension of f to $\partial D = Z \cup \partial E$, then for any $\epsilon > 0$ the restriction of $u + \epsilon v$ to D is a member of $\mathfrak{U}_g^D$. Hence, in particular, $U_g^D(p_0) \leq u(p_0) + \epsilon v(p_0)$ for all $\epsilon > 0$, which implies that $U_g^D(p_0) \leq u(p_0)$. Since u is arbitrary, we deduce that $U_g^D(p_0) \leq U_f^E(p_0)$. Conversely, given any function $w \in \mathfrak{U}_g^D$ and $\epsilon > 0$, the function w_ϵ defined by

$$w_\epsilon(p) = \begin{cases} w(p) + \epsilon v(p) & \text{if } p \in D, \\ +\infty & \text{if } p \in Z, \end{cases}$$

is a lower bounded hypertemperature on E such that $\liminf_{p \to q} w_\epsilon(p) \geq f(q)$ in the appropriate sense, for all revelant points q of ∂E. Thus $w_\epsilon \in \mathfrak{U}_f^E$, and hence $U_f^E(p_0) \leq w_\epsilon(p_0)$ for all $\epsilon > 0$. It follows that $U_f^E(p_0) \leq w(p_0)$, and therefore, by the arbitrariness of w, that $U_f^E(p_0) \leq U_g^D(p_0)$. Thus $U_f^E = U_g^D$ on D, so that the deletion of Z from E, or the addition of Z to D, makes no difference. Analogously, a similar change has no effect on the corresponding lower class of upper bounded hypotemperatures. Therefore, in our classification of boundary points, we always suppose that the boundary of our open set E does not contain any polar set whose union with E would give another open set.

We now classify the various types of boundary point of E, using the following notations for the upper and lower half-balls.

Given any point $p_0 = (x_0, t_0) \in \mathbb{R}^{n+1}$ and $r > 0$, we denote by $H(p_0, r)$ the

open lower half-ball
$$\Lambda(p_0; B(p_0, r)) = \{(x,t) : |x - x_0|^2 + (t - t_0)^2 < r^2, t < t_0\},$$
and by $H^*(p_0, r)$ the open upper half-ball
$$\Lambda^*(p_0; B(p_0, r)) = \{(x,t) : |x - x_0|^2 + (t - t_0)^2 < r^2, t > t_0\}.$$

For the remainder of this chapter, the boundary of a set is taken relative to the one-point compactification of $\mathbb{R}^{n+1}$. Thus ∂E contains the point at infinity if and only if E is unbounded.

DEFINITION 8.1. Let q be a boundary point of the open set E, including the point at infinity if E is unbounded. We call q a *normal* boundary point if either

(a) q is the point at infinity, or

(b) $q \in \mathbb{R}^{n+1}$ and every lower half-ball centred at q meets the complement of E; that is, for every $r > 0$, $H(q,r)\backslash E \neq \emptyset$.

Otherwise, we call q an *abnormal* boundary point. In this case, there is a positive number r_0 such that $H(q, r_0) \subseteq E$, and we can define $\Lambda(q; E)$ by putting $\Lambda(q; E) = \Lambda(q; E \cup B(q, r_0))$. The abnormal boundary points are of two kinds, according to whether they can be approached from above by points in E. If there is some $r_1 < r_0$ such that $H^*(q, r_1) \cap E = \emptyset$, then q is called a *singular* boundary point. In this case, $H(q, r_1) = B(q, r_1) \cap E$. On the other hand if, for every $r < r_0$, we have $H^*(q, r) \cap E \neq \emptyset$, then q is called a *semi-singular* boundary point.

The set of all normal boundary points of E is called the *normal boundary* of E, and is denoted by $\partial_n E$. The set of all abnormal boundary points of E is called the *abnormal boundary* of E, and is denoted by $\partial_a E$. The set of all singular points is called the *singular boundary* of E, and is denoted by $\partial_s E$. The set of all semi-singular points is called the *semi-singular boundary* of E, and is denoted by $\partial_{ss} E$. Thus $\partial E = \partial_n E \cup \partial_a E$ and $\partial_a E = \partial_s E \cup \partial_{ss} E$. The *essential boundary* $\partial_e E$ is defined by
$$\partial_e E = \partial_n E \cup \partial_{ss} E = \partial E \backslash \partial_s E.$$

It is convenient to present a version of the boundary point maximum principle in terms of our classification of boundary points.

THEOREM 8.2. *Let w be a hypotemperature on an open set E, and let Z be a polar subset of $\partial_e E$. Suppose that*
$$\limsup_{(x,t)\to(y,s)} w(x,t) \leq A$$
for all $(y, s) \in \partial_n E \backslash Z$, that
$$\limsup_{(x,t)\to(y,s)} w(x,t) < +\infty$$
for all $(y, s) \in \partial_n E \cap Z$, that
$$\limsup_{(x,t)\to(y,s+)} w(x,t) \leq A$$
for all $(y, s) \in \partial_{ss} E \backslash Z$, and that
$$\limsup_{(x,t)\to(y,s+)} w(x,t) < +\infty$$

for all $(y,s) \in \partial_{ss}E \cap Z$. Then $w(x,t) \leq A$ for all $(x,t) \in E$.

PROOF. This follows easily from Theorem 7.9. □

DEFINITION 8.3. Let E be an open set, and let $f \in C(\partial_e E)$. We say that a temperature u on E is a *classical solution of the Dirichlet problem for f* if both

$$\lim_{(x,t) \to (y,s)} u(x,t) = f(y,s) \quad \text{for all} \quad (y,s) \in \partial_n E,$$

and

$$\lim_{(x,t) \to (y,s+)} u(x,t) = f(y,s) \quad \text{for all} \quad (y,s) \in \partial_{ss} E.$$

LEMMA 8.4. *Let E be an open set, let $p_0 \in E$, and put $\Lambda = \Lambda(p_0; E)$. Then $\partial_e \Lambda \subseteq \partial_e E$ and $\partial_{ss} \Lambda \subseteq \partial_{ss} E$. If $q \in \partial_n \Lambda \cap \partial_a E$, there is an open half-ball $H(q, r_1)$ such that $H(q, r_1) \cap \Lambda = \emptyset$.*

PROOF. We first show that $\partial_s E \cap \partial \Lambda \subseteq \partial_s \Lambda$, then that $\partial_e \Lambda \cap E = \emptyset$. For if $\partial_s E \cap \partial \Lambda \subseteq \partial_s \Lambda$, then for any point $q \in \partial_e \Lambda = \partial \Lambda \backslash \partial_s \Lambda$, we have $q \notin \partial_s E$, so that $q \in \partial_e E \cup E$. If also $\partial_e \Lambda \cap E = \emptyset$, then $q \in \partial_e E$, and hence $\partial_e \Lambda \subseteq \partial_e E$.

Let $q \in \partial_s E \cap \partial \Lambda$, so that there is $r > 0$ such that $H(q,r) = B(q,r) \cap E$, and a sequence $\{q_k\}$ in Λ such that $q_k \to q$ as $k \to \infty$. Since $\Lambda \subseteq E$, we can assume that $q_k \in H(q,r)$ for all k. Since $\{q_k\}$ converges to q, we have

$$H(q,r) = \bigcup_{k=1}^{\infty} \Lambda(q_k; B(q,r)) \subseteq \bigcup_{k=1}^{\infty} \Lambda(q_k; E) \subseteq \Lambda,$$

so that

$$H(q,r) \subseteq \Lambda \cap B(q,r) \subseteq E \cap B(q,r) = H(q,r).$$

Hence $\Lambda \cap B(q,r) = H(q,r)$, and so $q \in \partial_s \Lambda$. Thus $\partial_s E \cap \partial \Lambda \subseteq \partial_s \Lambda$.

We now take any point $p \in \partial_e \Lambda \cap E$. We take a ball $B(p, \epsilon) \subseteq E$, and a sequence $\{p_k\}$ in Λ such that $p_k \to p$ as $k \to \infty$. We can assume that $p_k \in B(p, \epsilon)$ for all k. Since $p_k \in \Lambda$ and $B(p, \epsilon) \subseteq E$, we have $\Lambda(p_k; B(p, \epsilon)) \subseteq \Lambda$ for all k. Therefore, because $\{p_k\}$ converges to p, we have

$$H(p, \epsilon) \subseteq \bigcup_{k=1}^{\infty} \Lambda(p_k; B(p, \epsilon)) \subseteq \Lambda,$$

so that $p \in \partial_a \Lambda$ and hence $p \in \partial_{ss} \Lambda$. Therefore $H^*(p, \epsilon) \cap \Lambda \neq \emptyset$. If q_ϵ belongs to $H^*(p, \epsilon) \cap \Lambda$, then $p \in \Lambda(q_\epsilon; B(p, \epsilon)) \subseteq \Lambda(q_\epsilon; E)$ and $\Lambda(q_\epsilon; E) \subseteq \Lambda$. Hence $p \in \Lambda$, which contradicts our assumption that $p \in \partial \Lambda$, because Λ is open. Therefore $\partial_e \Lambda \cap E = \emptyset$, and it follows that $\partial_e \Lambda \subseteq \partial_e E$.

For the second part, if $q \in \partial_{ss} \Lambda$ then $q \in \partial_e \Lambda \subseteq \partial_e E = \partial_n E \cup \partial_{ss} E$ and $q \in \partial_a \Lambda \subseteq E \cup \partial_a E$. Hence $q \in \partial_{ss} E$.

For the last part, since $q \in \partial_a E$, there is an open half-ball $H(q, r_0) \subseteq E$. If there was a sequence $\{q_k\}$ in $H(q, r_0) \cap \Lambda$ such that $q_k \to q$ as $k \to \infty$, we would have

$$\Lambda(q_k; H(q, r_0)) \subseteq \Lambda(q_k; E) \subseteq \Lambda(p_0; E) = \Lambda$$

for all k, which implies that

$$H(q, r_0) = \bigcup_{k=1}^{\infty} \Lambda(q_k; H(q, r_0)) \subseteq \Lambda,$$

contrary to the hypothesis that $q \in \partial_n \Lambda$. Hence there is no such sequence, so that there is a positive number $r_1 < r_0$ such that $H(q, r_1) \cap \Lambda = \emptyset$. □

REMARK 8.5. In general, it is not true that $\partial_n \Lambda \subseteq \partial_n E$. For example, if we take $E = \mathbb{R}^{n+1} \backslash (\mathbb{R}^n \times \{0\})$ and $p_0 = (0, 1)$, then $\Lambda = \Lambda(p_0, E) = \mathbb{R}^n \times {]0, 1[}$ and $\mathbb{R}^n \times \{0\} \subseteq \partial_n \Lambda \cap \partial_{ss} E$.

LEMMA 8.6. *Let E be an open set, let $p_0 \in E$, and put $\Lambda = \Lambda(p_0; E)$. Then for any point $q \in \Lambda \cup \partial_a \Lambda$, we have $\Lambda(q; \Lambda) = \Lambda(q; E)$.*

PROOF. If $q \in \Lambda \cup \partial_a \Lambda$, then there is $r > 0$ such that $H(q, r) \subseteq \Lambda \subseteq E$. Hence $q \in E \cup \partial_a E$, so that $\Lambda(q; E)$ is defined. Since $\Lambda \subseteq E$, it follows that $\Lambda(q; \Lambda) \subseteq \Lambda(q; E)$.

To prove the opposite inclusion, we take any point $p \in \Lambda(q; E)$, and denote by γ a polygonal path in E joining q to p, along which the temporal variable is strictly decreasing. If the open lower half-ball $H(q, r) \subseteq \Lambda$, then $\gamma \cap H(q, s) \neq \emptyset$ whenever $0 < s \leq r$. For each such value of s, we choose a point $q_s \in \gamma \cap H(q, s)$. Then $q_s \in \Lambda$, so that all points on γ between q_s and p belong to Λ. This holds whenever $0 < s \leq r$, and so $\gamma \backslash \{q\} \subseteq \Lambda$. Thus $p \in \Lambda(q; \Lambda)$, so that $\Lambda(q; E) \subseteq \Lambda(q; \Lambda)$. □

8.2. Upper and Lower PWB Solutions

DEFINITION 8.7. An extended real-valued function w on an open set E is called a *hypertemperature* on E if $-w$ is a hypotemperature on E.

Note that if w is a hypertemperature on E, and $w(p) < +\infty$ for some point $p \in E$, then w is a supertemperature on $\Lambda(p; E)$, by Corollary 3.55.

DEFINITION 8.8. Let E be an open set, and let f be an extended real-valued function defined on $\partial_e E$. The *upper class* determined by f, denoted by $\mathfrak{U}_f^E$, consists of all lower bounded hypertemperatures w on E that satisfy

$$\liminf_{(x,t) \to (y,s)} w(x,t) \geq f(y,s) \quad \text{for all} \quad (y,s) \in \partial_n E,$$

and

$$\liminf_{(x,t) \to (y,s+)} w(x,t) \geq f(y,s) \quad \text{for all} \quad (y,s) \in \partial_{ss} E.$$

The *lower class* determined by f, denoted by $\mathfrak{L}_f^E$, consists of all upper bounded hypotemperatures w on E that satisfy

$$\limsup_{(x,t) \to (y,s)} w(x,t) \leq f(y,s) \quad \text{for all} \quad (y,s) \in \partial_n E,$$

and

$$\limsup_{(x,t) \to (y,s+)} w(x,t) \leq f(y,s) \quad \text{for all} \quad (y,s) \in \partial_{ss} E.$$

Note that $\mathfrak{U}_f^E$ contains the function which is identically $+\infty$ on E, and that $\mathfrak{L}_f^E$ contains that which is identically $-\infty$.

DEFINITION 8.9. The function $U_f^E = \inf\{w : w \in \mathfrak{U}_f^E\}$ is called the *upper solution* for f on E, and $L_f^E = \sup\{w : w \in \mathfrak{L}_f^E\}$ is called the *lower solution* for f on E.

If the temporal variable truly represents time, then we would expect the values of $f(y,s)$ for $s \geq a$ to have no effect on the values of the upper solution $U_f^E(x,t)$ and lower solution $L_f^E(x,t)$ for $t < a$. We now show that this is indeed the case.

LEMMA 8.10. *Let E be an open set, let $a \in \mathbb{R}$, and put $D = E \cap (\mathbb{R}^n \times\,]-\infty, a[)$. If f is an extended real-valued function defined on $\partial_e E$, then it is also defined on $\partial_e D$. Moreover, $U_f^E = U_f^D$ and $L_f^E = L_f^D$ on D.*

PROOF. Any points of ∂D that are not points of ∂E belong to E, and are therefore singular points of ∂D. Hence $\partial_n D \subseteq \partial_n E$, $\partial_{ss} D \subseteq \partial_{ss} E$, and f is defined on $\partial_e D$.

The result for the lower classes and lower solutions is the dual of that for the upper classes and upper solutions, so we give details only for the latter.

We define a function u on $\mathbb{R}^{n+1}$ by putting

$$u(x,t) = \begin{cases} \left(\frac{a}{a-t}\right)^{\frac{n}{2}} \exp\left(\frac{|x|^2}{4(a-t)}\right) & \text{if } t < a, \\ +\infty & \text{if } t \geq a. \end{cases}$$

Then u is a nonnegative, continuous hypertemperature on $\mathbb{R}^{n+1}$ (and a temperature on $\mathbb{R}^n \times\,]-\infty, a[$). Given any hypertemperature $w \in \mathfrak{U}_f^D$ and any number $\epsilon > 0$, we put

$$w_\epsilon(p) = \begin{cases} w(p) + \epsilon u(p) & \text{if } p \in D, \\ +\infty & \text{if } p \in E \setminus D. \end{cases}$$

Then $w_\epsilon \in \mathfrak{U}_f^E$, and so $U_f^E \leq w_\epsilon$ on E for all $\epsilon > 0$. Making $\epsilon \to 0$, we obtain $U_f^E \leq w$ on D, so that $U_f^E \leq U_f^D$ on D.

On the other hand, given any $v \in \mathfrak{U}_f^E$, the restriction of v to D belongs to $\mathfrak{U}_f^D$, so that $U_f^D \leq v$ on D. Hence $U_f^D \leq U_f^E$ on D, and so equality holds. □

Our next result is in a similar vein, but is more subtle. Let $p_0 \in E$, and put $\Lambda = \Lambda(p_0; E)$. Then $\partial_e \Lambda \subseteq \partial_e E$ by Lemma 8.4. Therefore, if f is an extended real-valued function on $\partial_e E$, the classes $\mathfrak{U}_f^\Lambda$ and $\mathfrak{L}_f^\Lambda$ are defined. Our next lemma shows that these classes are related in a natural and convenient way.

LEMMA 8.11. *Let E be an open set, let $p_0 \in E$, and put $\Lambda = \Lambda(p_0; E)$. If f is an extended real-valued function on $\partial_e E$, then $\mathfrak{U}_f^\Lambda$ is precisely the class of restrictions to Λ of the members of $\mathfrak{U}_f^E$, and $\mathfrak{L}_f^\Lambda$ is that of the restrictions to Λ of the members of $\mathfrak{L}_f^E$. Hence U_f^Λ is the restriction to Λ of U_f^E, and L_f^Λ is that of L_f^E.*

PROOF. The result for the lower classes and lower solutions is the dual of that for the upper classes and upper solutions, so we give details only for the latter.

Given any hypertemperature $w \in \mathfrak{U}_f^\Lambda$, we define a function $\bar{w}$ on E by putting

$$\bar{w}(p) = \begin{cases} w(p) & \text{if } p \in \Lambda, \\ +\infty & \text{if } p \in E \setminus \overline{\Lambda}, \\ \liminf_{q \to p,\, q \in \Lambda} w(q) & \text{if } p \in \partial \Lambda \cap E. \end{cases}$$

We claim that $\bar{w} \in \mathfrak{U}_f^E$. We show that $\bar{w}$ is a hypertemperature on E using Theorem 3.51. Clearly $\bar{w}$ is lower semicontinuous on E, and because w is lower bounded on

Λ, $\bar{w}$ is lower bounded on E. It remains to show that, given any point $p \in E$ and any $\epsilon > 0$, we can find a positive number $c < \epsilon$ such that the inequality $\bar{w}(p) \geq \mathcal{V}(\bar{w}; p; c)$ holds. Clearly this holds if $p \in E \backslash \partial \Lambda$, so suppose that $p \in E \cap \partial \Lambda$. Since $\partial_e \Lambda \subseteq \partial_e E$ by Lemma 8.4, $p \in \partial_s \Lambda$. Therefore we can find $r_0 > 0$ such that $H(p, 2r_0) = B(p, 2r_0) \cap \Lambda$. We now choose $c_0 > 0$ such that $\overline{\Omega}(q; c) \subseteq \Lambda$ whenever $q \in H(p, r_0)$ and $c \leq c_0$. Then, for any $c \leq c_0$, we have

$$\bar{w}(p) = \liminf_{q \to p, q \in \Lambda} w(q) \geq \liminf_{q \to p, q \in \Lambda} \mathcal{V}(w; q; c) = \liminf_{q \to p, q \in \Lambda} \mathcal{V}(\bar{w}; q; c) \geq \mathcal{V}(\bar{w}; p; c),$$

by Fatou's lemma. Hence $\bar{w}$ is a hypertemperature on E.

We now take any point $(y, s) \in \partial_e E$. If $(y, s) \notin \partial \Lambda$, then

$$\liminf_{(x,t) \to (y,s)} \bar{w}(x, t) = +\infty \geq f(y, s).$$

If $(y, s) \in \partial_s \Lambda$ then $(y, s) \in \partial_{ss} E$ and

$$\liminf_{(x,t) \to (y,s+)} \bar{w}(x, t) = +\infty \geq f(y, s).$$

If $(y, s) \in \partial_{ss} \Lambda$, then $(y, s) \in \partial_{ss} E$ by Lemma 8.4, and

$$\liminf_{(x,t) \to (y,s+)} \bar{w}(x, t) = \liminf_{(x,t) \to (y,s+)} w(x, t) \geq f(y, s).$$

If $(y, s) \in \partial_n \Lambda$, then

$$\liminf_{(x,t) \to (y,s)} \bar{w}(x, t) = \liminf_{(x,t) \to (y,s)} w(x, t) \geq f(y, s).$$

Hence $\bar{w} \in \mathfrak{U}_f^E$, and so w is the restriction to Λ of a function in $\mathfrak{U}_f^E$.

We now show that, given any hypertemperature $v \in \mathfrak{U}_f^E$, its restriction to Λ belongs to $\mathfrak{U}_f^\Lambda$. Obviously v is a lower bounded hypertemperature on Λ. Let $q = (y, s) \in \partial_e \Lambda$. Then either $q \in \partial_{ss} \Lambda \subseteq \partial_{ss} E$, or $q \in \partial_n \Lambda \subseteq \partial_e E$, by Lemma 8.4. In the former case, we have

$$\liminf_{(x,t) \to (y,s+), (x,t) \in \Lambda} v(x, t) \geq \liminf_{(x,t) \to (y,s+), (x,t) \in E} v(x, t) \geq f(y, s).$$

If $q \in \partial_n \Lambda \cap \partial_n E$, then

$$\liminf_{(x,t) \to (y,s), (x,t) \in \Lambda} v(x, t) \geq \liminf_{(x,t) \to (y,s), (x,t) \in E} v(x, t) \geq f(y, s).$$

Finally, if $q \in \partial_n \Lambda \cap \partial_{ss} E$, then there is $\delta > 0$ such that $H(q, \delta) \subseteq E$, but for all $r > 0$ we have $H(q, r) \backslash \Lambda \neq \emptyset$. If there is a sequence $\{q_k\}$ of points in $H(q, \delta) \cap \Lambda$ such that $q_k \to q$ as $k \to \infty$, then

$$\Lambda \supseteq \bigcup_{k=1}^{\infty} \Lambda(q_k; H(q, \delta)) = H(q, \delta),$$

a contradiction. There is therefore no such sequence, and hence there is a half-ball $H(q, \eta)$ contained in $\mathbb{R}^{n+1} \backslash \Lambda$. It follows that

$$\liminf_{(x,t) \to (y,s), (x,t) \in \Lambda} v(x, t) \geq \liminf_{(x,t) \to (y,s+), (x,t) \in E} v(x, t) \geq f(y, s).$$

Hence the restriction of v to Λ belongs to $\mathfrak{U}_f^\Lambda$. □

At the beginning of Section 8.1, we observed that the PWB method ignores relatively closed polar subsets of ∂E. In a similar way, it also ignores changes to the boundary function on a polar set.

LEMMA 8.12. *Let E be an open set, and suppose that f and g are extended real-valued functions on $\partial_e E$ such that $f = g$ except on a polar set. Then $U_f^E = U_g^E$ and $L_f^E = L_g^E$ on E.*

PROOF. Let p_0 be an arbitrary point of E, and let Z denote the polar subset of $\partial_e E$ where $f \neq g$. By Theorem 7.3, there is a heat potential v on $\mathbb{R}^{n+1}$ such that $v(p) = +\infty$ for all $p \in Z$, but $v(p_0) < +\infty$. If $u \in \mathfrak{U}_f^E$ and $\epsilon > 0$, then $u + \epsilon v \in \mathfrak{U}_g^E$. Therefore $U_g^E(p_0) \leq u(p_0) + \epsilon v(p_0)$ for every $\epsilon > 0$, and hence $U_g^E(p_0) \leq u(p_0)$. Since u is an arbitrary function from $\mathfrak{U}_f^E$, it follows that $U_g^E(p_0) \leq U_f^E(p_0)$. Interchanging f and g we obtain the reverse inequality, and hence equality. Thus $U_f^E = U_g^E$ on E. The proof that $L_f^E = L_g^E$ is similar. □

LEMMA 8.13. *Let E be an open set, and let f be an extended real-valued function defined on the essential boundary $\partial_e E$. If $u \in \mathfrak{L}_f^E$ and $v \in \mathfrak{U}_f^E$, then $u \leq v$ on E. Hence $L_f^E \leq U_f^E$.*

PROOF. Since u is a hypotemperature and v is a hypertemperature on E, $v - u$ is a hypertemperature, by Corollary 3.55 in open subsets where both are finite almost everywhere. Furthermore, if $(y, s) \in \partial_n E$ (and is possibly the point at infinity) and $f(y, s)$ is finite, then as $(x, t) \to (y, s)$ we have

$$\liminf(v - u)(x, t) \geq \liminf v(x, t) - \limsup u(x, t) \geq f(y, s) - f(y, s) = 0.$$

On the other hand, if $f(y, s) = +\infty$ then $\lim v(x, t) = +\infty$ and $\limsup u(x, t) < +\infty$ because u is upper bounded, so that $\liminf(v - u)(x, t) \geq 0$; and if $f(y, s) = -\infty$ then $\lim u(x, t) = -\infty$ and $\liminf v(x, t) > -\infty$, so that $\liminf(v - u)(x, t) \geq 0$. Moreover, if instead $(y, s) \in \partial_{ss}\Lambda$, then similarly $\liminf(v - u)(x, t) \geq 0$ if the limit is taken as $(x, t) \to (y, s+)$. Therefore, by Theorem 3.13, $v \geq u$ on E. It follows that $L_f^E \leq U_f^E$. □

LEMMA 8.14. *Let E be an open set, let f and g be extended real-valued functions on $\partial_e E$, and let $\alpha \in \mathbb{R}$.*
  (a) *Without further conditions, $U_{-f}^E = -L_f^E$.*
  (b) *If $\alpha > 0$, then $U_{\alpha f}^E = \alpha U_f^E$ and $L_{\alpha f}^E = \alpha L_f^E$.*
  (c) *If $f \leq g$, then $U_f^E \leq U_g^E$ and $L_f^E \leq L_g^E$.*
  (d) *Let $(f + g)(q)$ be defined arbitrarily at each point $q \in \partial_e E$ where $f(q) + g(q)$ is undefined. Then for each point $p \in E$,*

$$U_{f+g}^E(p) \leq U_f^E(p) + U_g^E(p)$$

provided that the sum on the right-hand side is defined, and

$$L_{f+g}^E(p) \geq L_f^E(p) + L_g^E(p)$$

with the same proviso.

PROOF. (a) Since $w \in \mathfrak{U}_{-f}^E$ if and only if $-w \in \mathfrak{L}_f^E$, we have

$$U_{-f}^E = \inf\{w : -w \in \mathfrak{L}_f^E\} = -\sup\{v : v \in \mathfrak{L}_f^E\} = -L_f^E.$$

(b) If $\alpha > 0$, then $w \in \mathfrak{U}_f^E$ if and only if $\alpha w \in \mathfrak{U}_{\alpha f}^E$. Therefore

$$U_{\alpha f}^E = \inf\{\alpha w : w \in \mathfrak{U}_f^E\} = \alpha U_f^E.$$

Similarly $L_{\alpha f}^E = \alpha L_f^E$.

(c) If $f \leq g$, then $\mathfrak{U}_g^E \subseteq \mathfrak{U}_f^E$ and $\mathfrak{L}_f^E \subseteq \mathfrak{L}_g^E$, so that $U_f^E \leq U_g^E$ and $L_f^E \leq L_g^E$.

(d) Let $v \in \mathfrak{U}_f^E$ and $w \in \mathfrak{U}_g^E$. Then $v + w$ is a lower bounded hypertemperature on E, and at all points $q \in \partial_e E$ where $f(q) + g(q)$ is well-defined, we have

$$\liminf_{p \to q}(v+w)(p) \geq \liminf_{p \to q} v(p) + \liminf_{p \to q} w(p) \geq f(q) + g(q),$$

where the limits are taken in the appropriate sense according to whether $q \in \partial_n E$ or $q \in \partial_{ss} E$. If q is a point in $\partial_e E$ where $f(q) + g(q)$ is undefined, then without loss of generality we take $f(q) = +\infty$ and $g(q) = -\infty$. Then $\lim_{p \to q} v(p) = +\infty$ (in the appropriate sense), and so $\lim_{p \to q}(v+w)(p) = +\infty$ because w is lower bounded. Thus, regardless of the value we assign to $(f+g)(q)$, we have

$$\lim_{p \to q}(v+w)(p) \geq (f+g)(q).$$

Hence $v + w \in \mathfrak{U}_{f+g}^E$, and so $v + w \geq U_{f+g}^E$. We now take any point $p \in E$. Clearly $U_{f+g}^E(p) \leq U_f^E(p) + U_g^E(p)$ if the sum on the right-hand side is defined and either term is $+\infty$. Since $U_f^E(p) = +\infty$ if and only if $v(p) = +\infty$ for all $v \in \mathfrak{U}_f^E$, it only remains to consider the case where $v(p) < +\infty$ and $w(p) < +\infty$ for some $v \in \mathfrak{U}_f^E$ and $w \in \mathfrak{U}_g^E$. In this case $U_f^E(p) + U_g^E(p)$ is defined, and since $U_{f+g}^E(p) \leq v(p) + w(p)$ we have $U_{f+g}^E(p) \leq U_f^E(p) + w(p)$, and hence the first result.

The proof for the lower solutions now follows easily from (a). □

LEMMA 8.15. *Let E be an open set, and let f be an extended real-valued function defined on the essential boundary $\partial_e E$. If there are points $p_0, q_0 \in E$ such that $q_0 \in \Lambda(p_0; E)$, $U_f^E(p_0) < +\infty$, and $U_f^E(q_0) > -\infty$, then U_f^E is a temperature on $\Lambda(q_0; E)$.*

PROOF. We put $\Lambda = \Lambda(p_0; E)$, and note that, by Lemma 8.11, we need to show that U_f^Λ is a temperature on $\Lambda(q_0; \Lambda) = \Lambda(q_0; E)$.

Since $U_f^E(p_0) < +\infty$, we can find a hypertemperature $w_0 \in \mathfrak{U}_f^E$ such that $w_0(p_0) < +\infty$. By Corollary 3.55, w_0 is a supertemperature on Λ. By Lemma 8.11, the restriction of w_0 to Λ belongs to $\mathfrak{U}_f^\Lambda$, and so we can write $U_f^\Lambda = \inf \mathcal{F}$, where $\mathcal{F}$ is the class of all supertemperatures that belong to $\mathfrak{U}_f^\Lambda$.

We show that $\mathcal{F}$ is a saturated family of supertemperatures on Λ, in order to apply Theorem 3.26. Let $u, v \in \mathcal{F}$. Then $u \wedge v$ is a lower bounded supertemperature on Λ. Moreover, whenever $(y, s) \in \partial_n \Lambda$ and $(x, t) \to (y, s)$ with $(x, t) \in \Lambda$, we have

$$\liminf(u \wedge v)(x, t) = \left(\liminf u(x, t)\right) \wedge \left(\liminf v(x, t)\right) \geq f(y, s);$$

and similarly, whenever $(y, s) \in \partial_{ss} \Lambda$ and $(x, t) \to (y, s+)$. Hence $u \wedge v \in \mathcal{F}$. We now take any function $w \in \mathcal{F}$, and any circular cylinder D such that $\overline{D} \subseteq \Lambda$. We denote by $\pi_D w$ the function defined in Theorem 3.21 (relative to Λ). Then $\pi_D w$ is a supertemperature on Λ, and is lower bounded by the same lower bound as w. Furthermore, since the compact set $\overline{D} \subseteq \Lambda$ and $\pi_D w = w$ on $\Lambda \backslash \overline{D}$, the boundary behaviour of $\pi_D w$ is the same as that of w. Therefore $\pi_D w \in \mathcal{F}$, and so $\mathcal{F}$ is a saturated family of supertemperatures on Λ.

Since $q_0 \in \Lambda$ and $U_f^\Lambda(q_0) > -\infty$, it follows from Theorem 3.26 that U_f^Λ is a temperature on $\Lambda(q_0; \Lambda) = \Lambda(q_0; E)$, as required. □

COROLLARY 8.16. *Let E be an open set, and let f be an extended real-valued function defined on the essential boundary $\partial_e E$. If there is a point $p_0 \in E$ such that $L_f^E(p_0)$ and $U_f^E(p_0)$ are both finite, then L_f^E and U_f^E are temperatures on $\Lambda(p_0; E)$.*

PROOF. Since $L_f^E(p_0) > -\infty$, we can find a hypotemperature $u \in \mathfrak{L}_f^E$ such that $u(p_0) > -\infty$. By Corollary 3.55, u is a subtemperature on $\Lambda(p_0; E)$, and in particular is finite on a dense subset F of $\Lambda(p_0; E)$. Therefore

$$-\infty < u(q) \leq L_f^E(q) \leq U_f^E(q)$$

for all $q \in F$. Since $U_f^E(p_0) < +\infty$, it follows from Lemma 8.15 that U_f^E is a temperature on the set

$$\bigcup_{q \in F} \Lambda(q; E) = \Lambda(p_0; E).$$

Applying this result to $-f$, and using Lemma 8.14(a), we obtain the result for L_f^E. □

LEMMA 8.17. *Suppose that f is the limit of an increasing sequence $\{f_j\}$ of extended real-valued functions on the essential boundary of the open set E, and that $U_{f_m}^E > -\infty$ on E for some m. If $p_0 \in E$ and satisfies $U_{f_j}^E(p_0) < +\infty$ for all j, then*

$$U_f^E = \lim_{j \to \infty} U_{f_j}^E$$

on $\Lambda(p_0; E)$.

PROOF. By Lemma 8.14(c), the sequence $\{U_{f_j}^E\}$ is increasing on E, and we have $U_{f_j}^E \leq U_f^E$ on E for all j. Therefore $\lim_{j \to \infty} U_{f_j}^E \leq U_f^E$ on E, and we may suppose that $U_{f_j}^E > -\infty$ on E for all j.

Suppose that $p_0 \in E$ and $U_{f_j}(p_0) < +\infty$ for all j. For each j, Lemma 8.15 and our supposition that $U_{f_j}^E > -\infty$ on E, imply that $U_{f_j}^E$ is a temperature on $\Lambda(p; E)$ for all $p \in \Lambda(p_0; E)$, and thus on $\Lambda(p_0; E)$ itself. We put $\Lambda = \Lambda(p_0; E)$, and note that by Lemma 8.11, $U_{f_j}^E = U_{f_j}^\Lambda$ on Λ. We now take any point $p_1 \in \Lambda$ and any positive number ϵ. For each j, we can find a hypertemperature $w_j \in \mathfrak{U}_{f_j}^\Lambda$ such that

$$w_j(p_1) - U_{f_j}^\Lambda(p_1) < 2^{-j}\epsilon.$$

Since each function $U_{f_j}^\Lambda$ is a temperature on Λ, $\lim_{j \to \infty} U_{f_j}^\Lambda$ is a hypertemperature on Λ, by Theorem 3.60. Moreover, since each function $w_j - U_{f_j}^\Lambda$ is a nonnegative hypertemperature on Λ, the same is true of $\sum_{j=1}^\infty (w_j - U_{f_j}^\Lambda)$, and hence of the function

$$v = \lim_{j \to \infty} U_{f_j}^\Lambda + \sum_{j=1}^\infty (w_j - U_{f_j}^\Lambda).$$

For each k, we have

$$v \geq U_{f_k}^\Lambda + (w_k - U_{f_k}^\Lambda) = w_k,$$

so that v is lower bounded on Λ and

$$\liminf_{p \to q} v(p) \geq f_k(q)$$

for all $q \in \partial_e \Lambda$, where the limits are taken in the appropriate sense according to whether $q \in \partial_n \Lambda$ or $q \in \partial_{ss} \Lambda$. It follows that (in the appropriate sense)

$$\liminf_{p \to q} v(p) \geq f(q)$$

for all $q \in \partial_e \Lambda$, so that $v \in \mathfrak{U}_f^\Lambda$ and hence $v \geq U_f^\Lambda$. In particular,

$$U_f^\Lambda(p_1) \leq v(p_1) \leq \lim_{j\to\infty} U_{f_j}^\Lambda(p_1) + \sum_{j=1}^\infty 2^{-j}\epsilon = \lim_{j\to\infty} U_{f_j}^\Lambda(p_1) + \epsilon.$$

This holds for all $\epsilon > 0$, so that

$$U_f^\Lambda(p_1) \leq \lim_{j\to\infty} U_{f_j}^\Lambda(p_1) \leq U_f^\Lambda(p_1).$$

Since p_1 is an arbitrary point of Λ, the result is established (in view of Lemma 8.11). □

8.3. Resolutivity and PWB Solutions

DEFINITION 8.18. We say that an extended real-valued function f on $\partial_e E$ is *resolutive* for E if $L_f^E = U_f^E$ and is a temperature on E. In this case, we define $S_f^E = L_f^E = U_f^E$ to be the *PWB Solution* for f on E.

LEMMA 8.19. *Let E be an open set, and let f be an extended real-valued function on $\partial_e E$. If, for each point $q \in E$, we can find a point $p \in \Lambda^*(q; E)$ such that $L_f^E(p)$ and $U_f^E(p)$ are equal and finite, then f is resolutive for E.*

PROOF. Because $L_f^E(p)$ and $U_f^E(p)$ are both finite, Corollary 8.16 shows that the functions L_f^E and U_f^E are temperatures on the neighbourhood $\Lambda(p; E)$ of the arbitrary point $q \in E$, and hence on the whole of E. Therefore the function $v = L_f^E - U_f^E$ is a nonpositive temperature on E, in view of Lemma 8.13. Since $v(p) = 0$, it follows from the strong maximum principle that $v = 0$ on $\Lambda(p; E)$, and hence $v(q) = 0$. Thus $v = 0$ on E, and so f is resolutive for E. □

It is an important fact that, if there is a classical solution for f, then the PWB solution for f exists and coincides with the classical solution.

THEOREM 8.20. *Let E be an open set, and let $f \in C(\partial_e E)$. If there is a classical solution h of the Dirichlet problem for f on E, then f is resolutive and $S_f = h$ on E.*

PROOF. Since $f \in C(\partial_e E)$ it is bounded, and therefore h is bounded, in view of the boundary point maximum principle of Theorem 8.2. Therefore, because of its boundary limits, h belongs to both $\mathfrak{U}_f^E$ and $\mathfrak{L}_f^E$. Hence $h \geq U_f^E$ and $h \leq L_f^E$, and so it follows from Lemma 8.13 that $h = U_f^E = L_f^E$. Since h is a temperature on E, this implies that f is resolutive and $S_f^E = h$ on E. □

It follows easily from Theorem 8.20 that, if $f(q) = \alpha \in \mathbb{R}$ for all $q \in \partial_e E$, then f is resolutive and $S_f = \alpha$ on E. Furthermore, in view of Lemma 8.14(c) and Lemma 8.13, if $g : \partial_e E \to [\alpha, \beta]$, then $\alpha \leq L_g^E \leq U_g^E \leq \beta$ on E.

EXAMPLE 8.21. Let E be an open set (whose boundary does not contain any polar set whose union with E would give another open set), let Z be a countable dense subset of $\partial_e E$, and let f be the characteristic function of Z on $\partial_e E$. Then f is nowhere continuous, but f is resolutive and $S_f = 0$ on E. This is because Z is polar and $f = 0$ except on Z, so that Lemma 8.12 shows that $U_f^E = U_0^E$ and

$L_f^E = L_0^E$ on E. Since Theorem 8.20 shows that $U_0^E = L_0^E = 0$, f is resolutive and $S_f^E = 0$ on E.

THEOREM 8.22. *Let E be an open set, let f and g be extended real-valued functions on $\partial_e E$, and let $\alpha \in \mathbb{R}$.*

(a) If f is resolutive, then αf is resolutive and $S_{\alpha f}^E = \alpha S_f^E$ on E.

(b) If f and g are both resolutive, and $(f+g)(q)$ is defined arbitrarily at each point $q \in \partial_e E$ where $f(q) + g(q)$ is undefined, then $f + g$ is resolutive and

$$S_{f+g}^E = S_f^E + S_g^E$$

on E.

PROOF. (a) If $\alpha = 0$, the result is trivial. If $\alpha > 0$, then Lemma 8.14(b) shows that $L_{\alpha f}^E = \alpha L_f^E = \alpha S_f^E = \alpha U_f^E = U_{\alpha f}^E$, because f is resolutive. The result follows. If $\alpha < 0$, then $-\alpha f$ is resolutive and, by Lemma 8.14(a),

$$L_{\alpha f}^E = -U_{-\alpha f}^E = -S_{-\alpha f}^E = -L_{-\alpha f}^E = U_{\alpha f}^E.$$

The result follows.

(b) If f and g are both resolutive, then Lemma 8.13 and Lemma 8.14(d) show that

$$S_f^E + S_g^E = L_f^E + L_g^E \leq L_{f+g}^E \leq U_{f+g}^E \leq U_f^E + U_g^E = S_f^E + S_g^E,$$

which implies the result. □

REMARK 8.23. Theorem 8.22(b) implies that, if f and g are both resolutive, then the set of points $q \in \partial_e E$ where $f(q) + g(q)$ is undefined cannot be too large.

THEOREM 8.24. *Let E be an open set, and let $\{f_j\}$ be a sequence of real-valued, resolutive functions on $\partial_e E$. If $\{f_j\}$ converges uniformly on $\partial_e E$ to a function f, then f is resolutive and $S_{f_j}^E \to S_f^E$ uniformly on E.*

PROOF. Given any $\epsilon > 0$, we choose a number k such that $|f_j - f| < \epsilon$ on $\partial_e E$ for all $j > k$. For such j, if $w \in \mathfrak{U}_{f_j}^E$ then $w + \epsilon \in \mathfrak{U}_f^E$. Therefore $U_f^E \leq w + \epsilon$, and it follows that $U_f^E \leq U_{f_j}^E + \epsilon$. Similarly $L_f^E \geq L_{f_j}^E - \epsilon$ for all $j > k$. It now follows from Lemma 8.13 and the resolutivity of the functions f_j that

$$S_{f_j}^E - \epsilon = L_{f_j}^E - \epsilon \leq L_f^E \leq U_f^E \leq U_{f_j}^E + \epsilon = S_{f_j}^E + \epsilon.$$

These inequalities show that $|S_{f_j}^E - L_f^E| < \epsilon$ and $|S_{f_j}^E - U_f^E| < \epsilon$ for all $j > k$, so that the sequence $\{S_{f_j}^E\}$ converges uniformly on E to both L_f^E and U_f^E. Therefore $L_f^E = U_f^E \in C(E)$, so that S_f^E exists and is a temperature on E, by Lemma 8.15. □

In our next theorem, we show that any function $f \in C(\partial_e E)$ is resolutive. The following lemma is used in the proof.

LEMMA 8.25. *Let E be an open set, let K be a compact subset of E, and let w be a function on $E \cup \partial_e E$ that is both a subtemperature on E and an element of $C((E \cup \partial_e E)\backslash K)$. Then the restriction of w to $\partial_e E$ is resolutive for E.*

PROOF. We denote by f the restriction of w to $\partial_e E$. Since $f \in C(\partial_e E)$ it is bounded, and so we can find real numbers α and β such that $\alpha \leq f \leq \beta$ on $\partial_e E$. Therefore $L_\alpha^E \leq L_f^E \leq L_\beta^E$ by Lemma 8.14(c), and so it follows from Theorem 8.20 that $\alpha \leq L_f^E \leq \beta$ on E. Hence L_f^E is a temperature on E, by Lemma 8.15.

For every point $q \in \partial_e E$, we have $\lim_{p \to q} w(p) = f(q) \leq \beta$, and so Theorem 8.2 shows that w is upper bounded. It follows that $w \in \mathfrak{L}_f^E$, and hence $w \leq L_f^E$ on E. Therefore
$$\liminf_{p \to q} L_f^E(p) \geq \lim_{p \to q} w(p) = f(q)$$
for all $q \in \partial_e E$, so that $L_f^E \in \mathfrak{U}_f^E$, and hence $L_f^E \geq U_f^E$ on E. Since $L_f^E \leq U_f^E$ by Lemma 8.13, equality holds and, because L_f^E is a temperature, f is resolutive. $\square$

THEOREM 8.26. *If E is any open set, and $f \in C(\partial_e E)$, then f is resolutive for E.*

PROOF. Let $\mathcal{G}$ denote the class of real-valued functions on $E \cup \partial_e E$ that are both supertemperatures on E and continuous on $(E \cup \partial_e E) \backslash K$ for some compact subset K of E. Let $\mathcal{D}$ denote the class of differences $u - v$ of functions in $\mathcal{G}$, and let $\mathcal{F}$ denote the class of restrictions to $\partial_e E$ of the functions in $\mathcal{D}$. Then $\mathcal{F}$ is a linear subspace of $C(\partial_e E)$ that contains the constant functions. By Lemma 8.25, the restrictions to $\partial_e E$ of the functions in $\mathcal{G}$ are resolutive, and so Theorem 8.22 shows that the functions in $\mathcal{F}$ are all resolutive. Furthermore, for any point $q_0 \notin \partial_e E$, the class $\mathcal{D}$ contains the function $G(\cdot; q_0) \wedge \alpha$ for every positive number α, and so $\mathcal{F}$ separates points of $\partial_e E$. Finally, if $u, v \in \mathcal{G}$ then Corollaries 3.18 and 3.19 imply that $u \wedge v, u + v \in \mathcal{G}$, so that if $u_1, u_2, v_1, v_2 \in \mathcal{G}$ the function
$$(u_1 - v_1) \vee (u_2 - v_2) = u_1 + u_2 - (u_2 + v_1) \wedge (u_1 + v_2) \in \mathcal{D}.$$
Thus $f \vee g \in \mathcal{F}$ whenever $f, g \in \mathcal{F}$. It now follows from the Stone-Weierstrass theorem for the one-point compactification of $\mathbb{R}^{n+1}$ that $\mathcal{F}$ is dense in $C(\partial_e E)$ with respect to the supremum norm. So every function in $C(\partial_e E)$ can be expressed as the uniform limit of a sequence in $\mathcal{F}$. Since every function in $\mathcal{F}$ is resolutive, it follows from Theorem 8.24 that every every function in $C(\partial_e E)$ is resolutive. $\square$

8.4. The Caloric Measure on the Essential Boundary

The caloric measure on the essential boundary of an arbitrary open set, arises in a similar way to that on the normal boundary of a circular cylinder (in Chapter 2). However, the proof is more complicated because the boundary behaviour of the PWB solution for an arbitrary open set is not as good as for a circular cylinder.

THEOREM 8.27. *Let E be an open set, and let $p \in E$. Then there is a unique nonnegative Borel measure μ_p^E on $\partial_e E$ such that the representation*
$$S_f^E(p) = \int_{\partial_e E} f \, d\mu_p^E$$
holds for every $f \in C(\partial_e E)$. Moreover $\mu_p^E(\partial_e E) = 1$.

PROOF. Any function $f \in C(\partial_e E)$ has a PWB-solution S_f^E on E, by Theorem 8.26. We show that the mapping $f \mapsto S_f^E(p)$ is a positive linear functional on the Banach space $C(\partial_e E)$ with the supremum norm. By Theorem 8.22, if $f, g \in C(\partial_e E)$ and $\alpha, \beta \in \mathbb{R}$, then
$$S_{\alpha f + \beta g}^E = S_{\alpha f}^E + S_{\beta g}^E = \alpha S_f^E + \beta S_g^E,$$

so that the mapping in question is a linear functional on $C(\partial_e E)$. Furthermore, if $f \geq 0$ and $w \in \mathfrak{U}_f^E$, then

$$\limsup_{(x,t)\to(y,s)} w(x,t) \geq f(y,s) \geq 0$$

for all $(y,s) \in \partial_n E$, and

$$\limsup_{(x,t)\to(y,s+)} w(x,t) \geq f(y,s) \geq 0$$

for all $(y,s) \in \partial_{ss} E$, so that $w \geq 0$ on E by Theorem 8.2. Hence $U_f^E \geq 0$ on E, so that the linear functional $f \mapsto S_f^E(p)$ is positive. It now follows from the Riesz Representation Theorem that there is a unique nonnegative Borel measure μ_p^E on $\partial_e E$ such that $S_f^E(p) = \int_{\partial_e E} f \, d\mu_p^E$ for every $f \in C(\partial_e E)$. In particular, if $f(q) = 1$ for all $q \in \partial_e E$, then $S_f^E = 1$ on E by Theorem 8.20, so that

$$1 = S_f^E(p) = \int_{\partial_e E} d\mu_p^E = \mu_p^E(\partial_e E).$$

$\square$

In Chapter 2, the treatment of the Dirichlet problem on a circular cylinder R involved only functions in $C(\partial_n R)$, and the caloric measure at a point $p \in R$ was defined to be the Borel measure in the representation theorem corresponding to Theorem 8.27. In this chapter, we shall consider much more general functions, so we shall define the caloric measure at $p \in E$ to be the completion of the measure in Theorem 8.27.

Recall that, if μ is a Borel measure on any set X, then the class of all subsets of X of the form $A \cup Y$, where A is a Borel set and Y is a subset of a Borel set Z with $\mu(Z) = 0$, is a σ-algebra that contains the Borel sets. We denote this σ-algebra by $\mathcal{B}_\mu$. The measure μ can be extended to a measure on $\mathcal{B}_\mu$, which we also denote by μ, by putting $\mu(A \cup Y) = \mu(A)$ whenever A and Y are as above. This extended measure is called the *completion* of μ.

DEFINITION 8.28. Let E be an open set, and let $p \in E$. Then the completion of the measure μ_p^E of Theorem 8.27, is called the *caloric measure* relative to E and p. It will also be denoted by μ_p^E. A function on $\partial_e E$ will be called μ_p^E-measurable if it is measurable with respect to the completed measure.

LEMMA 8.29. *Let E be an open set, let $p_0 \in E$, and put $\Lambda = \Lambda(p_0; E)$. Then for any point $p \in \Lambda$, the caloric measure μ_p^E is supported in $\partial_e \Lambda$, and μ_p^Λ is the restriction to $\partial_e \Lambda$ of μ_p^E.*

PROOF. Applying Theorem 8.27 on Λ, we obtain

$$S_f^\Lambda(p) = \int_{\partial_e \Lambda} f \, d\mu_p^\Lambda$$

for any point $p \in \Lambda$ and any function $f \in C(\partial_e \Lambda)$. Since we can extend any such f to a function $\bar{f} \in C(\partial_e E)$, an application of Theorem 8.27 on E also gives

$$S_{\bar{f}}^E(p) = \int_{\partial_e E} \bar{f} \, d\mu_p^E.$$

By Lemma 8.11, $S_{\bar{f}}^E(p) = S_f^\Lambda(p)$ for $p \in \Lambda$, and hence
$$S_f^\Lambda(p) = \int_{\partial_e E} \bar{f}\, d\mu_p^E.$$
This equality is independent of the choice of $\bar{f}$, so that $\mu_p^E(\partial_e E \backslash \partial_e \Lambda) = 0$ and
$$S_f^\Lambda(p) = \int_{\partial_e \Lambda} f\, d\mu_p^E.$$
The uniqueness of the caloric measure now gives the result. □

LEMMA 8.30. *Let E be an open set, and let f be a lower semicontinuous, lower finite function on $\partial_e E$. Then*
$$L_f^E(p) = U_f^E(p) = \int_{\partial_e E} f\, d\mu_p^E$$
for all $p \in E$, and if $U_f^E < +\infty$ on E then f is resolutive for E.

PROOF. By Lemma 3.6, there is an increasing sequence $\{f_j\}$ of functions in $C(\partial_e E)$ that converges pointwise to f on $\partial_e E$. By Theorem 8.26, each function f_j is resolutive for E so that, in particular, $S_{f_j}^E$ is finite-valued on E for all j. Therefore, by Lemma 8.17,
$$U_f^E = \lim_{j \to \infty} S_{f_j}^E$$
on E. Furthermore, Lemma 8.14(c) shows that $S_{f_j}^E \leq L_f^E$ on E for all j, so it follows that $U_f^E \leq L_f^E$ on E. Since Lemma 8.13 shows that $L_f^E \leq U_f^E$ on E, equality holds. By Theorem 8.27, for all $p \in E$ we have
$$L_f^E(p) = U_f^E(p) = \lim_{j \to \infty} S_{f_j}^E(p) = \lim_{j \to \infty} \int_{\partial_e E} f_j\, d\mu_p^E = \int_{\partial_e E} f\, d\mu_p^E,$$
by the Lebesgue monotone convergence theorem. Finally, since $U_f^E \geq S_{f_1}^E > -\infty$ on E, it follows from Lemma 8.15 that U_f^E is a temperature on E if it is upper finite, so that f is resolutive for E in this case. □

LEMMA 8.31. *Let E be an open set, let $p \in E$, and let f be an extended real-valued function on $\partial_e E$. Given any number $A > U_f^E(p)$, we can find a lower finite, lower semicontinuous function g on $\partial_e E$, such that $f \leq g$ on $\partial_e E$ and $U_g^E(p) < A$. Given any number $B < L_f^E(p)$, we can find an upper finite, upper semicontinuous function h on $\partial_e E$, such that $h \leq f$ on $\partial_e E$ and $L_h^E(p) > B$.*

PROOF. Since $U_f^E(p) < A$, we can find a function $w \in \mathfrak{U}_f^E$ such that $w(p) < A$. We define a function g on $\partial_e E$ by putting
$$g(y, s) = \liminf_{(x,t) \to (y,s)} w(x, t)$$
for all $(y, s) \in \partial_n E$, and
$$g(y, s) = \liminf_{(x,t) \to (y,s+)} w(x, t)$$
for all $(y, s) \in \partial_{ss} E$. Then g is lower bounded and lower semicontinuous on $\partial_e E$. Since $w \in \mathfrak{U}_f^E$, we also have $g \geq f$ on $\partial_e E$. Finally, we note that $w \in \mathfrak{U}_g^E$, which implies that $U_g^E(p) \leq w(p) < A$.

Given any $B < L_f^E(p)$, we have $-B > -L_f^E(p) = U_{-f}^E(p)$, by Lemma 8.14(a).

Therefore, by the part just proved, we can find a lower finite, lower semicontinuous function $-h$ on $\partial_e E$, such that $-f \leq -h$ on $\partial_e E$ and $U^E_{-h}(p) < -B$. So h is an upper finite, upper semicontinuous function on $\partial_e E$, such that $h \leq f$ on $\partial_e E$ and $L^E_h(p) = -U^E_{-h}(p) > B$. □

THEOREM 8.32. *Let E be an open set, let $p \in E$, and let f be an extended real-valued function on $\partial_e E$.*
(a) *If $\int_{\partial_e E} f \, d\mu^E_p$ exists, then*

(8.1) $$U^E_f(p) = L^E_f(p) = \int_{\partial_e E} f \, d\mu^E_p.$$

(b) *Conversely, if $U^E_f(p) = L^E_f(p)$ and is finite, then f is μ^E_p-integrable (and (8.1) holds).*

PROOF. (a) We establish (8.1) for increasingly general classes of functions.

If f is the characteristic function χ_A of a relatively open subset A of $\partial_e E$, then f is finite and lower semicontinuous on $\partial_e E$, so that (8.1) follows from Lemma 8.30.

We denote by $\mathcal{B}$ the σ-algebra of all Borel subsets of $\partial_e E$, and by $\mathcal{F}$ the class of all sets $A \in \mathcal{B}$ for which (8.1) holds when $f = \chi_A$. We prove that $\mathcal{F} = \mathcal{B}$. We know that $\mathcal{F}$ contains all the relatively open subsets of $\partial_e E$, so we can prove that $\mathcal{F} = \mathcal{B}$ by showing that $\mathcal{F}$ is a σ-algebra. Clearly $\partial_e E \in \mathcal{F}$. Suppose that $A \in \mathcal{F}$, so that

$$\mu^E_p(A) = \int_{\partial_e E} \chi_A \, d\mu^E_p = U^E_{\chi_A}(p) = L^E_{\chi_A}(p).$$

We denote by A^c the complement of A in $\partial_e E$. Then, using Theorem 8.27 and Lemma 8.14, we have

$$\mu^E_p(A^c) = 1 - \mu^E_p(A) = 1 - U^E_{\chi_A}(p) = L^E_1(p) + L^E_{-\chi_A}(p) \leq L^E_{\chi_{A^c}}(p) \leq U^E_{\chi_{A^c}}(p)$$
$$\leq U^E_1(p) + U^E_{-\chi_A}(p) = 1 - L^E_{\chi_A}(p) = 1 - \mu^E_p(A) = \mu^E_p(A^c).$$

Therefore equality holds throughout, and hence

$$L^E_{\chi_{A^c}}(p) = U^E_{\chi_{A^c}}(p) = \mu^E_p(A^c) = \int_{\partial_e E} \chi_{A^c} \, d\mu^E_p.$$

Thus $A^c \in \mathcal{F}$. We now let $\{F_j\}$ be an expanding sequence of sets in $\mathcal{F}$, and put $F = \bigcup_{j=1}^\infty F_j$. By Lemma 8.14, we have $1 = L^E_1 \geq L^E_{\chi_F} \geq L^E_{\chi_{F_{j+1}}} \geq L^E_{\chi_{F_j}} \geq L^E_0 = 0$ for all j. Since $U^E_{\chi_{F_j}}$ is finite on E for all j, it therefore follows from Lemma 8.17 that

$$L^E_{\chi_F}(p) \geq \lim_{j \to \infty} L^E_{\chi_{F_j}}(p) = \lim_{j \to \infty} U^E_{\chi_{F_j}}(p) = U^E_{\chi_F}(p) \geq L^E_{\chi_F}(p).$$

Hence

$$L^E_{\chi_F}(p) = U^E_{\chi_F}(p) = \lim_{j \to \infty} U^E_{\chi_{F_j}}(p) = \lim_{j \to \infty} \mu^E_p(F_j) = \mu^E_p(F) = \int_{\partial_e E} \chi_F \, d\mu^E_p,$$

so that $F \in \mathcal{F}$. It follows that $\mathcal{F}$ is a σ-algebra, and hence $\mathcal{F} = \mathcal{B}$.

Now we extend (8.1) to the characteristic functions of all μ^E_p-measurable sets. Let A be a μ^E_p-measurable set. Then we can write $A = F \cup Y$ for some Borel set F and some subset Y of a Borel set Z with $\mu^E_p(Z) = 0$. Then $\mu^E_p(A) = \mu^E_p(F)$, and

$$L^E_{\chi_F}(p) \leq L^E_{\chi_A}(p) \leq U^E_{\chi_A}(p) \leq U^E_{\chi_{F \cup Z}}(p),$$

8.4. THE CALORIC MEASURE ON THE ESSENTIAL BOUNDARY

by Lemmas 8.13 and 8.14. Since $U^E_{\chi_F}$ is finite on E, we can use Lemma 8.14 to obtain

$$U^E_{\chi_{F \cup Z}}(p) \leq U^E_{\chi_F}(p) + U^E_{\chi_Z}(p).$$

Since $Z, F \in \mathcal{B}$, we have

$$U^E_{\chi_Z}(p) = \int_{\partial_e E} \chi_Z \, d\mu^E_p = 0,$$

and (8.1) with $f = \chi_F$. Hence

$$L^E_{\chi_F}(p) \leq L^E_{\chi_A}(p) \leq U^E_{\chi_A}(p) \leq U^E_{\chi_F}(p) = \int_{\partial_e E} \chi_F \, d\mu^E_p = L^E_{\chi_F}(p).$$

Therefore equality holds throughout, and so

$$L^E_{\chi_A}(p) = U^E_{\chi_A}(p) = U^E_{\chi_F}(p) = \int_{\partial_e E} \chi_F \, d\mu^E_p = \mu^E_p(F) = \mu^E_p(A) = \int_{\partial_e E} \chi_A \, d\mu^E_p.$$

Thus (8.1) holds with $f = \chi_A$.

Our next step is to extend (8.1) to all nonnegative, μ^E_p-measurable, simple functions on $\partial_e E$. Suppose that f can be written in the form $f = \sum_{i=1}^k \alpha_i \chi_{A_i}$, for some positive numbers $\alpha_1, ..., \alpha_k$ and μ^E_p-measurable sets $A_1, ..., A_k$. Then (8.1) holds for each function χ_{A_i}, and therefore Lemmas 8.13 and 8.14 can be used to show that

$$\sum_{i=1}^k \alpha_i \mu^E_p(A_i) = \sum_{i=1}^k \alpha_i L^E_{\chi_{A_i}}(p) \leq L^E_f(p) \leq U^E_f(p) \leq \sum_{i=1}^k \alpha_i U^E_{\chi_{A_i}}(p) \leq \sum_{i=1}^k \alpha_i \mu^E_p(A_i).$$

Hence

$$L^E_f(p) = U^E_f(p) = \sum_{i=1}^k \alpha_i \mu^E_p(A_i) = \sum_{i=1}^k \alpha_i \int_{\partial_e E} \chi_{A_i} \, d\mu^E_p = \int_{\partial_e E} f \, d\mu^E_p,$$

so that (8.1) holds for f.

We now consider the case where f is an arbitrary nonnegative, μ^E_p-measurable function on $\partial_e E$. Let $\{g_j\}$ be an increasing sequence of nonnegative, μ^E_p-measurable, simple functions with limit f on $\partial_e E$. Since (8.1) holds for each function g_j, the Lebesgue monotone convergence theorem gives

$$L^E_{g_j}(p) = U^E_{g_j}(p) = \int_{\partial_e E} g_j \, d\mu^E_p \to \int_{\partial_e E} f \, d\mu^E_p.$$

Moreover, using Lemma 8.14 we obtain

$$L^E_f(p) \geq \lim_{j \to \infty} L^E_{g_j}(p) = \lim_{j \to \infty} U^E_{g_j}(p).$$

Each function g_j is bounded, so that each $U^E_{g_j}$ is also bounded, and hence Lemma 8.17 can be used to show that $\lim_{j \to \infty} U^E_{g_j}(p) = U^E_f(p)$. Since $U^E_f \geq L^E_f$, it follows that

$$L^E_f(p) = U^E_f(p) = \int_{\partial_e E} f \, d\mu^E_p,$$

as required.

Finally, we let f be an arbitrary μ^E_p-measurable function for which $\int_{\partial_e E} f \, d\mu^E_p$

exists. Then (8.1) holds for the positive and negative parts of f, so that Lemma 8.14 gives

$$\int_{\partial_e E} f \, d\mu_p^E = U_{f^+}^E(p) - L_{f^-}^E(p) = U_{f^+}^E(p) + U_{-f^-}^E(p) \geq U_f^E(p),$$

and also

$$\int_{\partial_e E} f \, d\mu_p^E = L_{f^+}^E(p) - U_{f^-}^E(p) = L_{f^+}^E(p) + L_{-f^-}^E(p) \leq L_f^E(p) \leq U_f^E(p),$$

with the help of Lemma 8.13. Now (8.1) follows.

(b) Since $U_f^E(p)$ is finite, it follows from Lemma 8.31 that, given any positive integer j, we can find a lower finite, lower semicontinuous function g_j on $\partial_e E$ such that $f \leq g_j$ on $\partial_e E$ and

$$U_{g_j}^E(p) < U_f^E(p) + \frac{1}{j}.$$

Furthermore, because $L_f^E(p)$ is finite, Lemma 8.31 also shows that we can find an upper finite, upper semicontinuous function h_j on $\partial_e E$ such that $h_j \leq f$ on $\partial_e E$ and

$$L_{h_j}^E(p) > L_f^E(p) - \frac{1}{j}.$$

We put

$$g = \inf_j g_j, \qquad h = \sup_j h_j,$$

and note that g, h are Borel measurable and satisfy $h \leq f \leq g$ on $\partial_e E$. By Lemma 8.30,

$$U_f^E(p) = \inf_j U_{g_j}^E(p) = \inf_j \int_{\partial_e E} g_j \, d\mu_p^E \geq \int_{\partial_e E} g \, d\mu_p^E.$$

Moreover, by Lemmas 8.14 and 8.30 we have

$$L_f^E(p) = -U_{-f}^E(p) = -\inf_j U_{-h_j}^E(p) = -\inf_j \int_{\partial_e E} (-h_j) \, d\mu_p^E$$

$$= \sup_j \int_{\partial_e E} h_j \, d\mu_p^E \leq \int_{\partial_e E} h \, d\mu_p^E.$$

Hence

$$L_f^E(p) \leq \int_{\partial_e E} h \, d\mu_p^E \leq \int_{\partial_e E} g \, d\mu_p^E \leq U_f^E(p) = L_f^E(p) \in \mathbb{R},$$

so that $h = g$ μ_p^E-almost everywhere on $\partial_e E$. Since g and h are Borel measurable, it follows that there is a Borel set Z such that $\mu_p^E(Z) = 0$ and $h = f = g$ on $(\partial_e E) \setminus Z$. All subsets of Z are μ_p^E-measurable, so that f is a μ_p^E-measurable function and

$$L_f^E(p) \leq \int_{\partial_e E} f \, d\mu_p^E \leq U_f^E(p) = L_f^E(p) \in \mathbb{R}.$$

Thus f is μ_p^E-integrable (and (8.1) holds). $\square$

COROLLARY 8.33. *Let E be an open set, and let f be a Borel measurable, extended real-valued function on $\partial_e E$. If both U_f^E and L_f^E are finite on E, then f is resolutive for E and*

$$S_f^E(p) = \int_{\partial_e E} f \, d\mu_p^E$$

for all $p \in E$.

PROOF. We choose any point $p \in E$. If f is Borel measurable, then f^+ is μ_p^E-measurable, so that Theorem 8.32(a) gives

$$U_{f^+}^E(p) = L_{f^+}^E(p) = \int_{\partial_e E} f^+ \, d\mu_p^E.$$

Since $U_f^E(p) < +\infty$, there is a hypertemperature $w \in \mathfrak{U}_f^E$ such that $w(p) < +\infty$, and since w is lower bounded on E, there is a number α such that $w + \alpha \in \mathfrak{U}_{f^+}^E$. Therefore $U_{f^+}^E(p) < +\infty$, and obviously $U_{f^+}^E(p) > -\infty$. Since p is an arbitrary point of E, Lemma 8.15 now shows that $U_{f^+}^E$ is a temperature on E, so that f^+ is resolutive for E, and

$$S_{f^+}^E(p) = \int_{\partial_e E} f^+ \, d\mu_p^E.$$

This result holds if f is replaced by $-f$ because, by Lemma 8.14, $U_{-f}^E = -L_f^E$ and $L_{-f}^E = -U_f^E$, which are finite, and so f^+ can be replaced by $(-f)^+ = f^-$. Therefore, by Theorem 8.22, the function $f = f^+ - f^-$ is resolutive and

$$S_f^E(p) = S_{f^+}^E(p) - S_{f^-}^E(p) = \int_{\partial_e E} f^+ \, d\mu_p^E - \int_{\partial_e E} f^- \, d\mu_p^E = \int_{\partial_e E} f \, d\mu_p^E.$$

$\square$

COROLLARY 8.34. *Let E be an open set, and let f be an extended real-valued function on $\partial_e E$. Then the following statements are equivalent:*

(a) f is resolutive for E;
(b) given any point $q \in E$, we can find a point $p \in \Lambda^*(q; E)$ such that f is μ_p^E-integrable;
(c) f is μ_p^E-integrable for all $p \in E$.
If these statements hold, then

$$S_f^E(p) = \int_{\partial_e E} f \, d\mu_p^E$$

for all $p \in E$.

PROOF. If statement (a) holds, then Theorem 8.32(b) shows that statement (c) holds also. If (c) holds, then obviously (b) holds too. Now suppose that (b) holds, and let $q \in E$. Then we can find a point $p \in \Lambda^*(q; E)$ such that f is μ_p^E-integrable, so that

$$L_f^E(p) = U_f^E(p) = \int_{\partial_e E} f \, d\mu_p^E$$

by Theorem 8.32(a), and the integral is finite. It now follows from Lemma 8.19 that (a) holds, and so the equivalence of the three statements is established.

Finally, if statement (a) holds, then

$$S_f^E(p) = \int_{\partial_e E} f \, d\mu_p^E$$

for all $p \in E$, by Theorem 8.32(b). $\square$

It follows from Corollary 8.34 that, if A is a subset of $\partial_e E$ which is μ_p^E-measurable for all $p \in E$, then its characteristic function χ_A is resolutive and $S_{\chi_A}^E(p) = \mu_p^E(A)$ for all $p \in E$. Therefore, if $\mu_{p_0}^E(A) = 0$ for some point $p_0 \in E$, then $\mu_p^E(A) = 0$ for all $p \in \Lambda(p_0; E)$, by the minimum principle.

8.5. Boundary Behaviour of PWB Solutions

In this section, we investigate the boundary behaviour of the PWB solution S_f^E, and see how it relates to the function f. In particular, if $f \in C(\partial_e E)$ it is desirable to have $S_f^E(p) \to f(q)$ as $p \to q$ in the appropriate sense, for all $q \in \partial_e E$. However, this does not always happen. Therefore we make the following definition.

DEFINITION 8.35. Let E be an open set. If $q \in \partial_n E$, then q is called *regular* if
$$\lim_{p \to q} S_f^E(p) = f(q)$$
for all $f \in C(\partial_e E)$. On the other hand, if $q = (y, s) \in \partial_{ss} E$, then q is called *regular* if
$$\lim_{(x,t) \to (y,s+)} S_f^E(x, t) = f(y, s)$$
for all $f \in C(\partial_e E)$. If q is not regular, then we say that q is *irregular*. The set E is called *regular* if every point $q \in \partial_e E$ is regular.

Theorems 2.3 and 8.20 together imply that any circular cylinder is regular. More generally, Theorem 3.39 gives a condition for a convex domain of revolution to be regular. On the other hand, Theorem 3.42 implies that some convex domains of revolution are not regular. In particular, we have the following example.

EXAMPLE 8.36. For any heat ball $\Omega = \Omega(q_0; c)$, the centre q_0 is irregular. Corollary 3.41 shows that every other boundary point is regular, so that if q_0 was also regular, the Dirichlet problem would be solvable for any function in $C(\partial\Omega)$. However, Theorem 3.42 shows that this is not the case.

The main result of this section is that finite regular points can be characterized in terms of functions called barriers, which we now define.

DEFINITION 8.37. Let E be an open set, and let $q = (y, s)$ be a finite point of $\partial_e E$. A function w is called a *barrier* at q if it is defined on $N \cap E$ for some open neighbourhood N of q, and possesses the following properties:
(a) w is a supertemperature on $N \cap E$;
(b) $w > 0$ on $N \cap E$;
(c) if $q \in \partial_n E$ then
$$\lim_{(x,t) \to (y,s)} w(x, t) = 0,$$
and if $q \in \partial_{ss} E$ then
$$\lim_{(x,t) \to (y,s+)} w(x, t) = 0.$$

In essence, there is a barrier at a point if a supertemperature can take a strict minimum value at that point. The use of a barrier is demonstrated in the proof of the following crucial lemma.

8.5. BOUNDARY BEHAVIOUR OF PWB SOLUTIONS

LEMMA 8.38. *Let D be a circular cylinder, let f be a bounded and resolutive function on $\partial_n D$, and let $q_0 = (y_0, s_0)$ be a point of continuity of f. Then the PWB solution for f on D satisfies $\lim_{p \to q_0} S_f^D(p) = f(q_0)$.*

PROOF. Outside D, we position a coheat ball
$$\Omega^* = \Omega^*(\eta_0, \sigma_0; c_0) = \{(x,t) : W(x - \eta_0, t - \sigma_0) > \tau(c_0)\},$$
with $\sigma_0 < s_0$, in such a way that $\partial \Omega^* \cap \partial D = \{q_0\}$. Then function w, defined on D by
$$w(x,t) = \tau(c_0) - W(x - \eta_0, t - \sigma_0),$$
is a positive temperature on D which satisfies $\lim_{p \to q_0} w(p) = 0$. Given any positive ϵ, we put $A = f(q_0) + \epsilon$, and choose a neighbourhood N of q_0 such that $f < A$ on $N \cap \partial_n D$. The barrier w satisfies $\inf_{D \setminus N} w > 0$, and so we can find a positive number α such that $\alpha \inf_{D \setminus N} w > \sup_{\partial_n D} f - A$. We now put $u = A + \alpha w$ on D, and note that u is a lower bounded temperature on D. Moreover, if $q \in (\partial_n D) \setminus N$ we have
$$\liminf_{p \to q} u(p) \geq A + \alpha \inf_{D \setminus N} w > \sup_{\partial_n D} f \geq f(q).$$
On the other hand, if $q \in (\partial_n D) \cap N$ we have
$$\liminf_{p \to q} u(p) \geq A > f(q).$$
Therefore $u \in \mathfrak{U}_f^D$. Hence $u \geq S_f^D$, so that
$$\limsup_{p \to q_0} S_f^D(p) \leq \limsup_{p \to q_0} u(p) = A + \alpha \lim_{p \to q_0} w(p) = A.$$
It follows that
$$\limsup_{p \to q_0} S_f^D(p) \leq f(q_0).$$
A similar inequality holds with f replaced by $-f$, and so it follows from Lemma 8.14(a) that
$$\liminf_{p \to q_0} S_f^D(p) = -\limsup_{p \to q_0} S_{-f}^D(p) \geq f(q_0).$$
Hence $S_f^D(p) \to f(q_0)$ as $p \to q_0$. □

The barrier in the above proof satisfies the additional condition $\inf_{D \setminus N} w > 0$, which is crucial to the reasoning. We proceed to show that, whenever there is a barrier at q_0, there is also a barrier w that satisfies this additional condition.

LEMMA 8.39. *Let E be an open set, let $D = B \times \,]c,d[$ be a circular cylinder such that $E \cap D \neq \emptyset$, let $\partial_l D = \partial B \times [c,d]$, and let $\partial_b D = B \times \{c\}$. If $q \in \partial_e(E \cap D)$, then either*
  (a) $q \in E \cap \partial_n D$, or
  (b) $q \in (\overline{D} \cap \partial_n E) \cup (\partial_b D \cap \partial_{ss} E) \subseteq \partial_n(E \cap D)$, or
  (c) $q \in D \cap \partial_{ss} E \subseteq \partial_{ss}(E \cap D)$, or
  (d) $q \in \partial_l D \cap \partial_a E$.

PROOF. We start from the inclusion
$$\partial(E \cap D) \subseteq (\overline{D} \cap \partial E) \cup (E \cap \partial_n D) \cup (E \cap \partial_s D).$$

If $q \in E \cap \partial_s D$, then there is $\delta > 0$ such that $B(q,\delta) \subseteq E$, $H(q,\delta) \subseteq D$, and $H^*(q,\delta) \subseteq \mathbb{R}^{n+1}\backslash D$, so that $H(q,\delta) \subseteq E \cap D$ and $H^*(q,\delta) \subseteq \mathbb{R}^{n+1}\backslash(E \cap D)$; hence $q \in \partial_s(E \cap D)$. Therefore

(8.2) $$\partial_e(E \cap D) \subseteq (\overline{D} \cap \partial E) \cup (E \cap \partial_n D).$$

Next,

(8.3) $$\overline{D} \cap \partial E = (\overline{D} \cap \partial_n E) \cup (\partial_l D \cap \partial_a E) \cup ((\partial_b D \cup D \cup \partial_s D) \cap \partial_a E).$$

We deal first with the last bracket in this union. If $q \in \partial_b D \cap \partial_s E$, then there is $\delta > 0$ such that $H(q,\delta) \subseteq E\backslash D$ and $H^*(q,\delta) \subseteq D\backslash E$, so that $q \notin \partial(E \cap D)$. If $q \in \partial_b D \cap \partial_{ss} E$, then there is $\delta > 0$ such that $H(q,\delta) \subseteq E\backslash D$ and $H^*(q,\delta) \subseteq D$; moreover, for all $\epsilon > 0$ we have $H^*(q,\epsilon) \cap E \neq \emptyset$, and so $H^*(q,\epsilon) \cap (E \cap D) \neq \emptyset$; hence $q \in \partial_n(E \cap D)$. If $q \in D \cap \partial_s E$, then there is $\delta > 0$ such that $B(q,\delta) \subseteq D$, $H(q,\delta) \subseteq E$, and $H^*(q,\delta) \subseteq \mathbb{R}^{n+1}\backslash E$; it follows that $H(q,\delta) \subseteq E \cap D$ and $H^*(q,\delta) \subseteq \mathbb{R}^{n+1}\backslash(E \cap D)$, so that $q \in \partial_s(E \cap D)$. If $q \in D \cap \partial_{ss} E$, then there is $\delta > 0$ such that $B(q,\delta) \subseteq D$ and $H(q,\delta) \subseteq E$; moreover, for all $\epsilon > 0$ we have $H^*(q,\epsilon) \cap E \neq \emptyset$, so that $H(q,\delta) \subseteq E \cap D$ and for all $\epsilon > 0$ $H^*(q,\epsilon) \cap (E \cap D) \neq \emptyset$, and hence $q \in \partial ss(E \cap D)$. Finally, if $q \in \partial_s D \cap \partial_a E$, then there is $\delta > 0$ such that $H(q,\delta) \subseteq E \cap D$ and $H^*(q,\delta) \subseteq \mathbb{R}^{n+1}\backslash D \subseteq \mathbb{R}^{n+1}\backslash(E \cap D)$, so that $q \in \partial_s(E \cap D)$. It now follows from (8.2) and (8.3) that

$$\partial_e(E \cap D) \subseteq (E \cap \partial_n D) \cup (\overline{D} \cap \partial_n E) \cup (\partial_l D \cap \partial_a E) \cup (\partial_b D \cap \partial_{ss} E) \cup (D \cap \partial_{ss} E).$$

Moreover, if $q \in \overline{D} \cap \partial_n E$, then for all $\epsilon > 0$ we have $H(q,\epsilon)\backslash E \neq \emptyset$, so that $H(q,\epsilon)\backslash(E \cap D) \neq \emptyset$, and hence $q \in \partial_n(E \cap D)$ provided that $q \in \partial(E \cap D)$. $\square$

THEOREM 8.40. *Given an open set E, there is a sequence of hyperplanes of the form $\mathbb{R}^n \times \{t\}$ which covers $\partial_a E$.*

PROOF. We prove the result with the additional assumption that E is bounded. The general case then follows from the facts that $E = \bigcup_{j=1}^{\infty}(E \cap B(0,j))$ and a union of countably many countable sets is itself countable.

Given any point $q \in \partial_a E$, we choose an open half-ball $H(q) = H(q,\epsilon)$ contained in E, and note that $H(q) \cap \partial_a E = \emptyset$. We put $U = \bigcup_{q \in \partial_a E} H(q)$. By Lindelöf's theorem, there is a countable subset C of $\partial_a E$ such that $U = \bigcup_{q \in C} H(q)$. We show that the temporal coordinate of each point of $\partial_a E$ coincides with that of some point in C.

We take any point $q_0 = (y_0, s_0) \in \partial_a E$, and denote by $2r_0$ the radius of the associated half-ball $H(q_0)$. Since $H(q_0) \subseteq U$, the line segment $L = \{y_0\} \times [s_0-r_0, s_0[$ lies in U, and hence is covered by some of the half-balls $H(q)$ for $q \in C$. If there was a half-ball $H(\bar{q})$, with $\bar{q} = (\bar{y}, \bar{s}) \in C$, such that $\bar{s} > s_0$ and $H(\bar{q}) \cap L \neq \emptyset$, then we would have $q_0 \in H(\bar{q})$, contrary to the fact that $H(\bar{q}) \cap \partial_a E = \emptyset$. Therefore, if $H(q) \cap L \neq \emptyset$ and $q = (y, s) \in C$, then $s \leq s_0$.

Suppose that $s < s_0$ whenever $H(q) \cap L \neq \emptyset$ and $q = (y, s) \in C$. Then there are infinitely many such half-balls $H(q)$. Moreover, because the distance between L and $(\mathbb{R}^n \times\,]-\infty, s_0[)\backslash H(q_0)$ is r_0, the radius of those half-balls must exceed r_0. Given any point $p_1 = (y_0, t_1) \in L$, we choose a half-ball $H(q_1)$ such that $p_1 \in H(q_1)$ and $q_1 = (y_1, s_1) \in C$. Then $t_1 < s_1$. We now choose a point $p_2 = (y_0, t_2) \in L\backslash H(q_1)$ such that $t_2 > t_1$, and a half-ball $H(q_2)$ such that $p_2 \in H(q_2)$ and $q_2 \in C$. If $q_2 = (y_2, s_2)$, then $s_1 < t_2 < s_2$. Furthermore, because $q_1 \in C \subseteq \partial_a E$, we know that $q_1 \notin H(q_2)$, and so the fact that $s_2 > s_1$ implies that the distance between

q_1 and q_2 is greater than the radius of $H(q_2)$, and therefore exceeds r_0. We now choose a point $p_3 = (y_0, t_3) \in L \backslash H(q_2)$ such that $t_3 > t_2$, and a half-ball $H(q_3)$ such that $p_3 \in H(q_3)$ and $q_3 = (y_3, s_3) \in C$. As before, the radius of $H(q_3)$ exceeds r_0, $s_3 > s_2 > s_1$, $q_1 \notin H(q_3)$, and $q_2 \notin H(q_3)$. Therefore the distance between q_3 and $\{q_1, q_2\}$ exceeds r_0. Continuing in this way, we obtain a sequence $\{q_k\}$ of isolated points in ∂E. However, E is bounded so ∂E is compact, and hence there can be no such sequence. It follows that our supposition that $s < s_0$ whenever $H(q) \cap L \neq \emptyset$ is untenable.

Hence there is a point $\hat{q} = (\hat{y}, \hat{s}) \in C$ such that $\hat{s} = s_0$. $\square$

COROLLARY 8.41. *Let E be an open set, let $D = B \times \,]c, d[$ be a circular cylinder such that $E \cap D \neq \emptyset$, and let $\partial_l D = \partial B \times [c, d]$. Then $\partial_l D \cap \partial_a E$ is a polar set.*

PROOF. By Theorem 8.40, there is a sequence of numbers $\{t_k\}$ such that the union $\bigcup_{k=1}^{\infty}(\mathbb{R}^n \times \{t_k\})$ covers $\partial_a E$, so that

$$\partial_l D \cap \partial_a E \subseteq \bigcup_{k=1}^{\infty}(\partial B \times \{t_k\}).$$

The following capacities are relative to $\mathbb{R}^{n+1}$. By Theorems 7.45(c) and 7.55, we have

$$\mathcal{C}_+(\partial_l D \cap \partial_a E) \leq \sum_{k=1}^{\infty} \mathcal{C}(\partial B \times \{t_k\}) = \sum_{k=1}^{\infty} m_n(\partial B) = 0,$$

so that $\partial_l D \cap \partial_a E$ is polar, by Theorem 7.46. $\square$

THEOREM 8.42. *Let E be a bounded open set, and let $q = (y, s)$ be a point of $\partial_e E$. If there is a barrier u at q, then there is a barrier w at q such that w is a temperature on the whole of E and $\inf_{E \backslash N} w > 0$ for each neighbourhood N of q.*

PROOF. We choose a number $\alpha > 1$ such that $\overline{E} \subseteq \mathbb{R}^n \times \,]-\frac{1}{2}n\alpha, \frac{1}{2}n\alpha[$, and define a function ψ on $\mathbb{R}^{n+1}$ by putting $\psi(x, t) = \alpha|x - y|^2 + (t - s)^2$. Since the restriction of ψ to $\partial_e E$ belongs to $C(\partial_e E)$, Theorem 8.26 shows that it is resolutive for E. We show that the temperature S_ψ^E has the required properties. For all $(x, t) \in E$, we have $\Theta\psi(x, t) = 2(n\alpha - t + s) > 0$, so that ψ is a subtemperature on E, by Corollary 3.49. Therefore $\psi \in \mathfrak{L}_\psi^E$ and $\psi \leq S_\psi^E$ on E. Hence $\inf_{E \backslash N} S_\psi^E > 0$ for each neighbourhood N of q.

We now put $w = S_\psi^E$. It remains to prove that, if $q \in \partial_n E$ then

(8.4) $$\lim_{(x,t) \to (y,s)} w(x, t) = 0,$$

and if $q \in \partial_{ss} E$ then

(8.5) $$\lim_{(x,t) \to (y,s+)} w(x, t) = 0.$$

Since u is a barrier at q, there is a ball $B(q, \rho)$ such that $u > 0$ on $B(q, \rho) \cap E$ and either (8.4) or (8.5) holds, as appropriate, with w replaced by u. Let D be a circular cylinder such that $q \in D$ and $\overline{D} \subseteq B(q, \rho)$. The set D is, of course, the cartesian product of an n-dimensional ball and a bounded interval; by reducing the radius of the ball while keeping the length of the interval unchanged, we can ensure that $E \cap \partial_n D \neq \emptyset$. We put $M = \max_{\partial_e E} \psi$, and choose a closed set $F \subseteq E \cap \partial_n D$ such that the caloric measure $\mu_q^D((E \cap \partial_n D) \backslash F) < \alpha\rho^2/M$. We define a function f on $\partial_n D$ by putting $f = M\chi_A$, where χ_A is the characteristic function of the set

$A = (E \cap \partial_n D) \backslash F$, and note that f is resolutive for D with $S_f^D(q) = M\mu_q^D(A)$, by Corollary 8.33. We put $m = \inf_F u > 0$, and for any hypotemperature $v \in \mathfrak{L}_\psi^E$ we consider the function $\bar{v}$, defined on $E \cap D$ by

$$\bar{v} = v^+ - \alpha\rho^2 - \frac{M}{m}u - S_f^D.$$

Corollary 3.18 shows that v^+ is a subtemperature on E, and hence Corollary 3.57 shows that $\bar{v}$ is a subtemperature on $E \cap D$.

We show that $\bar{v} \leq 0$ on $E \cap D$, using the maximum principle of Theorem 8.2. Note that, because $\psi \geq 0$ we have $v^+ \in \mathfrak{L}_\psi^E \subseteq \mathfrak{L}_M^E$, so that $v^+ \leq M$ on E. We put $Z = \partial_l D \cap \partial_a E$, which Corollary 8.41 shows is polar. Lemma 8.39 shows that $\partial_e(E \cap D)$ is contained in the union of the sets $E \cap \partial_n D$,

$$X = \left((\overline{D} \cap \partial_n E) \cup (\partial_b D \cap \partial_{ss} E)\right) \subseteq \partial_n(E \cap D),$$

$$Y = D \cap \partial_{ss} E \subseteq \partial_{ss}(E \cap D),$$

and Z. At all points $q' \in A$, the function f is continuous, and so

$$\lim_{p \to q'} S_f^D(p) = f(q') = M,$$

by Lemma 8.38. At all points $q' \in F$, the barrier u is lower semicontinuous, so that

$$\liminf_{p \to q', p \in E \cap D} u(p) \geq u(q') \geq m.$$

At all points $q' \in X (\subseteq \partial_e E)$, since $v^+ \in \mathfrak{L}_\psi^E$ we have

$$\limsup_{p \to q', p \in E \cap D} v^+(p) \leq \psi(q') \leq \alpha|q' - q|^2 \leq \alpha\rho^2.$$

At all points $(y', s') \in Y$, we have

$$\limsup_{(x,t) \to (y',s'+), (x,t) \in E \cap D} v^+(x,t) \leq \psi(y', s') \leq \alpha\rho^2.$$

Using these deductions, and the facts that $S_f^D \geq 0$, $u > 0$ and $v^+ \leq M$ on $E \cap D$, it follows, by considering separately points in F, A, X and Y, that

$$\limsup_{p \to q', p \in E \cap D} \bar{v}(p) \leq 0$$

for all $q' \in \partial_n(E \cap D) \backslash Z$, and

$$\limsup_{(x,t) \to (y',s'+), (x,t) \in E \cap D} \bar{v}(x,t) \leq 0$$

for all points $(y', s') \in \partial_{ss}(E \cap D) \backslash Z$. Now the facts that v is upper bounded on E, that S_f^D is bounded on D, and that $u > 0$ on E, together imply that $\bar{v}$ is upper bounded on $E \cap D$. It therefore follows from Theorem 8.2 that $\bar{v} \leq 0$, and hence

$$v^+ \leq \alpha\rho^2 + \frac{M}{m}u + S_f^D,$$

on $E \cap D$. Since $\psi \geq 0$, we have $w(= S_\psi^E) = \sup\{v^+ : v \in \mathfrak{L}_\psi^E\}$ on E, and so it follows that

(8.6) $$w \leq \alpha\rho^2 + \frac{M}{m}u + S_f^D$$

on $E \cap D$. We chose F to satisfy $\mu_q^D(A) < \alpha\rho^2/M$, and f to be $M\chi_A$. These choices imply that

$$\lim_{p \to q} S_f^D(p) = S_f^D(q) = \int_{\partial_n D} M\chi_A \, d\mu_q^D = M\mu_q^D(A) < \alpha\rho^2.$$

This inequality, together with (8.6) and the fact that (8.4) and (8.5) hold with w replaced by u, shows that

$$\limsup_{(x,t) \to (y,s)} w(x,t) \leq 2\alpha\rho^2$$

if $q \in \partial_n E$, and

$$\limsup_{(x,t) \to (y,s+)} w(x,t) \leq 2\alpha\rho^2$$

if $q \in \partial_{ss} E$. Since ρ can be taken arbitrarily small, and $w > 0$, it follows that (8.4) and (8.5) hold. $\square$

COROLLARY 8.43. *Let E be an open set, and let q be a finite point of $\partial_e E$. If there is a barrier u at q, then there is also a barrier v at q such that v is a supertemperature on the whole of E and $\inf_{E \setminus N} v > 0$ for each neighbourhood N of q.*

PROOF. We choose any $r > 0$, and put $E' = E \cap B(q, 2r)$. We let w denote a barrier at q for E', that is defined on the whole of E' and satisfies the condition $\inf_{E' \setminus N} w > 0$ for every open superset N of $\{q\}$, whose existence is guaranteed by Theorem 8.42. We put

$$\alpha = \inf_{E' \setminus B(q,r)} w,$$

and define a function v on E by

$$v = \begin{cases} w \wedge \alpha & \text{on} \quad E \cap B(q,r) \\ \alpha & \text{on} \quad E \setminus B(q,r). \end{cases}$$

Since α can be regarded as a supertemperature on E, and w is one on $E \cap B(q,r)$, Lemma 7.20 (with $V = E \cap B(q,r)$) shows that v is a supertemperature on E. Clearly v has the other properties we require. $\square$

THEOREM 8.44. *Let E be an open set, and let (y,s) be a finite point of $\partial_e E$. If f is an upper bounded function on $\partial_e E$, and there is a barrier at (y,s), then*

$$\limsup_{(x,t) \to (y,s)} U_f^E(x,t) \leq \limsup_{(y',s') \to (y,s)} f(y',s')$$

if $(y,s) \in \partial_n E$, and

$$\limsup_{(x,t) \to (y,s+)} U_f^E(x,t) \leq \limsup_{(y',s') \to (y,s)} f(y',s')$$

if $(y,s) \in \partial_{ss} E$.

PROOF. Corollary 8.43 shows that there is a barrier v at (y,s) such that v is a supertemperature on the whole of E and $\inf_{E \setminus N} v > 0$ for each neighbourhood N of (y,s). We put $L = \limsup_{(y',s') \to (y,s)} f(y',s')$, and note that $L < +\infty$ because f is upper bounded. Given any number $M > L$, we can find a neighbourhood V of

(y, s) such that $f(y', s') < M$ for all points $(y', s') \in V \cap \partial_e E$. Since $\inf_{E \setminus V} v > 0$, we can choose a positive number c such that

$$M + c \inf_{E \setminus V} v > \sup_{\partial_e E} f.$$

We now put $u = M + cv$, and note that u is a lower bounded supertemperature on E. For all points $(y', s') \in (\partial_e E) \setminus V$, we have

$$\liminf_{(x,t) \to (y',s')} u(x,t) \geq M + c \inf_{E \setminus V} v > \sup_{\partial_e E} f \geq f(y', s').$$

For all points $(y', s') \in (\partial_e E) \cap V$, we have $f(y', s') < M$, so that

$$\liminf_{(x,t) \to (y',s')} u(x,t) \geq M > f(y', s').$$

It follows that $u \in \mathfrak{U}_f^E$, so that $u \geq U_f^E$ on E. Hence

$$\limsup_{(x,t) \to (y,s)} U_f^E(x,t) \leq \limsup_{(x,t) \to (y,s)} u(x,t) \leq M + c \lim_{(x,t) \to (y,s)} v(x,t) = M$$

if $(y, s) \in \partial_n E$, and similarly

$$\limsup_{(x,t) \to (y,s+)} U_f^E(x,t) \leq M$$

if $(y, s) \in \partial_{ss} E$. Since M is arbitrary, the result follows. $\square$

COROLLARY 8.45. *Let E be an open set, and let $q = (y, s)$ be a finite point of $\partial_e E$. If f is a bounded function on $\partial_e E$ which is continuous at (y, s), and there is a barrier at (y, s), then*

$$\lim_{(x,t) \to (y,s)} L_f^E(x,t) = \lim_{(x,t) \to (y,s)} U_f^E(x,t) = \lim_{(y',s') \to (y,s)} f(y', s')$$

if $(y, s) \in \partial_n E$, and

$$\lim_{(x,t) \to (y,s+)} L_f^E(x,t) = \lim_{(x,t) \to (y,s+)} U_f^E(x,t) = \lim_{(y',s') \to (y,s)} f(y', s')$$

if $(y, s) \in \partial_{ss} E$.

PROOF. If $(y, s) \in \partial_n E$, then Theorem 8.44 shows that

$$\limsup_{(x,t) \to (y,s)} U_f^E(x,t) \leq \limsup_{(y',s') \to (y,s)} f(y', s')$$

and that

$$\limsup_{(x,t) \to (y,s)} U_{-f}^E(x,t) \leq \limsup_{(y',s') \to (y,s)} -f(y', s'),$$

which implies that

$$\liminf_{(x,t) \to (y,s)} L_f^E(x,t) \geq \liminf_{(y',s') \to (y,s)} f(y', s'),$$

in view of Lemma 8.14(a). Therefore, if f is continuous at (y, s), we have

$$\lim_{(y',s')\to(y,s)} f(y', s') \leq \liminf_{(x,t)\to(y,s)} L_f^E(x,t)$$
$$\leq \big(\limsup_{(x,t)\to(y,s)} L_f^E(x,t)\big) \wedge \big(\liminf_{(x,t)\to(y,s)} U_f^E(x,t)\big)$$
$$\leq \big(\limsup_{(x,t)\to(y,s)} L_f^E(x,t)\big) \vee \big(\liminf_{(x,t)\to(y,s)} U_f^E(x,t)\big)$$
$$\leq \limsup_{(x,t)\to(y,s)} U_f^E(x,t)$$
$$\leq \lim_{(y',s')\to(y,s)} f(y', s'),$$

in view of Lemma 8.13. This proves the result in this case, and the other case has a similar proof. $\square$

THEOREM 8.46. *Let E be an open set.*
(a) *If $q = (y, s)$ is a finite point of $\partial_e E$, then q is regular if and only if there is a barrier at q.*
(b) *If E is unbounded, then the point at infinity is regular.*

PROOF. (a) If there is a barrier at q and $f \in C(\partial_e E)$, then Corollary 8.45 shows that
$$\lim_{p\to q} S_f^E(p) = f(q)$$
if $q \in \partial_n E$, and that
$$\lim_{(x,t)\to(y,s+)} S_f^E(x,t) = f(y, s)$$
if $q \in \partial_{ss} E$. Hence q is regular.

Conversely, if q is regular, we put $N = E \cap B(q, \frac{1}{2})$, and define a function ψ on $\mathbb{R}^{n+1}$ by putting $\psi(p) = |p-q|^2$. The restriction of ψ to $\partial_e N$ belongs to $C(\partial_e N)$, and so Theorem 8.26 shows that it is resolutive for N. We show that the temperature S_ψ^N is a barrier at q. For all $(x, t) \in N$, we have $\Theta\psi(x,t) = 2(n - t + s) > 0$, so that ψ is a subtemperature on N. Therefore $\psi \in \mathfrak{L}_\psi^N$, so that $S_\psi^N \geq \psi > 0$ on N. Finally, since q is regular and $\psi(q) = 0$, we have
$$\lim_{p\to q} S_\psi^N(p) = 0$$
if $q \in \partial_n E$ (so that $q \in \partial_n N$), and
$$\lim_{(x,t)\to(y,s+)} S_\psi^N(x,t) = 0$$
if $q \in \partial_{ss} E$.

(b) Let $f \in C(\partial_e E)$ and take the value λ at infinity. Then f is resolutive, so that the function $g = f - \lambda$ is resolutive with $S_g^E = S_f^E - \lambda$, by Theorem 8.22. Hence $S_g^E(p) \to 0$ as $|p| \to \infty$ if and only if $S_f^E(p) \to \lambda$ as $|p| \to \infty$. Let $M = \max_{\partial_e E} |g|$. Given any positive number ϵ, we choose $R > 0$ such that $|g| < \epsilon$ on $\{p \in \partial_e E : |p| > R\}$. Then we choose a coheat ball $\Omega^*(p_0; c_0)$ that contains $\{p : |p| \leq R\}$, so that $(4\pi c_0)^{n/2} M G(\cdot; p_0) \geq |g|$ on $\{p \in \partial_e E : |p| \leq R\}$. It follows that the function $u_\epsilon = \epsilon + (4\pi c_0)^{n/2} M G(\cdot; p_0)$ belongs to the upper class $\mathfrak{U}_g^E$, and that $-u_\epsilon \in \mathfrak{L}_g^E$, so that $|S_g^E| \leq u_\epsilon$ on E. Since $u_\epsilon(p) \to \epsilon$ as $|p| \to \infty$, and ϵ is arbitrary, it follows that $S_g^E(p) \to 0$ as $|p| \to \infty$, as required. $\square$

COROLLARY 8.47. *Let E be an open set, let q be a finite point of $\partial_e E$, and let V be an open subset of E. If q is a regular point of $\partial_e E$, and either $q \in \partial_{ss} V$ or $q \in \partial_n V \cap \partial_n E$, then q is a regular point of $\partial_e V$.*

PROOF. If $q \in \partial_{ss} V$, then there is a half-ball $H(q, \rho) \subseteq V \subseteq E$, and hence $q \in \partial_{ss} E$. We use the same argument in the two cases $q \in \partial_{ss} V \cap \partial_{ss} E$ and $q \in \partial_n V \cap \partial_n E$. By Theorem 8.46, there is a barrier w at q relative to E, defined on $N \cap E$ for some open neighbourhood N of q, and the restriction of w to $N \cap V$ is a barrier at q relative to V. Now Theorem 8.46 shows that q is a regular point of $\partial_e V$. □

EXAMPLE 8.48. In the context of the above corollary, if $q \in \partial_n V \cap \partial_{ss} E$, then q may be a regular point of $\partial_e E$ but an irregular point of $\partial_e V$. For example, let $E = D_1 \cup D_2$ where $D_1 = B(x_0, 1) \times \,]0, 1[$ and $D_2 = B(x_0, 1) \times \,]1, 2[$, let $q = (x_0, 1)$, and let V be any heat ball $\Omega(q; c) \subseteq E$. Then q is a regular point of $\partial_e D_2$, and hence of $\partial_e E$, but an irregular point of $\partial_e V$ by Example 8.36.

8.6. Geometric Tests for Regularity

In this section, we give two geometric sufficient conditions for regularity, and a consequence of one of them. The first test was anticipated in the proofs of Theorem 3.39 and Lemma 8.38.

THEOREM 8.49. *Let E be an open set, and let q be a finite point of $\partial_n E$. If there exist a point $p \neq q$ and a positive number c such that $\overline{\Omega}^*(p; c) \cap \overline{E} = \{q\}$, then q is regular.*

PROOF. The function $\tau(c) - G(\cdot; p)$ is a temperature on $\mathbb{R}^{n+1} \backslash \{p\}$, is zero on $\partial \Omega^*(p; c)$, and is positive on $\Lambda^*(p; \mathbb{R}^{n+1}) \backslash \overline{\Omega}^*(p; c) \supseteq E$. It is therefore a barrier at q, so that q is regular by Theorem 8.46. □

THEOREM 8.50. *Any open set E can be written as a union of a sequence $\{E_k\}$ of bounded open sets such that, for each k,*
 (a) $\overline{E}_k \subseteq E_{k+1}$,
 (b) $\partial_s E_k = \emptyset$,
 (c) $\partial_{ss} E_k$ *has only finitely many points, all of which are irregular, and*
 (d) *every point of $\partial_n E_k$ is regular.*

PROOF. We use the following notation. Given a point $p = (x, t)$ and a positive number c, we put $p^c = (x, t - c/e)$. Note that $p \in \Omega^*(p^c; c)$.

We first take a bounded open set U, and show that the result holds for U. We define a sequence of open sets $\{V_k\}$ inductively, as follows. We choose a positive number c_1 such that $U \backslash \bigcup_{p \in \partial U} \Omega^*(p^{c_1}; c_1) \neq \emptyset$. Since the compact set $\partial U \subseteq \bigcup_{p \in \partial U} \Omega^*(p^{c_1}; c_1)$, we can find a finite subset $\{p_{1,1}, ..., p_{1,m(1)}\}$ of ∂U such that $\partial U \subseteq \bigcup_{i=1}^{m(1)} \Omega^*(p_{1,i}^{c_1}; c_1)$. We put

$$V_1 = U \backslash \bigcup_{i=1}^{m(1)} \overline{\Omega}^*(p_{i,1}^{c_1}; c_1).$$

We now suppose that $k \in \mathbb{N}$, that we have a positive number c_k and a finite subset $\{p_{k,1}, ..., p_{k,m(k)}\}$ of ∂U such that $\partial U \subseteq \bigcup_{i=1}^{m(k)} \Omega^*(p_{k,i}^{c_k}; c_k)$, and that we have

defined
$$V_k = U \setminus \bigcup_{i=1}^{m(k)} \overline{\Omega}^*(p_{k,i}^{c_k}; c_k).$$

Since the compact set ∂U is contained in $\bigcup_{i=1}^{m(k)} \Omega^*(p_{k,i}^{c_k}; c_k)$, we can find a positive number c_{k+1} such that $2c_{k+1} < c_k$ and $\Omega^*(p^{2c_{k+1}}; 2c_{k+1}) \subseteq \bigcup_{i=1}^{m(k)} \Omega^*(p_{k,i}^{c_k}; c_k)$ for all $p \in \partial U$. Because ∂U is contained in $\bigcup_{p \in \partial U} \Omega^*(p^{c_{k+1}}; c_{k+1})$, we can find a finite subset $\{p_{k+1,1}, ..., p_{k+1,m(k+1)}\}$ of ∂U such that
$$\partial U \subseteq \bigcup_{i=1}^{m(k+1)} \Omega^*(p_{k+1,i}^{c_{k+1}}; c_{k+1}).$$

We now define
$$V_{k+1} = U \setminus \bigcup_{i=1}^{m(k+1)} \overline{\Omega}^*(p_{k+1,i}^{c_{k+1}}; c_{k+1}) = \overline{U} \setminus \bigcup_{i=1}^{m(k+1)} \overline{\Omega}^*(p_{k+1,i}^{c_{k+1}}; c_{k+1}).$$

The sequence of open subsets $\{V_k\}$ of U, thus defined, satisfies
$$\overline{V}_k \subseteq \overline{U} \setminus \bigcup_{i=1}^{m(k)} \Omega^*(p_{k,i}^{c_k}; c_k) \subseteq \overline{U} \setminus \bigcup_{i=1}^{m(k+1)} \overline{\Omega}^*(p_{k+1,i}^{c_{k+1}}; c_{k+1}) = V_{k+1}$$

for all k. Moreover, each point $q \in \partial V_k$ also belongs to $\partial \Omega^*(p_{k,i}^{c_k}; c_k)$ for some i. Hence, if $q \neq p_{k,i}^{c_k}$, it is a normal boundary point of V_k, and is regular by Theorem 8.49. On the otherhand, if $q = p_{k,i}^{c_k}$ for some i, it may again be normal and regular if it belongs to $\partial \Omega^*(p_{k,j}^{c_k}; c_k)$ for some $j \neq i$, or it may be an isolated semi-singular boundary point, and thus irregular. This establishes the result for a bounded open set.

Given now an arbitrary open set E, we choose a sequence of bounded open sets $\{U_j\}$ such that $\overline{U}_j \subseteq U_{j+1}$ for all j, and $\bigcup_{j=1}^{\infty} U_j = E$. As above, for each j we can find a sequence of open sets $\{V_{j,k}\}$ such that, for each k, $\overline{V}_{j,k} \subseteq V_{j,k+1}$, $\partial_s V_{j,k} = \emptyset$, $\partial_{ss} V_{j,k}$ has only finitely many points (all of which are irregular), and every point of $\partial_n V_{j,k}$ is regular. For each $j \geq 2$, we choose a set V_{j,k_j} from the sequence $\{V_{j,k}\}$ such that $\overline{U}_{j-1} \subseteq V_{j,k_j}$. Putting $E_j = V_{j,k_j}$ for all j, we obtain the result of the theorem. $\square$

DEFINITION 8.51. Let $q = (y, s)$, let $r < s$, and let B be a closed n-dimensional ball. Given any point $x \in B$, we denote by γ_x the parabolic curve with vertex at q that passes through the point (x, r). The set
$$\Gamma = \bigcup_{x \in B} \gamma_x$$
is called a *parabolic tusk with vertex at* q.

THEOREM 8.52. *Let E be an open set, let $q_0 = (y_0, s_0) \in \partial_e E$, and suppose that there is a parabolic tusk Γ with vertex at q_0 such that*
$$\Gamma \cap (\mathbb{R}^n \times]r_0, s_0[) \subseteq \mathbb{R}^{n+1} \setminus E$$
for some $r_0 < s_0$. Then q_0 is a regular point of $\partial_n E$.

PROOF. By translating the set E, we can assume that $q_0 = 0$. By using a parabolic dilation, we can also assume that $r_0 = -1$. Given these assumptions, we choose a positive number ρ such that $\Gamma \cap (\mathbb{R}^n \times \{-1\}) \subseteq B(0,\rho) \times \{-1\}$, and let D denote the circular cylinder $B(0,\rho) \times]-1,1[$. If we produce a barrier at $q_0 = 0$ for $D\backslash\Gamma$, then its restriction to $E \cap (D\backslash\Gamma)$ will be a barrier for E.

We put $U = D\backslash\Gamma$, and define a function f on $\partial_e U = \partial_n U$ by putting

$$f(x,t) = \begin{cases} 1 & \text{if } (x,t) \in \partial_n U \backslash \Gamma, \\ -t & \text{if } (x,t) \in \partial_n U \cap \Gamma. \end{cases}$$

Since $\Gamma \cap (\mathbb{R}^n \times \{-1\}) \subseteq B(0,\rho) \times \{-1\}$, we see that $f \in C(\partial_e U)$, so that f is resolutive. Because $0 \leq f \leq 1$ on $\partial_e U$, we have $0 \leq S_f^U \leq 1$ on U. Furthermore, all points of $\partial_e U \backslash \{0\}$ are regular points of $\partial_n U$ by Theorem 8.49, so that $S_f^U(p) \to f(q)$ as $p \to q$ for all $q \in \partial_e U \backslash \{0\}$. If there was a point $p_0 \in U$ with $S_f^U(p_0) \subseteq \{0,1\}$, then the strong maximum principle would imply that $S_f^U(p) = S_f^U(p_0)$ for every $p \in \Lambda(p_0; U)$, which would contradict, for some points $q \in \Gamma \cap \partial_e U$, the fact that $S_f^U(p) \to f(q)$ as $p \to q$. Therefore $0 < S_f^U < 1$ on U, and it only remains to prove that $S_f^U(p) \to 0$ as $p \to 0$.

We now let C denote the circular cylinder $B(0, \frac{1}{2}\rho) \times]-\frac{1}{4}, \frac{1}{4}[$, and put $V = C\backslash\Gamma$. We define a function g on $\partial_e V = \partial_n V$ by putting

$$g = \begin{cases} f & \text{on } \Gamma \cap \partial_n V, \\ S_f^U & \text{on } U \cap \partial_n V. \end{cases}$$

Since all the points of $\partial_n U \backslash \{0\}$ are regular for U, the function g belongs to $C(\partial_e V)$, and hence is resolutive. Moreover, $0 \leq g < 1$ on $\partial_e V$. The restriction to V of S_f^U has continuous boundary values on $\partial_e V \backslash \{0\}$, and these coincide with those of S_g^V. Since $\{0\}$ is a polar set and the functions are bounded, it follows from Theorem 8.2 that $S_g^V = S_f^U$ on V. Therefore

$$\sup_V S_f^U \leq \sup_{\partial_e V} g.$$

Now $g \in C(\partial_e V)$, and $\partial_e V$ is compact, so that g has a maximum value. Hence $\sup_V S_f^U < 1$.

We now put

$$\alpha = \left(\frac{1}{4}\right) \vee \left(\sup_V S_f^U\right),$$

and define a function v on V by putting

$$v(x,t) = S_f^U(x,t) - \alpha S_f^U(2x, 4t).$$

Then v is a bounded temperature on V. At each point $(y,s) \in \partial_n V \backslash \Gamma$, we have

$$\limsup_{(x,t) \to (y,s)} v(x,t) = S_f^U(y,s) - \alpha \leq 0,$$

and at each point $(y,s) \neq (0,0)$ on $\partial_n V \cap \Gamma$ we have

$$\limsup_{(x,t) \to (y,s)} v(x,t) = -s - \alpha(-4s) = -s(1 - 4\alpha) \leq 0.$$

Therefore, by the maximum principle, $v \leq 0$ on V. It follows that

$$0 \leq \limsup_{(x,t) \to (0,0)} S_f^U(x,t) \leq \alpha \limsup_{(x,t) \to (0,0)} S_f^U(x,t) < +\infty.$$

Since $0 < \alpha < 1$, this implies that $\limsup_{(x,t)\to(0,0)} S_f^U(x,t) = 0$. Thus $S_f^U(p) \to 0$ as $p \to 0$, so that S_f^U is a barrier at 0, and hence 0 is regular by Theorem 8.46. □

8.7. Green Functions, Heat Potentials, and Thermal Capacity

In this section, we show that the Green function for an open set E, with pole at a point q, has a limit zero at every regular point of $\partial_n E$, and a one-sided limit zero at every regular point of $\partial_{ss} E$. So it is a barrier at any point $q_0 \in \partial_e E$, provided that it is positive on $E \cap N$ for some neighbourhood N of q_0. Theorem 8.53(c) gives a more general criterion for regularity that involves sets of such functions. Theorem 8.55 strengthens the zero limit property by showing that it holds locally uniformly, and Corollary 8.56 extends the property to heat potentials of measures with compact supports in E. The proof of Theorem 8.58 uses that last result to prove a characterization of the thermal capacity of a compact subset of E, subject to a condition on E.

Theorem 8.53(a) generalizes Theorem 6.5 from convex domains of revolution to arbitrary open sets.

THEOREM 8.53. *Let E be an open set.*

(a) The Green function for E has the representation $G_E(\cdot; q) = G(\cdot; q) - S_{G(\cdot;q)}^E$ for each point $q \in E$.

(b) For each point $(y, s) \in E$, as (x, t) tends to any regular point of $\partial_n E$ we have

$$\lim G_E(x, t; y, s) = 0; \tag{8.7}$$

and if (y_0, s_0) is any regular point of $\partial_{ss} E$, we have

$$\lim_{(x,t)\to(y_0,s_0+)} G_E(x, t; y, s) = 0. \tag{8.8}$$

(c) Conversely, given a finite point $q_0 = (y_0, s_0) \in \partial_e E$, and a countable set $\{p_k : k \in I\}$ of points in E such that

$$\bigcup_{k \in I} \Lambda^*(p_k; E) \supseteq E \cap N \tag{8.9}$$

for some neighbourhood N of q_0, then if $q_0 \in \partial_n E$ and (8.7) holds with $(y, s) = p_k$ for every $k \in I$, or if $q_0 \in \partial_{ss} E$ and (8.8) holds with $(y, s) = p_k$ for every $k \in I$, then q_0 is regular for E.

PROOF. (a) We fix a point $q \in E$, and observe that the restriction of $G(\cdot; q)$ to $\partial_e E$ belongs to $C(\partial_e E)$, and so is resolutive. Furthermore $G(\cdot; q) \in \mathfrak{U}_{G(\cdot;q)}^E$, so that $G(\cdot; q) \geq S_{G(\cdot;q)}^E$ on E. Hence, if h_q denotes the greatest thermic minorant of $G(\cdot; q)$ on E, we have $h_q \geq S_{G(\cdot;q)}^E$ on E. On the other hand, since $h_q \leq G(\cdot; q)$ on E, we have $h_q \in \mathfrak{L}_{G(\cdot;q)}^E$, so that $h_q \leq S_{G(\cdot;q)}^E$ on E. Therefore $h_q = S_{G(\cdot;q)}^E$ on E, and the required representation follows from the definition of $G_E(\cdot; q)$.

(b) If (y_0, s_0) is a finite regular point of $\partial_n E$, then the representation in (a) shows that

$$\lim_{(x,t)\to(y_0,s_0)} G_E(x, t; y, s) = G(y_0, s_0; y, s) - \lim_{(x,t)\to(y_0,s_0)} S_{G(\cdot,\cdot;y,s)}^E(x, t) = 0.$$

If E is unbounded, then an analogous calculation works for the point at infinity, in view of Theorem 8.46(b). Similarly, if (y_0, s_0) is a regular point of $\partial_{ss} E$, then (a) implies that (8.8) holds.

(c) We may suppose that I has the form $I = [1, m] \cap \mathbb{N}$ for some $m \in \mathbb{N} \cup \{+\infty\}$. For each $k \in I$, we put $\Lambda_k^* = \Lambda^*(p_k; E)$. We define a function u on E by putting

$$u = \sum_{k=1}^{m} \left(G_E(\cdot; p_k) \wedge 2^{-k-1} \right),$$

and show that u is a barrier at q_0. By Theorem 6.7, $G_E(\cdot; p_k) > 0$ on Λ_k^* for each k, so that condition (8.9) implies that $u > 0$ on $E \cap N$. The function u is a supertemperature on E, by Corollary 3.57 if $m < +\infty$, and by Corollary 3.57 and Theorem 3.60 if $m = +\infty$. Let ϵ be a given positive number.

If q_0 is a finite point of $\partial_n E$, and (8.7) holds with $(y, s) = p_k$ for every $k \in I$, then for each k we can find a ball $B_k = B(q_0, r_k)$ such that $G_E(\cdot; p_k) < 2^{-k-1}\epsilon$ on $B_k \cap E$. For any integer $j \in I$ such that $2^{-j} < \epsilon$, we put $N_j = \bigcap_{k=1}^{j} B_k$. Then on $N_j \cap E$ we have

$$u \leq \sum_{k=1}^{m \wedge j} 2^{-k-1} \epsilon + \sum_{k=(m \wedge j)+1}^{m} 2^{-k-1} \leq \frac{\epsilon}{2} + 2^{-j-1} < \epsilon.$$

Thus $\lim_{p \to q_0} u(p) = 0$, so that u is a barrier for E at q_0, and hence q_0 is regular for E, by Theorem 8.46.

On the other hand, if $q_0 \in \partial_{ss} E$, and (8.8) holds with $(y, s) = p_k$ for every k, then for each k we can find an open upper half-ball $H_k^* = H^*(q_0, r_k)$ such that $G_E(\cdot; p_k) < 2^{-k-1}\epsilon$ on $H_k^* \cap E$. Now an argument similar to that for the previous case shows that q_0 is regular for E. □

REMARK 8.54. In Theorem 8.53(c), if q_0 is a finite point of $\partial_n E$, then every open lower half-ball centred at q_0 meets $\mathbb{R}^{n+1} \backslash E$, so that every such half-ball meets $\mathbb{R}^{n+1} \backslash \Lambda_k^*$ for all k. Hence $q_0 \in \partial_n \Lambda_k^*$ whenever $q_0 \in \partial \Lambda_k^*$. Thus condition (8.7) is appropriate.

On the other hand, if $q_0 \in \partial_{ss} E$, then each of the four possibilities $q_0 \in \partial_s \Lambda_k^*$, $q_0 \notin \partial \Lambda_k^*$, $q_0 \in \partial_n \Lambda_k^*$, or $q_0 \in \partial ss \Lambda_k^*$ may occur. Consider, for example, the cases $E = \mathbb{R}^{n+1} \backslash (B(y_0, 1) \times \{s_0\})$ with $p_k = (y_0, s_0 - 1)$,

$$E = \mathbb{R}^{n+1} \backslash ((B(y_0, 1) \times \{s_0\}) \cup (\partial B(y_0, 1) \times [s_0 - 1, s_0]))$$

with $p_k = (y_0, s_0 - 1)$ and with $p_k = (z_0, s_0 - 1)$ for $|z_0 - y_0| > 1$. However, in all cases only the one-sided approach to q_0, as in (8.8), is appropriate.

We now use Bauer's form of the Harnack inequality to show that the result of Theorem 8.53(b) holds uniformly on compact subsets of E.

THEOREM 8.55. *If K is a compact subset of the open set E, then as p tends to any regular point of $\partial_n E$ we have $\lim G_E(p; \cdot) = 0$, uniformly on K, and for any regular point (y_0, s_0) of $\partial_{ss} E$ we have $\lim_{(x,t) \to (y_0, s_0+)} G_E(x, t; \cdot, \cdot) = 0$ uniformly on K.*

PROOF. Let q_0 denote any regular point of $\partial_n E$, and let U be a bounded open superset of K such that $\overline{U} \subseteq E$. The compact set K is at a positive distance from ∂U, so that for any point $q \in K$ we can find a point $q' \in U \backslash K$ such that

$q \in \Lambda^*(q'; U)$, and hence

$$\bigcup_{q' \in U \setminus K} \Lambda^*(q'; U) = \bigcup_{q' \in U} \Lambda^*(q'; U) = U.$$

Since K is compact, we can therefore find a finite set set $S = \{q'_1, ..., q'_k\} \subseteq U \setminus K$ such that

$$K \subseteq \bigcup_{i=1}^k \Lambda^*(q'_i; U).$$

We now define a measure ν on U, supported by S, by putting $\nu(\{q'_i\}) = 1$ for all i. Then K is a compact subset of U such that for each point $r \in K$ there is a point $r' \in S$ such that $r \in \Lambda^*(r'; U)$. Therefore the dual of Theorem 1.32 for the adjoint equation shows that there is a constant κ such that

$$\max_K u \leq \kappa \int_S u \, d\nu = \kappa \sum_{i=1}^k u(q'_i)$$

for every nonnegative cotemperature u on U. In particular,

$$\max_K G_E(p; \cdot) \leq \kappa \sum_{i=1}^k G_E(p; q'_i)$$

for all $p \in E \setminus U$. By Theorem 8.53(b), we have $\lim_{p \to q_0} G_E(p; q'_i) = 0$ for each point q'_i, so that $\lim_{p \to q_0} \max_K G_E(p; \cdot) = 0$; that is, $\lim_{p \to q_0} G_E(p; \cdot) = 0$ uniformly on K.

The proof for the case where (y_0, s_0) is a regular point of $\partial_{ss} E$ is similar. $\square$

COROLLARY 8.56. *Let E be an open set, and let μ be a nonnegative measure whose support is a compact subset of E. Then as p tends to any regular point of $\partial_n E$ we have $\lim G_E \mu(p) = 0$, and for any regular point (y_0, s_0) of $\partial_{ss} E$ we have*

$$\lim_{(x,t) \to (y_0, s_0+)} G_E \mu(x, t) = 0.$$

PROOF. If q_0 denotes any regular point of $\partial_n E$, then it follows from Theorem 8.55 that

$$\lim_{p \to q_0} G_E \mu(p) = \int_K \lim_{p \to q_0} G_E(p; \cdot) \, d\mu = 0.$$

The proof for a regular point of $\partial_{ss} E$ is similar. $\square$

We use Corollary 8.56 to prove a characterization of the thermal capacity of compact sets, relative to certain open sets which we define below. In Lemma 7.54, this characterization was given for a very restricted class of sets.

DEFINITION 8.57. *An open set E is called quasi-regular if the set of irregular points of $\partial_e E$ is polar.*

It follows from Corollary 3.41 and Theorem 7.55 that every convex domain of revolution is quasi-regular.

THEOREM 8.58. *Let E be a quasi-regular open set, let K be a compact subset of E, and let $M^+(K)$ denote the collection of all nonnegative measures on E whose supports are contained in K. Then the thermal capacity of K relative to E is given by*

$$\mathcal{C}(K) = \max\{\mu(K) : \mu \in M^+(K),\ G_E\mu \leq 1 \text{ on } E\}.$$

PROOF. Let $\mu \in M^+(K)$ and satisfy $G_E\mu \leq 1$ on E. Let v be a nonnegative supertemperature on E such that $v \geq 1$ on K. Then for each point $q \in \partial K$, we have

$$\liminf_{p \to q,\ p \notin K}(v(p) - G_E\mu(p)) \geq \liminf_{p \to q} v(p) - \limsup_{p \to q} G_E\mu(p) \geq v(q) - 1 \geq 0.$$

Furthermore, whenever q is a regular point of $\partial_n E$, we have

$$\liminf_{p \to q}(v(p) - G_E\mu(p)) \geq 0 - \lim_{p \to q} G_E\mu(p) = 0,$$

by Corollary 8.56. Similarly, whenever (y,s) is a regular point of $\partial_{ss} E$, we have

$$\liminf_{(x,t) \to (y,s+)} (v(x,t) - G_E\mu(x,t)) \geq 0.$$

By Corollary 6.22, the heat potential $G_E\mu$ is a temperature on $E\backslash K$, so that $v - G_E\mu$ is a supertemperature there, and clearly $v - G_E\mu \geq -1$ on E. Since E is quasi-regular, it now follows from the boundary point minimum principle of Theorem 8.2 that $v - G_E\mu \geq 0$ on $E\backslash K$, and clearly $v \geq 1 \geq G_E\mu$ on K. Thus $v \geq G_E\mu$ on E, which implies that $R_1^K \geq G_E\mu$ on E, and hence that $\widehat{R}_1^K \geq G_E\mu$ on E because $G_E\mu$ is lower semicontinuous on E. Writing $\widehat{R}_1^K = G_E\omega_K$, we obtain $G_E\omega_K \geq G_E\mu$ on E, so that Lemma 7.37(a) shows that $\omega_K(E) \geq \mu(E)$. Since $\omega_K \in M^+(K)$, we have $\mathcal{C}(K) = \omega_K(K) \geq \mu(K)$, as required. □

8.8. Notes and Comments

This version of the Dirichlet problem, in which the prescribed function is defined only on the essential boundary and only one-sided regularity is sought at point of the semi-singular boundary, is taken from Watson [**72, 90**]. It seems to be the most natural extension to arbitrary open sets of the versions adopted by classical authors, who considered only special open sets. It acknowledges that the temporal variable behaves differently from the spatial variables, so that prescribed values can rarely by taken at the latest time. It does, however, contrast with the version adopted by several other authors, including Bauer [**5**], Constantinescu & Cornea [**12**], and Doob [**14**], in which the prescribed function is defined on the whole of the boundary and full regularity is sought at every point. Their approach has the advantage that elliptic and parabolic equations can be treated together in the same axiomatic system. However, rarely does an axiomatic system do justice to any particular case that it covers. Gains in generality are inevitably accompanied by losses in precision. The approach adopted here is more closely aligned with the maximum principle, gives a wider class of regular sets, gives information about one-sided limits at semi-singular boundary points, and allows a systematic treatment of caloric measure for arbitrary open sets.

Caloric measure, sometimes called harmonic measure for the heat equation, sometimes parabolic measure, has been studied by many authors for particular open sets, including Bousch & Heurteaux [**8**], Fabes & Salsa [**19**], Fabes, Garofalo & Salsa [**20**], Heurteaux [**34**], Hofmann, Lewis & Nyström [**35**], Kaufman & Wu

[**38, 39, 40, 41**], Kemper [**42**], Lewis & Silver [**50**], Nyström [**57**], and Wu [**92, 93, 94**]. Moreover, assuming that there exists a caloric measure on the whole boundary of an arbitrary open set, Suzuki [**64**] has proved that it must be supported by the essential boundary.

The regularity of boundary points for particular domains has a long history. Some notable early works are those of Petrowsky [**58**], Sternberg [**63**], and Pini [**59**]. Effros & Kazdan [**15**] discuss the topic from a different perspective and include several examples to illustrate the complexity of the problem. The parabolic tusk test, given in Theorem 8.52, comes from Effros & Kazdan [**16**]. More recently, Abdulla [**1**] gave a sufficient condition for the regularity of a finite normal boundary point of any open set, which is different to, but resembles, the parabolic tusk test. There is also a necessary and sufficient condition for the regularity of the boundary point (x_0, b) of the domain of revolution $\{(x,t) : |x - x_0| < \rho(t), a < t < b\}$, in Abdulla [**2**].

Theorems 8.50 and 8.55 appear to be new, although the result of Corollary 8.56 is given in Watson [**73**] for bounded open sets. A different form of Theorem 8.58 is also given in Watson [**73**].

The Wiener criterion is a very important test for the regularity of normal boundary points, but is not included in the text because of the length of its proof. It states that a point $p_0 \in \partial_n E$ is regular if and only if

$$\sum_{k=1}^{\infty} 2^{kn/2} \mathcal{C}(\overline{A}(p_0; 2^{-k-1}, 2^{-k}) \backslash E) = +\infty,$$

where $\mathcal{C}$ denotes thermal capacity and $A(p; b, c)$ denotes the heat annulus of centre p, inner radius b and outer radius c. The necessity of this condition for regularity was proved by Lanconelli [**46**], and the sufficiency by Evans & Gariepy [**18**]. A different necessary and sufficient condition for regularity is given by Landis [**48**].

Watson [**82**] considered a different kind of regularity for semi-singular boundary points, in which limits in the cothermal fine topology replace the one-sided limits considered here. The main result is that the set of 'cofine' irregular points of the semi-singular boundary is polar, for any open set.

Taylor & Watson [**67**] give examples to show that the set of irregular normal boundary points is not generally polar.

CHAPTER 9

The Thermal Fine Topology

The thermal fine topology is an excellent tool for studying the continuity traits of supertemperatures. In this topology every supertemperature is continuous, and it is, in a sense, the minimal such topology. It is thus a natural topology for the study of heat potential theory. It enables us to improve some of the results about reductions given in Chapter 7, some by obtaining similar conclusions under weaker hypotheses, others by sharpening the conclusions. The main result, Theorem 9.27, characterizes those sets where the infimum of a family of supertemperatures differs from its lower semicontinuous smoothing. This result has several applications to reductions and, conversely, reductions are used to study the thermal fine topology. In Section 9.5 we give a characterization, in terms of the thermal fine topology, of those normal boundary points of an open set which are regular for the Dirichlet problem. Subsequently, we discuss the relationship between the thermal fine limits of functions and their Euclidean limits. Finally, we show that the thermal fine topology has a property which is close to the well known Lindelöf property that every collection of open sets has a countable subcollection with the same union.

9.1. Definitions and Basic Properties

Recall that the topology of a topological space is the class of all open subsets of the space.

DEFINITION 9.1. If $\mathcal{T}_1$ and $\mathcal{T}_2$ are topologies on the same set, and $\mathcal{T}_2 \subseteq \mathcal{T}_1$, then we say that $\mathcal{T}_1$ is *finer* than $\mathcal{T}_2$, and that $\mathcal{T}_2$ is *coarser* than $\mathcal{T}_1$.

For any family of extended real-valued functions on a space, there is a coarsest topology on the space that makes every member continuous. It is the intersection of all topologies that make every member continuous. In heat potential theory there are two fine topologies, the thermal fine topology which emanates from the heat operator, and the cothermal fine topology which emanates from the adjoint operator. The cothermal fine topology is just the dual of the thermal one, so we give a detailed discussion only for the latter.

DEFINITION 9.2. The *Thermal Fine Topology* is the coarsest topology on $\mathbb{R}^{n+1}$ that makes every supertemperature on $\mathbb{R}^{n+1}$ continuous.

Thus sets of the form $w^{-1}(D)$, where w is a supertemperature on $\mathbb{R}^{n+1}$ and D an open subset of the extended real line, are open in the thermal fine topology. One subbase for this topology consists of all sets of the forms $\{p : w(p) < a\}$ and $\{p : w(p) > a\}$, where w is a supertemperature on $\mathbb{R}^{n+1}$ and $a \in \mathbb{R}$. Sets of the latter form are open in the Euclidean topology.

Concepts relative to the thermal fine topology will be prefixed with "$\Theta - f$"; for example, $\Theta - f\lim$, $\Theta - f\limsup$. Concepts with no prefix will refer to the Euclidean topology.

Any coheat ball $\Omega^*(q;c) = \{p : G(p;q) > \tau(c)\}$ is open in the thermal fine topology. Any (Euclidean) open set can be written as a union of coheat balls, and is therefore open in the thermal fine topology. Thus the thermal fine topology is at least as fine as the Euclidean topology. Since $G(\cdot;q)$ is a supertemperature discontinuous at q, the thermal fine topology is strictly finer.

More strikingly, if $A = \mathbb{R}^n \times\,]a,+\infty[$ for some $a \in \mathbb{R}$, then its characteristic function χ_A is a supertemperature, and hence thermal fine continuous. Therefore no point of the hyperplane $\mathbb{R}^n \times \{a\}$ is a thermal fine limit point of A, and in fact the half-space $\mathbb{R}^n \times\,]-\infty,a] = \{p : \chi_A(p) < 1\}$ is a thermal fine open set. It follows that the set

$$(\mathbb{R}^n \times\,]-\infty,a]) \cap B((x,a),r) = \{(y,s) : |x-y|^2 + (a-s)^2 < r^2,\, s \leq a\}$$

is a thermal fine open neighbourhood of (x,a), for any $x \in \mathbb{R}^n$ and any positive number r. Hence the closed half-ball $\overline{H}(q,r)$ is a thermal fine neighbourhood of q. Therefore, if $D = B\times\,]c,d[$ is an open circular cylinder, then the set $\overline{D}\backslash \partial_n D$ is thermal fine open.

LEMMA 9.3. *If w is a supertemperature on an open set E, then w is thermal fine continuous on E.*

PROOF. Let μ denote the Riesz measure associated with w on E. Let B be an open ball such that $\overline{B} \subseteq E$. The restriction μ_B of μ to B is finite, so that $G\mu_B$ is a heat potential on $\mathbb{R}^{n+1}$, by Theorem 6.18. Therefore, by Theorem 6.24, there is a nonnegative temperature u on B such that $G\mu_B = G_B\mu_B + u$ on B. The heat potential $G\mu_B$ is thermal fine continuous on $\mathbb{R}^{n+1}$, and differs from $G_B\mu_B$ by a continuous function on B. Therefore $G_B\mu_B$ is thermal fine continuous on B. Since w is lower bounded on B, the Riesz decomposition theorem shows that $w = G_B\mu_B + h$, where h is a temperature on B. It follows that w is thermal fine continuous on B, and hence, due to the arbitrary nature of B, on E. □

LEMMA 9.4. *A polar set has no thermal fine limit point.*

PROOF. Let Z be a polar set, and let $p_0 \in \mathbb{R}^{n+1}$. We can assume that $p_0 \notin Z$. By Theorem 7.3, we can find a heat potential v on $\mathbb{R}^{n+1}$ such that $v(p) = +\infty$ for all $p \in Z$ and $v(p_0) < +\infty$. The set $V = \{q : v(q) < v(p_0) + 1\}$ is thermal fine open and contains p_0, but does not meet Z. Thus the arbitrary point p_0 is not a thermal fine limit point of Z. □

REMARK 9.5. Example 9.21 below shows that the converse of Lemma 9.4 is false, by exhibiting a set with no thermal fine limit point that is not polar.

LEMMA 9.6. *The thermal fine topology has a neighbourhood base consisting of (Euclidean) compact sets.*

PROOF. Let $p_0 \in \mathbb{R}^{n+1}$, and let N be a thermal fine neighbourhood of p_0. Then there exist supertemperatures $w_1,...,w_m$ on $\mathbb{R}^{n+1}$, and real numbers $a_1,...,a_m$, such

that
$$p_0 \in \left(\bigcap_{i=1}^{k}\{q : w_i(q) < a_i\}\right) \cap \left(\bigcap_{i=k+1}^{m}\{q : w_i(q) > a_i\}\right) \subseteq N.$$

The sets $\{q : w_i(q) > a_i\}$ are (Euclidean) open, and so we can find a coheat ball $\Omega^*(p;c)$ such that
$$p_0 \in \Omega^*(p;c) \subseteq \overline{\Omega}^*(p;c) \subseteq \bigcap_{i=k+1}^{m}\{q : w_i(q) > a_i\}.$$

We choose numbers $b_1, ..., b_k$ such that $w_i(p_0) < b_i < a_i$ for all $i \leq k$. Then the set
$$\left(\bigcap_{i=1}^{k}\{q : w_i(q) \leq b_i\}\right) \cap \overline{\Omega}^*(p;c)$$

is a thermal fine neighbourhood of p_0, is (Euclidean) compact, and is contained in N. $\square$

THEOREM 9.7. *The set $\mathbb{R}^{n+1}$, endowed with the thermal fine topology, is a Baire space. That is, if $\{D_i\}$ is a sequence of thermal fine open sets, each of which is thermal fine dense in $\mathbb{R}^{n+1}$, then $\bigcap_{i=1}^{\infty} D_i$ is also thermal fine dense in $\mathbb{R}^{n+1}$.*

PROOF. Let D be a nonempty, thermal fine open subset of $\mathbb{R}^{n+1}$. Then, for each i, $D \cap D_i \neq \emptyset$. By Lemma 9.6, there is a compact set K_1, with nonempty thermal fine interior, such that $K_1 \subseteq D \cap D_1$. Similarly, there is a compact set K_2, with nonempty thermal fine interior, such that $K_2 \subseteq K_1 \cap D_2 (\subseteq D \cap (D_1 \cap D_2))$. We proceed inductively. Given compact sets $K_1, ..., K_m$ ($m \geq 2$), with nonempty thermal fine interiors, such that $K_{i+1} \subseteq K_i \cap D_{i+1} (\subseteq D \cap (\bigcap_{j=1}^{i+1} D_j))$ whenever $i \leq m-1$, we can find a compact set K_{m+1}, with nonempty thermal fine interior, such that $K_{m+1} \subseteq K_m \cap D_{m+1} (\subseteq D \cap (\bigcap_{j=1}^{m+1} D_j))$. Since $K_i \neq \emptyset$ for any i, the nested sequence of compact sets $\{K_i\}$ has nonempty intersection, so that
$$\emptyset \neq \bigcap_{i=1}^{\infty} K_i \subseteq D \cap \left(\bigcap_{i=1}^{\infty} D_i\right).$$

Thus $\bigcap_{i=1}^{\infty} D_i$ meets the arbitrary nonempty thermal fine open set D, and hence is thermal fine dense in $\mathbb{R}^{n+1}$. $\square$

DEFINITION 9.8. A set S is said to be *thermally thin* at a point if that point is not a thermal fine limit point of S.

Thus S is thermally thin at p if there is a thermal fine neighbourhood of p that does not meet $S\setminus\{p\}$.

Clearly, if S is thermally thin at p then every subset of S is thermally thin at p. If $m \in \mathbb{N}$ and $S_1, ..., S_m$ are thermally thin at p, then so is $\bigcup_{i=1}^{m} S_m$, because the intersection of finitely many thermal fine neighbourhoods of p is itself a thermal fine neighbourhood of p.

EXAMPLE 9.9. By Lemma 9.4, any polar set is thermally thin at every point of $\mathbb{R}^{n+1}$.

EXAMPLE 9.10. If S is a set, and q is a point such that $\overline{H}(q,r) \cap S = \emptyset$ for some $r > 0$, then S is thermally thin at q because the closed half-ball is a thermal fine neighbourhood of q. Thus, if E is an open set and $q \in E \cup \partial_a E$, then $\mathbb{R}^{n+1} \backslash \overline{E}$ is thermally thin at q.

THEOREM 9.11. *Let q be a limit point of the set S. If there is a supertemperature w on an open neighbourhood of q such that*

(9.1) $$\liminf_{p \to q, \, p \in S} w(p) > w(q),$$

then S is thermally thin at q.

Conversely, if S is thermally thin at q, then there is a supertemperature w on $\mathbb{R}^{n+1}$ such that (9.1) holds.

PROOF. If S is not thermally thin at q, and w is a supertemperature on an open neighbourhood of q, then the thermal fine continuity of w at q implies that

$$\liminf_{p \to q, \, p \in S} w(p) \leq \Theta - f \lim_{p \to q, \, p \in S} w(p) = w(q),$$

so that (9.1) does not hold.

Conversely, if S is thermally thin at q, we can suppose that $q \notin S$. Then there is a thermal fine neighbourhood N of q such that $N \cap S = \emptyset$. There exist supertemperatures $w_1, ..., w_m$ on $\mathbb{R}^{n+1}$, and real numbers $a_1, ..., a_m$, such that

$$q \in \left(\bigcap_{i=1}^{k} \{p : w_i(p) < a_i\} \right) \cap \left(\bigcap_{i=k+1}^{m} \{p : w_i(p) > a_i\} \right) \subseteq N,$$

and hence an open set V such that

$$q \in \left(\bigcap_{i=1}^{k} \{p : w_i(p) < a_i\} \right) \cap V \subseteq N.$$

Our hypothesis that q is a limit point of S implies that $V \cap S \neq \emptyset$, so that $\bigcap_{i=1}^{k}$ is not vacuous. Since $w_i(q) < a_i$ for all $i \leq k$, we can choose a positive number ϵ such that $\epsilon < a_i - w_i(q)$ for all such i. Then the set

$$U = \bigcap_{i=1}^{k} \{p : w_i(p) < w_i(q) + \epsilon\} \cap V$$

is a thermal fine neighbourhood of q and a subset of N. We put $w = \sum_{i=1}^{k} w_i$, which is a supertemperature on $\mathbb{R}^{n+1}$. Each function w_i is lower semicontinuous at q, and hence we can find an open neighbourhood X of q such that $X \subseteq V$ and

$$w_i(p) > w_i(q) - \frac{\epsilon}{k}$$

for all $i \leq k$ and $p \in X$. Let $p \in S \cap X$. Since $S \cap U \subseteq S \cap N = \emptyset$ we know that $p \notin U$, so that $w_j(p) \geq w_j(q) + \epsilon$ for some $j \leq k$. Hence

$$w(p) = \sum_{i=1}^{k} w_i(p) > \sum_{i=1}^{k} w_i(q) - \left(\frac{k-1}{k}\right)\epsilon + \epsilon = w(q) + \frac{\epsilon}{k}.$$

The inequality (9.1) follows. □

COROLLARY 9.12. *If a set S is thermally thin at one of its limit points q, then there is an open superset of $S \backslash \{q\}$ that is also thermally thin at q.*

PROOF. By Theorem 9.11, there is a supertemperature w on $\mathbb{R}^{n+1}$ such that (9.1) holds. We put $l = \liminf_{p \to q,\, p \in S} w(p)$, and choose a number m such that $l > m > w(q)$. Then we can find a closed neighbourhood N of q such that $w(p) > m$ for all $p \in (S \cap N) \backslash \{q\}$. The set $T = \{p : w(p) > m\} \cup (\mathbb{R}^{n+1} \backslash N)$ is open and contains $S \backslash \{q\}$. Moreover, because $\liminf_{p \to q,\, p \in T} w(p) \geq m > w(q)$, Theorem 9.11 shows that T is thermally thin at q. □

COROLLARY 9.13. *If L is a thermal fine open set and $q \in L$, then q is a thermal fine limit point of L.*

PROOF. If $q = (y, s)$ and $S = \mathbb{R}^n \times\,]-\infty, s[$, then Lemma 3.16 shows that $\liminf_{p \to q,\, p \in S} w(p) = w(q)$ for every supertemperature w on $\mathbb{R}^{n+1}$. Hence q is a thermal fine limit point of S, by Theorem 9.11. It follows that, for each positive number r, the thermal fine neighbourhood $L \cap B(q, r)$ of q contains a point other than q. Thus q is a thermal fine limit point of L. □

If S is thermally thin at a limit point q, we can improve upon Theorem 9.11 by showing that there is a heat potential w on $\mathbb{R}^{n+1}$ such that (9.1) holds with the lower limit equal to $+\infty$. We illustrate this with the following example, which is required for the proof of the general result.

EXAMPLE 9.14. If $A = \mathbb{R}^n \times\,]a, +\infty[$ for some $a \in \mathbb{R}$, then we know that A is thermally thin at every point of $\mathbb{R}^n \times \{a\}$. Let $q = (y, a)$ for some $y \in \mathbb{R}^n$. By Theorem 7.57 with $Z = \{q\}$, there is a positive temperature u on A such that

$$\lim_{p \to q,\, p \in A} u(p) = +\infty.$$

By Theorem 4.18, u is the Gauss-Weierstrass integral of a nonnegative measure on $\mathbb{R}^n$. Therefore, by Example 6.14, u can be extended by 0 to a heat potential v on $\mathbb{R}^{n+1}$. Then

$$\lim_{p \to q,\, p \in A} v(p) = +\infty > 0 = v(q).$$

THEOREM 9.15. *Let q be a limit point of the set S. If S is thermally thin at q, then there is a heat potential $G\mu$ on $\mathbb{R}^{n+1}$ such that*

(9.2) $$+\infty = \lim_{p \to q,\, p \in S} G\mu(p) > G\mu(q).$$

PROOF. Let $q = (y, a)$. It suffices to prove the result with S replaced by the set $T = \{(x, t) \in S : t \leq a\}$. For if we can find a heat potential u on $\mathbb{R}^{n+1}$ such that

$$+\infty = \lim_{p \to q,\, p \in T} u(p) > u(q),$$

and v is the heat potential in Example 9.14 above, then the heat potential $w = u + v$ satisfies (9.2).

Suppose that T is thermally thin at q. Then Theorem 9.11 shows that there is a supertemperature w on $\mathbb{R}^{n+1}$ such that

$$\liminf_{p \to q,\, p \in T} w(p) > w(q).$$

Let D be a bounded open set that contains q. Corollary 6.37, shows that there is a temperature h such that $w = G\nu + h$ on D, where ν is the restriction to D of the

Riesz measure associated with w. Since
$$\liminf_{p \to q,\, p \in T} G\nu(p) = \liminf_{p \to q,\, p \in T} w(p) - \lim_{p \to q} h(p) > w(q) - h(q) = G\nu(q),$$
we can replace w with $G\nu$. We can also assume that $\nu(\{q\}) = 0$, because the replacement of ν by its restriction to $\mathbb{R}^{n+1}\backslash\{q\}$ would not change the values of the heat potential on $T \cup \{q\}$.

Given any positive number r, we denote by ν_r the restriction of ν to the open ball $B(q,r)$. By Corollary 6.37, for each r there is a temperature h_r such that $G\nu = G\nu_r + h_r$ on $B(q,r)$. Therefore, if
$$\delta = \liminf_{p \to q,\, p \in T} G\nu(p) - G\nu(q),$$
then for all r we have

(9.3) $\quad \liminf_{p \to q,\, p \in T} G\nu_r(p) = \liminf_{p \to q,\, p \in T} G\nu(p) - h_r(q) = \delta + G\nu_r(q) \geq \delta > 0.$

Since $0 = \nu(\{q\}) = \lim_{r \to 0+} \nu(B(q,r))$, we can find a null sequence $\{r_i\}$ such that $\sum_{i=1}^{\infty} \nu(B(q,r_i)) < +\infty$. For each i, we put $\mu_i = \nu_{r_i}$ and note that
$$\sum_{i=1}^{\infty} \mu_i(\mathbb{R}^{n+1}) = \sum_{i=1}^{\infty} \nu(B(q,r_i)) < +\infty.$$

We now put $\mu = \sum_{i=1}^{\infty} \mu_i$. Then $G\mu$ is a heat potential on $\mathbb{R}^{n+1}$, by Theorem 6.18. Furthermore, because $G\nu(q) < +\infty$, we have $\lim_{i \to \infty} G\mu_i(q) = 0$. Therefore, by switching to a subsequence if necessary, we can arrange to have $G\mu_i(q) < 2^{-i}$ for all i, so that $G\mu(q) < 1$. Then (9.3) implies that
$$\liminf_{p \to q,\, p \in T} G\mu(p) \geq \sum_{i=1}^{\infty} \liminf_{p \to q,\, p \in T} G\mu_i(p) = +\infty > 1 > G\mu(q),$$
as required. $\qquad\square$

COROLLARY 9.16. *If a set S is thermally thin at one of its limit points q, and $0 < a < b < +\infty$, then there is a heat potential $G\mu$ on $\mathbb{R}^{n+1}$, and a neighbourhood N of q, such that $G\mu(p) > b$ for all $p \in S \cap N\backslash\{q\}$ but $G\mu(q) < a$.*

PROOF. By Theorem 9.15, there is a heat potential $G\mu$ on $\mathbb{R}^{n+1}$ such that (9.2) holds. If $G\mu(q) = 0$, then the result follows immediately. If $G\mu(q) > 0$, then multiplication of $G\mu$ by a positive constant gives a heat potential v such that $v(q) < a$, and the result follows. $\qquad\square$

COROLLARY 9.17. *If a Borel set S is thermally thin at one of its limit points q, then*
$$\lim_{c \to 0+} \mathcal{L}(\chi_S; q; c) = \lim_{c \to 0+} \mathcal{M}(\chi_S; q; c) = 0.$$

PROOF. By Theorem 9.15, there is a heat potential $G\mu$ on $\mathbb{R}^{n+1}$ such that (9.2) holds. Because of the equality in (9.2), we have both $\inf_{S \cap \partial_n \Delta(q;c)} G\mu(p) \to +\infty$ and $\inf_{S \cap \partial \Omega(q;c)} G\mu(p) \to +\infty$ as $c \to 0+$. Moreover, for any $c > 0$ we have
$$+\infty > G\mu(q) \geq \mathcal{L}(G\mu; q; c) \geq \left(\inf_{S \cap \partial_n \Delta(q;c)} G\mu\right) \mathcal{L}(\chi_S; q; c)$$

by Theorem 3.17, and similarly

$$+\infty > \left(\inf_{S \cap \partial\Omega(q;c)} G\mu\right) \mathcal{M}(\chi_S; q; c)$$

by Theorem 3.48. The result follows. □

9.2. Further Properties of Reductions

In this section, we give properties of reductions in addition to those presented in Chapter 7, and also prove some extensions of those earlier results. These will be applied in later sections to prove further properties of the thermal fine topology.

The first theorem of this section is a variant of Theorem 7.31(d), in which the replacement of open sets by thermal fine open sets permits a similar conclusion without the smoothness hypothesis on the supertemperature.

THEOREM 9.18. *Let v be a nonnegative supertemperature on an open set E, and let $L \subseteq E$. If v is finite-valued on L, then*

$$R_v^L = \inf\{R_v^C : C \text{ is a thermal fine open superset of } L\}.$$

PROOF. Since v is thermal fine continuous on E, the set of point where $v = +\infty$ is thermal fine closed. Therefore, since v is finite valued on L, it is finite-valued on a thermal fine open superset D of L. Let w be a nonnegative supertemperature on E such that $w \geq v$ on L, and let $\epsilon > 0$. Since v and w are thermal fine continuous on E, and v is finite-valued on D, the set

$$V = \{p \in D : w(p) - v(p) > -\epsilon\}$$

is a thermal fine open superset of L. Because $w + \epsilon \geq v$ on V, we have $w + \epsilon \geq R_v^V$ on E, so that

$$w + \epsilon \geq \inf\{R_v^C : C \text{ is a thermal fine open superset of } L\}.$$

Therefore

$$R_v^L + \epsilon \geq \inf\{R_v^C : C \text{ is a thermal fine open superset of } L\}$$

for every $\epsilon > 0$, so that

$$R_v^L \geq \inf\{R_v^C : C \text{ is a thermal fine open superset of } L\}.$$

The reverse inequality follows from Theorem 7.27(b). □

Theorem 9.19 below is required for the proof of the important Theorem 9.27. After the latter has been proved, we will be able to remove the conditions that the supertemperatures in Theorem 9.19 belong to $C(E)$. See Theorem 9.33 for details.

THEOREM 9.19. *Suppose that $\{L_j\}$ is an expanding sequence of subsets of the open set E, that $L = \bigcup_{j=1}^{\infty} L_j$, that $\{v_j\}$ is an increasing sequence of nonnegative supertemperatures in $C(E)$, and that $v = \lim_{j \to \infty} v_j$ also belongs to $C(E)$. Then*

$$\lim_{j \to \infty} R_{v_j}^{L_j} = R_v^L \quad \text{and} \quad \lim_{j \to \infty} \widehat{R}_{v_j}^{L_j} = \widehat{R}_v^L$$

on E.

PROOF. We first consider the case where each set L_j is open, and hence L is open. Then each of the functions $R_{v_j}^{L_j}$, R_v^L, is a supertemperature on E and the two equalities coincide, by Theorem 7.31(a). Since $\{L_j\}$ is expanding and $\{v_j\}$ is increasing, Theorem 7.27(a),(b) show that the sequence $\{R_{v_j}^{L_j}\}$ is also increasing, and therefore tends to a limit $\bar{v}$. Since $R_{v_j}^{L_j} \leq R_v^L$ for all j, we have $\bar{v} \leq R_v^L$, which implies that $\bar{v}$ is a supertemperature on E, by Theorem 3.60. Each function $R_{v_j}^{L_j}$ is equal to v_j on L_j, so that $\bar{v} = v$ on L, and hence $\bar{v} \geq R_v^L$. Thus the assertions hold if each set L_j is open.

To prove the general case, we choose any point $p_0 \in E$, and any numbers $\epsilon > 0$ and $\alpha > 1$. We define $L_j^\alpha = \{p \in L_j : v(p) < \alpha v_j(p)\}$. The sequence $\{L_j^\alpha\}$ is expanding, and its union is the set $L_+ = \{p \in L : v(p) > 0\}$. Using Theorem 7.31(d), for each j we choose an open superset M_j of L_j^α such that

$$R_v^{M_j}(p_0) < R_v^{L_j^\alpha}(p_0) + 2^{-j}\epsilon,$$

and define the set

$$M = \bigcup_{j=1}^\infty M_j \supseteq \bigcup_{j=1}^\infty L_j^\alpha = L_+.$$

For all j we have $L_j^\alpha \subseteq M_j$ and $L_j^\alpha \subseteq L_{j+1}^\alpha \subseteq M_{j+1}$, and hence

$$L_j^\alpha \subseteq \left(\bigcup_{i=1}^j M_i\right) \cap M_{j+1}.$$

Putting $N_j = \bigcup_{i=1}^j M_i$ for all j, and using the strong subadditivity property of Theorem 7.32 repeatedly, we therefore obtain

$$R_v^{N_k} + \sum_{j=1}^k R_v^{L_j^\alpha} \leq R_v^{N_{k-1} \cup M_k} + \sum_{j=1}^{k-1} R_v^{N_j \cap M_{j+1}} + R_v^{L_k^\alpha}$$

$$\leq R_v^{M_k} + R_v^{N_{k-1}} + \sum_{j=1}^{k-2} R_v^{N_j \cap M_{j+1}} + R_v^{L_k^\alpha}$$

$$\leq \ldots$$

$$\leq R_v^{M_k} + \ldots + R_v^{M_1} + R_v^{L_k^\alpha}.$$

Hence

$$R_v^{N_k}(p_0) \leq R_v^{L_k^\alpha}(p_0) + \sum_{j=1}^k R_v^{M_j}(p_0) - \sum_{j=1}^k R_v^{L_j^\alpha}(p_0) \leq R_v^{L_k^\alpha}(p_0) + \epsilon,$$

by our choice of the sets M_j. We have proved the result for open sets above, and so if we make $k \to \infty$ we obtain

$$R_v^{L_+}(p_0) \leq R_v^M(p_0) = \lim_{k \to \infty} R_v^{N_k}(p_0) \leq \lim_{k \to \infty} R_v^{L_k^\alpha}(p_0) + \epsilon$$

$$\leq \lim_{k \to \infty} R_{\alpha v_k}^{L_k^\alpha}(p_0) + \epsilon = \alpha \lim_{k \to \infty} R_{v_k}^{L_k^\alpha}(p_0) + \epsilon,$$

using Theorem 7.27(c). Since Theorem 7.32 shows that $R_v^L(p_0) = R_v^{L_+}(p_0)$, it follows from the arbitrariness of ϵ and α that

$$R_v^L(p_0) \leq \lim_{k \to \infty} R_{v_k}^{L_k^\alpha}(p_0) \leq \lim_{k \to \infty} R_{v_k}^{L_k}(p_0) \leq R_v^L(p_0),$$

since $L_k \subseteq L$ and $v_k \leq v$ for all k. This proves the first equality in the general case.

It follows from the result just proved, and Theorem 7.13, that the equality for smoothed reductions holds almost everywhere. Since both sides of the equality are supertemperatures, it follows from Theorem 3.59 that the equality holds everywhere on E. □

We denote by L^Θ the set of thermal fine limit points of a set L.

By Corollary 9.13, every point of a thermal fine open set L is a limit point of L; that is, $L \subseteq L^\Theta$.

The following Theorem 9.20 is also necessary for the proof of Theorem 9.27. We shall prove below, in Corollary 9.28 and Corollary 9.34, that the hypothesis $v \in C(E)$ can be removed from Theorem 9.20.

THEOREM 9.20. *Let v be a nonnegative supertemperature on an open set E, and let L be a subset of E. Then $R_v^L = v$ on $L^\Theta \cap E$.*

If, in addition, $v \in C(E)$, then $R_v^L = \widehat{R}_v^L$ on $E \backslash L$, and $R_v^L = \widehat{R}_v^L = v$ on $L^\Theta \cap E$.

PROOF. If u is a nonnegative supertemperature on E such that $u \geq v$ on L, then the thermal fine continuity of u and v shows that $u \geq v$ on $L^\Theta \cap E$. Hence $R_v^L = v$ on $L^\Theta \cap E$.

Suppose that $v \in C(E)$. Given a point $p_0 \in E \backslash L$, we choose a contracting sequence of open balls $\{B_j\}$ with intersection $\{p_0\}$, and put $L_j = L \backslash B_j$ for all j. Then $\{L_j\}$ is an expanding sequence with union $L \backslash \{p_0\} = L$. For each j, Theorem 7.27(d) shows that $R_v^{L_j}(p_0) = \widehat{R}_v^{L_j}(p_0)$. Therefore, making $j \to \infty$ we deduce from Theorem 9.19 that $R_v^L(p_0) = \widehat{R}_v^L(p_0)$.

To prove the last part, we take any point $p_1 \in L^\Theta \cap E$. Then the first part of this theorem shows that $R_v^L(p_1) = v(p_1)$, and also that $R_v^{L \backslash \{p_1\}}(p_1) = v(p_1)$ because $p_1 \in (L \backslash \{p_1\})^\Theta \cap E$. Moreover, by the second part, $R_v^{L \backslash \{p_1\}}(p_1) = \widehat{R}_v^{L \backslash \{p_1\}}(p_1)$. Furthermore, because $\{p_1\}$ is a polar set, Theorem 7.27(e) shows that $\widehat{R}_v^{L \backslash \{p_1\}} = \widehat{R}_v^L$ on E. Hence

$$R_v^L(p_1) = v(p_1) = R_v^{L \backslash \{p_1\}}(p_1) = \widehat{R}_v^{L \backslash \{p_1\}}(p_1) = \widehat{R}_v^L(p_1),$$

as required. □

Using Theorem 9.20, we can give an example of a set which is thermally thin at every point of $\mathbb{R}^{n+1}$ but is not polar.

EXAMPLE 9.21. Let $v = 1$ on $\mathbb{R}^{n+1}$, and let $L = \mathbb{R}^n \times \{a\}$ for some real number a. We can calculate R_v^L and $\widehat{R}_v^L$ explicitly. If u is a nonnegative supertemperature on $\mathbb{R}^{n+1}$ such that $u \geq 1$ on L, then $u \geq 1$ on the set $A = \mathbb{R}^n \times \,]a, +\infty[$ by the strong minimum principle. Hence $R_v^L \geq 1$ on $\overline{A}$. Moveover, if $-\infty < b < a$ and B is $\mathbb{R}^n \times \,]b, +\infty[$, then the characteristic function χ_B is a nonnegative supertemperature on $\mathbb{R}^{n+1}$ that majorises v on L, so that $\chi_B \geq R_v^L$. It follows that $R_v^L = \chi_{\overline{A}}$, and hence that $\widehat{R}_v^L = \chi_A$. Since $v \in C(\mathbb{R}^{n+1})$, Theorem 9.20 shows that $\widehat{R}_v^L = 1$ on L^Θ. However $\widehat{R}_v^L = 0$ on L, so that no point of L is a thermal fine limit point of L. Thus for each point of L there is a thermal fine neighbourhood that includes no other point of L. Since we already know that the set $\{(y, s) : |x-y|^2 + (a-s)^2 < r^2, \, s \leq a\}$

is a thermal fine neighbourhood of (x,a), it follows that the open half-ball $H(q,r)$ is a deleted thermal fine neighbourhood of q. Moreover, since L is a closed set, it thermally thin at every point of $\mathbb{R}^{n+1}$. Thus the converse of Lemma 9.4 is false, because L is not polar.

Example 9.21 motivates the following definition.

DEFINITION 9.22. Given an open set E, a set $L \subseteq E$ is called a *semipolar subset of E* if it can be written in the form $L = \bigcup_{i=1}^{\infty} L_i$, where each set L_i has no thermal fine limit point in E.

Note that the sets L_i may have thermal fine limit points outside E, and that L may have thermal fine limit points anywhere.

EXAMPLE 9.23. For any open set E, the set $\partial_a E$ of abnormal boundary points of E is a semipolar subset of $\mathbb{R}^{n+1}$. For, by Theorem 8.40, there is a sequence of hyperplanes of the form $\mathbb{R}^n \times \{t\}$ which covers $\partial_a E$, and by Example 9.21 each such hyperplane is thermally thin at every point of $\mathbb{R}^{n+1}$.

LEMMA 9.24. *If L is a semipolar subset of $\mathbb{R}^{n+1}$, then L is thermal fine nowhere dense in $\mathbb{R}^{n+1}$.*

PROOF. Let $L = \bigcup_{i=1}^{\infty} L_i$, where each set L_i has no thermal fine limit point in $\mathbb{R}^{n+1}$, and hence is thermal fine closed. Then each L_i has no thermal fine interior point, because such a point would be a thermal fine limit point of the thermal fine interior of L_i by Corollary 9.13, and hence also of L_i. It follows that each complement $\mathbb{R}^{n+1}\backslash L_i$ is thermal fine open and thermal fine dense in $\mathbb{R}^{n+1}$, so that $\bigcap_{i=1}^{\infty}(\mathbb{R}^{n+1}\backslash L_i)$ is itself thermal fine dense in $\mathbb{R}^{n+1}$, by Theorem 9.7. Thus L is thermal fine nowhere dense. $\square$

In the next section, we give a refinement of Theorem 7.13 that characterizes the semipolar subsets of E as being the sets where a locally lower bounded infimum of a family of supertemperatures differs from its lower semicontinuous smoothing.

9.3. The Fundamental Convergence Theorem

In order to prove the theorem, we require a particular criterion for a set to be thermally thin at a limit point. This is given in the following lemma, and will also be used in the proof of Theorem 9.46 below.

LEMMA 9.25. *Let E be an open set, let v be a nonnegative supertemperature in the class $C(E)$, let $q \in E$, and let L be a subset of E that is thermally thin at q. Then*

$$\lim_{r \to 0+} \widehat{R}_v^{L \cap B(q,r)}(q) = 0.$$

PROOF. We can assume that $q \notin L$, because if $q \in L$ then we could replace L with $L\backslash\{q\}$ without affecting the hypothesis that L is thermally thin at q, and without affecting the smoothed reductions (in view of Theorem 7.27(e)) because $\{q\}$ is a polar set. We can also assume that q is a limit point of L, because otherwise we would have $\widehat{R}_v^{L\cap B(q,r)} = 0$ on E for all sufficiently small values of r, which makes the result trivial. The result is also trivial if $v(q) = 0$, because $0 \le \widehat{R}_v^S(q) \le v(q)$ for any subset S of E. We therefore suppose that $v(q) > 0$.

By Theorem 9.15, there is a heat potential w on $\mathbb{R}^{n+1}$ such that

(9.4) $$+\infty = \lim_{p \to q,\, p \in L} w(p) > w(q).$$

Since we can multiply w by a positive constant without affecting (9.4), we may suppose that $w(q) < v(q)$. Moreover, because $v \in C(E)$ and $v(q) > 0$, we can find a positive number r_0 such that $0 < v(p) < 2v(q)$ for all $p \in B(q, r_0)$. For all $r < r_0$, we write
$$\alpha(r) = \frac{\inf_{L \cap B(q,r)} w}{2v(q)}.$$
Then $w \ge \alpha(r)v$ on $L \cap B(q,r)$, so that
$$+\infty > w(q) \ge \widehat{R}_w^{L\cap B(q,r)}(q) \ge \widehat{R}_{\alpha(r)v}^{L\cap B(q,r)}(q) = \alpha(r)\widehat{R}_v^{L\cap B(q,r)}(q),$$
by Theorem 7.27(a),(c). As $r \to 0+$ we have $\alpha(r) \to +\infty$ by (9.4), which implies the result. $\square$

REMARK 9.26. Under the hypotheses on v in Lemma 9.25, we know from Theorem 9.20 that $\widehat{R}_v^{L\cap B(q,r)}(q) = v(q)$ for all $r > 0$ if L is not thermally thin at q.

THEOREM 9.27. *Let $\mathcal{F} = \{u_\alpha : \alpha \in I\}$ be a family of supertemperatures on an open set E, and let $u = \inf \mathcal{F}$. If u is locally lower bounded on E, and $\widehat{u}$ is its lower semicontinuous smoothing, then for rational numbers r_1 and r_2, the set*
$$S_{r_1,r_2} = \{p \in E : \widehat{u}(p) < r_1 < r_2 < u(p)\}$$
has no thermal fine limit point in E, so that $\widehat{u} = u$ except on the semipolar subset $\bigcup_{r_1,r_2 \in \mathbb{Q}} S_{r_1,r_2}$ of E. Moreover,

(9.5) $$\widehat{u}(q) = \Theta - f\lim_{p \to q} u(p)$$

for all $q \in E$.

Conversely, if L is a semipolar subset of E, then there is a decreasing sequence $\{v_j\}$ of nonnegative supertemperatures on E whose pointwise limit v satisfies $v > \widehat{v}$ on L. Moreover, L has Lebesgue measure zero.

PROOF. We denote by $\mathcal{G}$ the family of pointwise minima $u_{\alpha_1} \wedge ... \wedge u_{\alpha_l}$ ($l \ge 1$) that can be formed using finitely many elements of $\mathcal{F}$. Corollary 3.18 shows that $\mathcal{G}$ is a family of supertemperatures on E, and clearly $u = \inf \mathcal{G}$. By Lemma 7.12, there is a sequence $\{w_k\}$ of functions in $\mathcal{G}$ whose infimum $w(\ge u)$ has lower semicontinuous smoothing $\widehat{u}$. We can assume that this sequence is decreasing, for otherwise we could replace it by the sequence $\{w_1 \wedge ... \wedge w_j\}$. Let $\{D_j\}$ be an expanding sequence of bounded open sets such that $\overline{D}_j \subseteq E$ for all j and $\bigcup_{j=1}^\infty D_j = E$. Given a pair of rational numbers (r_1, r_2) such that $r_1 < r_2$, for each j we put
$$S_{r_1,r_2}^{D_j} = \{p \in D_j : \widehat{u}(p) < r_1 < r_2 < w(p)\}.$$
Then the expanding sequence of sets $\{S_{r_1,r_2}^{D_j}\}$ has a union T_{r_1,r_2} that contains S_{r_1,r_2}. If S_{r_1,r_2} has a thermal fine limit point $p_0 \in E$, then p_0 belongs to the open set D_l for some l, so that every thermal fine neighbourhood of p_0 meets $T_{r_1,r_2} \cap D_l = S_{r_1,r_2}^{D_l}$, and hence p_0 is a thermal fine limit point of $S_{r_1,r_2}^{D_l}$ in D_l. Therefore, if we prove

that $S_{r_1,r_2}^{D_j}$ has no thermal fine limit point in D_j, it will follow that S_{r_1,r_2} has no such limit point in E. Since $\overline{D}_j$ is compact, there is a real number m_j such that $w \geq m_j$ on D_j, so if we replace each function w_k by $w_k - m_j$, we can assume the functions in the sequence are nonnegative on D_j.

Thus it suffices to prove the direct part of the result for a decreasing sequence $\{u_j\}$ of nonnegative supertemperatures on E. Given a pair of rational numbers (r_1, r_2) such that $r_1 < r_2$, we put $S = S_{r_1,r_2}$, take any point $q \in E$, and show that S is thermally thin at q. If q is not a limit point of S, then the assertion is trivial. If q is a limit point of S, then the lower semicontinuity of $\widehat{u}$ implies that we can find a neighbourhood N of q such that

$$\widehat{u}(p) > \widehat{u}(q) - \frac{r_2 - r_1}{2}$$

for all $p \in N$. For all $p \in S$ we have $u(p) - \widehat{u}(p) > r_2 - r_1$, so that if $p \in N$ also we have

$$u(p) > \widehat{u}(p) + r_2 - r_1 > \widehat{u}(q) + \frac{r_2 - r_1}{2} = \beta,$$

say. Hence $u_j \geq u > \beta$ on $S \cap N$. Therefore $u_j \geq R_\beta^{S \cap N}$ for all j, so that $u \geq R_\beta^{S \cap N}$, and hence $\widehat{u} \geq \widehat{R}_\beta^{S \cap N}$ on E. By Theorem 9.20, $\widehat{R}_\beta^{S \cap N} = \beta$ on $(S \cap N)^\Theta \cap E$, so that the inequality $\widehat{u}(q) < \beta$ implies that $q \notin S^\Theta$. Since q is an arbitrary point of E, the set $S = S_{r_1,r_2}$ has no thermal fine limit point in E. Hence the union of the sets S_{r_1,r_2} over all pairs (r_1, r_2), is a semipolar subset of E, and clearly $\widehat{u} = u$ outside this union.

To prove (9.5), we first note that for any point $q \in E$, we have

$$\Theta - f \limsup_{p \to q} u(p) \geq \Theta - f \liminf_{p \to q} u(p) \geq \liminf_{p \to q} u(p) = \widehat{u}(q),$$

by Theorem 7.13. If there was a point q such that $\Theta - f \limsup_{p \to q} u(p) > \widehat{u}(q)$, then there would be rational numbers r_1 and r_2 such that

$$\Theta - f \limsup_{p \to q} u(p) > r_2 > r_1 > \Theta - f \lim_{p \to q} \widehat{u}(p),$$

because $\widehat{u}$ is thermal fine continuous. This implies the existence of a deleted thermal fine neighbourhood of q on which $\widehat{u} < r_1$, and also that u has a thermal fine cluster value strictly greater than r_2 at q. The corresponding set S_{r_1,r_2} would have a thermal fine limit point at q, contrary to what we have just proved. Hence there is no such point q, and (9.5) holds for all $q \in E$.

For the converse part, we write $L = \bigcup_{k=1}^\infty L_k$, where each of the sets L_k has no thermal fine limit point in E. We take a sequence $\{B_m\}$ of open subsets of E that form a base for the Euclidean topology on E. For all k and m, we define the sets

$$L_{k,m} = L_k \cap B_m \cap \{p \in E : \widehat{R}_1^{L_k \cap B_m}(p) < 1\}.$$

For each point $p \in L_{k,m}$, we have $\widehat{R}_1^{L_{k,m}}(p) \leq \widehat{R}_1^{L_k \cap B_m}(p) < 1 = R_1^{L_{k,m}}(p)$, so that

(9.6) $$L_{k,m} \subseteq \{p \in E : \widehat{R}_1^{L_{k,m}}(p) < R_1^{L_{k,m}}(p)\}.$$

Moreover, given any of the sets L_k, for each $p \in E$ Lemma 9.25 implies that

$$\lim_{r \to 0+} \widehat{R}_1^{L_k \cap B(p,r)}(p) = 0,$$

and so we can find $\rho > 0$ such that $\widehat{R}_1^{L_k \cap B(p,\rho)}(p) < 1$. We can then find one of the sets B_m, say B_μ, such that $p \in B_\mu \subseteq B(p,\rho)$, so that

$$\widehat{R}_1^{L_k \cap B_\mu}(p) \leq \widehat{R}_1^{L_k \cap B(p,\rho)}(p) < 1.$$

Thus the arbitrary point $p \in E$ belongs to a set B_μ such that $\widehat{R}_1^{L_k \cap B_\mu}(p) < 1$. Hence

$$(9.7) \quad \bigcup_{m=1}^\infty L_{k,m} = L_k \cap \left(\bigcup_{m=1}^\infty \left(B_m \cap \{p \in E : \widehat{R}_1^{L_k \cap B_m}(p) < 1\} \right) \right) = L_k.$$

We now denote by $\mathcal{H}_{k,m}$ the set of nonnegative supertemperatures on E that majorize 1 on $L_{k,m}$. Since $w \wedge 1 \in \mathcal{H}_{k,m}$ whenever $w \in \mathcal{H}_{k,m}$, we can suppose that $w \leq 1$ for all such w. By Lemma 7.12, there is a sequence $\{w_j^{k,m}\}$ of functions in $\mathcal{H}_{k,m}$ whose infimum $w^{k,m}$ has lower semicontinuous smoothing $\widehat{R}_1^{L_{k,m}}$. We can assume that this sequence is decreasing, for otherwise we could replace it by the sequence $\{w_1^{k,m} \wedge ... \wedge w_i^{k,m}\}$. We now put

$$v_j = \sum_{k,m=1}^\infty w_j^{k,m} 2^{-k-m}.$$

Clearly $v_j \geq 0$ for all j, and because $0 \leq w_j^{k,m} \leq 1$ for all j, k and m, it follows from Theorem 3.60 that each v_j is a supertemperature on E. Moreover, since each sequence $\{w_j^{k,m}\}$ is decreasing, so is $\{v_j\}$. We put

$$v = \lim_{j \to \infty} v_j = \sum_{k,m=1}^\infty w^{k,m} 2^{-k-m}.$$

By Theorem 7.13, there is a Lebesgue null subset Z of E such that $v = \widehat{v}$ and $w^{k,m} = \widehat{R}_1^{L_{k,m}}$ on $E \backslash Z$. Hence

$$\widehat{v} = \sum_{k,m=1}^\infty \widehat{R}_1^{L_{k,m}} 2^{-k-m}$$

on $E \backslash Z$, and since both sides of this equation are supertemperatures, the equation holds throughout E, by Theorem 3.59. For every k and m we have $w^{k,m} \geq R_1^{L_{k,m}}$ on E, and also $R_1^{L_{k,m}} > \widehat{R}_1^{L_{k,m}}$ on $L_{k,m}$ by (9.6), so that $v > \widehat{v}$ on $L_{k,m}$. Hence, by (9.7), $v > \widehat{v}$ on L_k for all k, and thus on L.

The fact that L has Lebesgue measure zero now follows from Theorem 7.13. $\square$

We now use Theorem 9.27 to obtain part of the conclusion of Theorem 9.20 without the hypothesis that $v \in C(E)$.

COROLLARY 9.28. *Let v be a nonnegative supertemperature on an open set E, and let L be a subset of E. Then $R_v^L = \widehat{R}_v^L = v$ on $L^\Theta \cap E$.*

PROOF. Since $R_v^L = v$ on L, it follows from (9.5) and the thermal fine continuity of v that, if $q \in L^\Theta \cap E$,

$$\widehat{R}_v^L(q) = \Theta - f \lim_{p \to q} R_v^L(p) = \Theta - f \lim_{p \to q} v(p) = v(q).$$

The result follows because $\widehat{R}_v^L \leq R_v^L \leq v$ on E. $\square$

COROLLARY 9.29. *If L is a semipolar subset of the open set E, then it is a subset of a Borel semipolar subset M of E, and M can be expressed as a countable union of Borel sets each of which has no thermal fine limit point in E.*

PROOF. By Theorem 9.27, there is a decreasing sequence $\{v_j\}$ of nonnegative supertemperatures on E whose limit v satisfies $v > \widehat{v}$ on L. Thus L is a subset of the Borel set $M = \{p \in E : v(p) > \widehat{v}(p)\}$. Moreover, the equality

$$M = \bigcup_{r_1, r_2 \in \mathbb{Q}} \{p \in E : \widehat{v}(p) < r_1 < r_2 < v(p)\}$$

expresses M as a countable union of Borel sets, each of which has no thermal fine limit point in E by Theorem 9.27. □

The following combination of results from Theorems 7.13 and 9.27 is known as the *Fundamental Convergence Theorem*.

THEOREM 9.30. *Let $\mathcal{F}$ be a family of supertemperatures on an open set E, and let $u = \inf \mathcal{F}$. If u is locally lower bounded on E, then its lower semicontinuous smoothing $\widehat{u}$ is a supertemperature on E, is equal to u except on a semipolar subset of E, and satisfies*

$$\widehat{u}(q) = \liminf_{p \to q} u(p) = \Theta - f \lim_{p \to q} u(p)$$

for all $q \in E$.

9.4. Applications of the Fundamental Convergence Theorem to Reductions

The first theorem of this section is an extension of Theorem 7.31(a),(b),(c) from open sets to thermal fine open sets.

THEOREM 9.31. *Let u and v be nonnegative supertemperatures on an open set E, and let L be a thermal fine open subset of E.*
(a) The equality $R_u^L = \widehat{R}_u^L$ holds on E, and so R_u^L is a supertemperature on E.
(b) The equality $R_{u+v}^L = R_u^L + R_v^L$ holds on E.
(c) If $L \subseteq M \subseteq E$, then

$$\widehat{R}_{\widehat{R}_v^L}^M = \widehat{R}_v^L$$

on E. If M is also thermal fine open, then

$$\widehat{R}_{\widehat{R}_v^M}^L = \widehat{R}_v^L$$

on E.

PROOF. Since L is thermal fine open, we have $L \subseteq L^\Theta \cap E$ by Corollary 9.13, so that $\widehat{R}_u^L = u$ on L by Corollary 9.28. Since $\widehat{R}_u^L$ is a nonnegative supertemperature on E, it follows that $\widehat{R}_u^L \geq R_u^L$ on E. The reverse inequality is always true, and so (a) holds.

Since $u = R_u^L$ and $v = R_v^L$ on L, the nonnegative supertemperature $R_u^L + R_v^L$ majorizes $u + v$ on L, and so $R_u^L + R_v^L \geq R_{u+v}^L$ on E. Therefore, by the natural order decomposition (Theorem 7.21), there are nonnegative supertemperatures u^* and v^* on E such that

$$u^* \leq R_u^L, \quad v^* \leq R_v^L, \quad R_{u+v}^L = u^* + v^*.$$

On L, we have $u = R_u^L \geq u^*$, $v = R_v^L \geq v^*$, and $u + v = R_{u+v}^L = u^* + v^*$, so that $u^* = u$ and $v^* = v$. Hence $u^* \geq R_u^L$ and $v^* \geq R_v^L$ on E, and so equality holds. This proves (b).

Since L is thermal fine open, we have $\widehat{R}_v^L = R_v^L$ on E, by part (a). Since $R_v^L = v$ on L, a subtemperature w majorizes v on L if and only if it majorizes R_v^L on L, and so

(9.8) $$R_{R_v^L}^L = R_v^L$$

on E. By Theorem 7.27(b), we have

$$R_{R_v^L}^L \leq R_{R_v^L}^M \leq R_v^L,$$

and so the first assertion follows from (9.8). If M is also thermal fine open, so that $R_v^M = \widehat{R}_v^M$ also, then Theorem 7.27(a),(b) show that

$$R_{R_v^L}^L \leq R_{R_v^M}^L \leq R_v^L,$$

and so the second assertion also follows from (9.8). □

Our next result extends the strong subadditivity property of reductions and smoothed reductions, given in Theorem 7.32 for nonnegative supertemperatures in the class $C(E)$.

THEOREM 9.32. *Let L and M be subsets of the open set E, and let v be a nonnegative supertemperature on E. Then*

(9.9) $$R_v^{L \cup M} + R_v^{L \cap M} \leq R_v^L + R_v^M,$$

and

(9.10) $$\widehat{R}_v^{L \cup M} + \widehat{R}_v^{L \cap M} \leq \widehat{R}_v^L + \widehat{R}_v^M$$

on E. Moreover, if $L_+ = \{p \in L : v(p) > 0\}$, then $R_v^L = R_v^{L_+}$ on E.

PROOF. We first consider the case of thermal fine open subsets of E, with a view to using Theorem 9.18. Let A and B be thermal fine open subsets of E, and put $w = R_v^A \wedge R_v^B$. On A, we have $R_v^A + R_v^B = v + R_v^B$ and $R_v^A = v \geq R_v^B$. Similarly, on B we have $R_v^A + R_v^B = R_v^A + v$ and $R_v^B = v \geq R_v^A$. Hence, on $A \cup B$ we have $R_v^A + R_v^B = v + R_v^A \wedge R_v^B = v + w$. It therefore follows from Theorem 9.31 that

(9.11) $$R_v^{A \cup B} + R_w^{A \cup B} = R_{v+w}^{A \cup B} = R_{R_v^A + R_v^B}^{A \cup B} = R_{R_v^A}^{A \cup B} + R_{R_v^B}^{A \cup B} = R_v^A + R_v^B$$

on E. Furthermore, on $A \cap B$ we have $R_w^{A \cup B} = w = R_v^A \wedge R_v^B = v$. Therefore, using Theorem 9.31(a) again, we have $R_w^{A \cup B} \geq R_v^{A \cap B}$ on E. Now (9.11) implies that

$$R_v^{A \cup B} + R_v^{A \cap B} \leq R_v^A + R_v^B,$$

for thermal fine open sets A and B.

We now consider the case where v is finite-valued on $L \cup M$. If $A \supseteq L$ and $B \supseteq M$, it now follows from Theorem 7.27(b) that

$$R_v^{L \cup M} + R_v^{L \cap M} \leq R_v^{A \cup B} + R_v^{A \cap B} \leq R_v^A + R_v^B,$$

so that

$$R_v^{L \cup M} + R_v^{L \cap M} \leq R_v^A + \inf\{R_v^C : C \text{ is a thermal fine open superset of } M\}$$
$$= R_v^A + R_v^M,$$

by Theorem 9.18. Hence

$$R_v^{L\cup M} + R_v^{L\cap M} \leq \inf\{R_v^C : C \text{ is a thermal fine open superset of } L\} + R_v^M$$
$$= R_v^L + R_v^M,$$

again by Theorem 9.18.

For the general case, we put $Z = \{p \in L \cup M : v(p) = +\infty\}$. Then, by the case just proved, we have

$$R_v^{(L\cup M)\setminus Z} + R_v^{(L\cap M)\setminus Z} = R_v^{(L\setminus Z)\cup(M\setminus Z)} + R_v^{(L\setminus Z)\cap(M\setminus Z)} \leq R_v^{L\setminus Z} + R_v^{M\setminus Z}$$

on E. Since Z is a polar set, Theorem 7.27(e) shows that, for any subset S of E, we have $R_v^{S\setminus Z} = R_v^S$ on $E\setminus Z$. Hence (9.9) holds on $E\setminus Z$. If $p \in Z$, then $p \in L \cup M$ and $R_v^{L\cup M}(p) = v(p) = +\infty$; and either $p \in L$ and $R_v^L(p) = v(p) = +\infty$, or $p \in M$ and $R_v^M(p) = v(p) = +\infty$. In both cases, (9.9) holds with equality. Hence (9.9) holds on E.

It follows that (9.10) holds almost everywhere on E, and hence everywhere on E because all the functions in (9.10) are supertemperatures.

For the last part, by Theorem 7.27(b) and (9.9), we have

$$R_v^{L_+} \leq R_v^L \leq R_v^{L_+} + R_v^{L\setminus L_+} = R_v^{L_+} + R_0^{L\setminus L_+} = R_v^{L_+}.$$

□

We can now improve upon Theorem 9.19 by obtaining the same conclusions without the continuity and finiteness hypotheses.

THEOREM 9.33. *Suppose that $\{L_j\}$ is an expanding sequence of subsets of the open set E, that $L = \bigcup_{j=1}^\infty L_j$, that $\{v_j\}$ is an increasing sequence of nonnegative supertemperatures on E, and that $v = \lim_{j\to\infty} v_j$ is also a supertemperature on E. Then*

$$\lim_{j\to\infty} R_{v_j}^{L_j} = R_v^L \quad \text{and} \quad \lim_{j\to\infty} \widehat{R}_{v_j}^{L_j} = \widehat{R}_v^L$$

on E.

PROOF. We first consider the case where each of the sets L_j is thermal fine open, and hence L is thermal fine open. Then each of the functions $R_{v_j}^{L_j}$, R_v^L, is a supertemperature on E and the two equalities coincide, by Theorem 9.31(a). Since $\{L_j\}$ is expanding and $\{v_j\}$ is increasing, Theorem 7.27(a),(b) show that the sequence $\{R_{v_j}^{L_j}\}$ is also increasing, and therefore tends to a limit $\bar{v}$. Since $R_{v_j}^{L_j} \leq R_v^L$ for all j, we have $\bar{v} \leq R_v^L$, which implies that $\bar{v}$ is a supertemperature on E, by Theorem 3.60. Each function $R_{v_j}^{L_j}$ is equal to v_j on L_j, so that $\bar{v} = v$ on L, and hence $\bar{v} \geq R_v^L$. Thus the assertions hold if each set L_j is thermal fine open.

We now consider the case where the function v is finite-valued on L. We choose any point $p_0 \in E$, and any numbers $\epsilon > 0$ and $\alpha > 1$. We define

$$L_j^\alpha = \{p \in L_j : v(p) < \alpha v_j(p)\}.$$

The sequence $\{L_j^\alpha\}$ is expanding, and its union is the set $L_+ = \{p \in L : v(p) > 0\}$. Using Theorem 9.18, for each j we choose a thermal fine open superset M_j of L_j^α such that

$$R_v^{M_j}(p_0) < R_v^{L_j^\alpha}(p_0) + 2^{-j}\epsilon.$$

9.4. APPLICATIONS TO REDUCTIONS

We define the set
$$M = \bigcup_{j=1}^{\infty} M_j \supseteq \bigcup_{j=1}^{\infty} L_j^\alpha = L_+.$$
For all j we have $L_j^\alpha \subseteq M_j$ and $L_j^\alpha \subseteq L_{j+1}^\alpha \subseteq M_{j+1}$, and hence
$$L_j^\alpha \subseteq \left(\bigcup_{i=1}^{j} M_i\right) \cap M_{j+1}.$$
Putting $N_j = \bigcup_{i=1}^{j} M_i$ for all j, and using the strong subadditivity property of Theorem 9.32 repeatedly, we therefore obtain
$$R_v^{N_k} + \sum_{j=1}^{k} R_v^{L_j^\alpha} \leq R_v^{N_{k-1} \cup M_k} + \sum_{j=1}^{k-1} R_v^{N_j \cap M_{j+1}} + R_v^{L_k^\alpha}$$
$$\leq R_v^{M_k} + R_v^{N_{k-1}} + \sum_{j=1}^{k-2} R_v^{N_j \cap M_{j+1}} + R_v^{L_k^\alpha}$$
$$\leq \ldots$$
$$\leq R_v^{M_k} + \ldots + R_v^{M_1} + R_v^{L_k^\alpha}.$$
Hence
$$R_v^{N_k}(p_0) \leq R_v^{L_k^\alpha}(p_0) + \sum_{j=1}^{k} R_v^{M_j}(p_0) - \sum_{j=1}^{k} R_v^{L_j^\alpha}(p_0) \leq R_v^{L_k^\alpha}(p_0) + \epsilon,$$
by our choice of the sets M_j. We have proved the result for thermal fine open sets above, and so if we make $k \to \infty$ we obtain
$$R_v^{L_+}(p_0) \leq R_v^M(p_0) = \lim_{k \to \infty} R_v^{N_k}(p_0) \leq \lim_{k \to \infty} R_v^{L_k^\alpha}(p_0) + \epsilon$$
$$\leq \lim_{k \to \infty} R_{\alpha v_k}^{L_k^\alpha}(p_0) + \epsilon = \alpha \lim_{k \to \infty} R_{v_k}^{L_k^\alpha}(p_0) + \epsilon,$$
using Theorem 7.27(c). By Theorem 9.32, we have $R_v^L(p_0) = R_v^{L_+}(p_0)$, so it follows from the arbitrariness of ϵ and α that
$$R_v^L(p_0) \leq \lim_{k \to \infty} R_{v_k}^{L_k^\alpha}(p_0) \leq \lim_{k \to \infty} R_{v_k}^{L_k}(p_0) \leq R_v^L(p_0),$$
since $L_k \subseteq L$ and $v_k \leq v$ for all k. This proves that the first of the two equalities holds in this case.

We now consider the general case. Putting $F = \{p \in E : v(p) < +\infty\}$, we have $\lim_{j \to \infty} R_{v_j}^{L_j \cap F} = R_v^{L \cap F}$ on E by the case just proved. Since the set $L \setminus F$ is polar, it follows from Theorem 7.27(e) that $R_{v_j}^{L_j} = R_{v_j}^{L_j \cap F}$ and $R_v^L = R_v^{L \cap F}$ except on $L \setminus F$. Therefore $\lim_{j \to \infty} R_{v_j}^{L_j} = R_v^L$ except on $L \setminus F$. If $p \in L \setminus F$, then there is an integer j_0 such that $p \in L_j$ for all $j \geq j_0$, and so $R_v^L(p) \geq R_v^{L_j}(p) = v(p) = +\infty$ for all $j \geq j_0$, and the first of the two equalities holds trivially at p.

It follows from the result just proved that the equality for smoothed reductions holds almost everywhere on E, in view of Theorem 7.13. Since both sides of the equality are supertemperatures, it follows from Theorem 3.59 that the equality holds everywhere on E. $\square$

Using Theorem 9.33, we are able to obtain part of Theorem 9.20 without the continuity and finiteness hypotheses, as follows.

COROLLARY 9.34. *If v be a nonnegative supertemperature on an open set E, and $L \subseteq E$, then $R_v^L = \widehat{R}_v^L$ on $E \backslash L$.*

PROOF. Given any point $p \in E \backslash L$, we choose a contracting sequence of open balls $\{B_j\}$ with intersection $\{p\}$, and put $L_j = L \backslash B_j$ for all j. Then $\{L_j\}$ is an expanding sequence with union L. For each j we have $R_v^{L_j}(p) = \widehat{R}_v^{L_j}(p)$, by Theorem 7.27(d). Therefore, making $j \to \infty$, we deduce from Theorem 9.33 that $R_v^L(p) = \widehat{R}_v^L(p)$. □

THEOREM 9.35. *If $\{v_j\}$ is a sequence of nonnegative supertemperatures on an open set E, such that the function $v = \sum_{j=1}^{\infty} v_j$ is also a supertemperature on E, and L is a subset of E, then*

$$R_v^L = \sum_{j=1}^{\infty} R_{v_j}^L \quad \text{and} \quad \widehat{R}_v^L = \sum_{j=1}^{\infty} \widehat{R}_{v_j}^L$$

on E.

PROOF. We first show that, if u and w are nonnegative supertemperatures on E, then

(9.12) $$R_{u+w}^L = R_u^L + R_w^L$$

on E. By Theorem 9.31(b), this equality holds if L is thermal fine open.

We first consider the case where L is arbitrary but u and w are finite-valued on L. Given any point $p_0 \in E$ and positive number ϵ, Theorem 9.18 shows that we can find a thermal fine open set M such that $L \subseteq M \subseteq E$, $R_u^M(p_0) \leq R_u^L(p_0) + \epsilon$, $R_w^M(p_0) \leq R_w^L(p_0) + \epsilon$, and $R_{u+w}^M(p_0) \leq R_{u+w}^L(p_0) + \epsilon$. Therefore, since (9.12) holds with L replaced by M, we have

$$-\epsilon = \left(R_{u+w}^M(p_0) - \epsilon\right) - R_u^M(p_0) - R_w^M(p_0)$$
$$\leq R_{u+w}^L(p_0) - R_u^L(p_0) - R_w^L(p_0)$$
$$\leq R_{u+w}^M(p_0) - \left(R_u^M(p_0) - \epsilon\right) - \left(R_w^M(p_0) - \epsilon\right)$$
$$= 2\epsilon.$$

Since ϵ and p_0 are arbitrary, it follows that (9.12) holds in this case.

For the general case, we put $F = \{p \in E : u(p) + w(p) < +\infty\}$. Then $R_{u+w}^{L \cap F} = R_u^{L \cap F} + R_w^{L \cap F}$ by the case just proved. Since $L \backslash F$ is polar, it follows from Theorem 7.27(e) that $R_u^L = R_u^{L \cap F}$, $R_w^L = R_w^{L \cap F}$ and $R_{u+w}^L = R_{u+w}^{L \cap F}$ except on $L \backslash F$. Thus (9.12) holds except on $L \backslash F$. If $p \in L \backslash F$, then $R_{u+w}^L(p) = u(p) + w(p) = +\infty$, and either $R_u^L(p) = u(p) = +\infty$ or $R_w^L(p) = w(p) = +\infty$. Thus (9.12) holds on $L \backslash F$ also, and hence on the whole of E.

Therefore the equality $\widehat{R}_{u+w}^L = \widehat{R}_u^L + \widehat{R}_w^L$ holds almost everywhere on E, by Theorem 7.13. Since both sides of this equality are supertemperatures, equality holds everywhere on E, by Theorem 3.59.

Now we let $\{v_j\}$ and v be as in the statement of the theorem, and for each positive integer k put $w_k = \sum_{j=1}^{k} v_j$. It follows from the results just proved that $R_{w_k}^L = \sum_{j=1}^{k} R_{v_j}^L$ and $\widehat{R}_{w_k}^L = \sum_{j=1}^{k} \widehat{R}_{v_j}^L$ on E, for all k. Applying Theorem 9.33 to

$\{w_k\}$, we obtain $R_v^L = \lim_{k\to\infty} R_{w_k}^L = \sum_{j=1}^{\infty} R_{v_j}^L$, and similarly for the smoothed reductions. □

9.5. Thermal Thinness and the Regularity of Normal Boundary Points

Regularity of finite normal boundary points, relative to the Dirichlet problem, can be characterized in terms of thermal thinness. To show this, we use a test for the thermal thinness of a set at a point which involves the smoothed reduction of nonnegative supertemperatures which peak at a specified point. We now make this terminology precise.

DEFINITION 9.36. Let u be an extended real-valued function on an open set E, and let q be a point in E. We say that u *peaks at* q if
$$\sup_{E \backslash B(q,r)} u < u(q)$$
for every positive number r such that $E \backslash B(q,r) \neq \emptyset$.

Elementary examples of such functions can easily be constructed. Here is one which we shall use in the proof of Theorem 9.39 below.

EXAMPLE 9.37. If $q = (y,s)$, and u is defined on $\mathbb{R}^{n+1}$ by
$$u(x,t) = -(t-s)^2 - |x-y|^2,$$
then $\Theta u(x,t) = -2n + 2(t-s)$, so that $\Theta u(x,t) < 0$ if and only if $t < s + n$. Thus u is a supertemperature on $\mathbb{R}^n \times \,]-\infty, s+n[$, and clearly u peaks at q.

THEOREM 9.38. *Let E be an open set, let $q \in E$, let $L \subseteq E$, and let u be a nonnegative supertemperature on E that peaks at q. Then L is thermally thin at q if and only if $\widehat{R}_u^L(q) < u(q)$.*

PROOF. If L is not thermally thin at q, then $\widehat{R}_u^L(q) = u(q)$ by Corollary 9.28.

We now suppose, conversely, that L is thermally thin at q. This is equivalent to supposing that $L \backslash \{q\}$ is thermally thin at q. Also, $\widehat{R}_u^L = \widehat{R}_u^{L \backslash \{q\}}$ on E by Theorem 7.27(e). Therefore we can assume that $q \notin L$, so that $\widehat{R}_u^L(q) = R_u^L(q)$ by Corollary 9.34.

If q is not a limit point of L, and r is chosen so that $B(q,r) \subseteq E \backslash L$, then putting $\delta = \sup_{E \backslash B(q,r)} u$ we have $R_u^L \leq R_\delta^L \leq \delta < u(q)$ on E, because u peaks at q. In particular $R_u^L(q) < u(q)$.

On the other hand, if q is a limit point of L then there is a heat potential w on $\mathbb{R}^{n+1}$ such that
$$w(q) < \liminf_{p \to q,\, p \in L} w(p),$$
by Theorem 9.11. We can assume that w is upper bounded, because we could replace it with $w \wedge (w(q)+1)$ if necessary. We now let α be a number such that
$$w(q) < \alpha < \liminf_{p \to q,\, p \in L} w(p),$$
and for each positive number λ we put
$$w_\lambda(p) = u(q) + \lambda(w(p) - \alpha)$$

for all $p \in E$. We choose a ball $B = B(q, \rho)$ such that $\overline{B} \subseteq E$ and $w(p) > \alpha$ for all $p \in B \cap L$. Because u peaks at q, for all $p \in E \backslash B$ we have $u(p) \leq \sup_{E \backslash B} u < u(q)$, so that if $\beta = u(q) - \sup_{E \backslash B} u$ then $\beta > 0$ and $u(p) \leq u(q) - \beta$. Since w is bounded, we can find a positive number γ such that $\gamma |w(p) - \alpha| \leq \beta$ for all $p \in E \backslash B$. We then have
$$w_\gamma(p) = u(q) + \gamma(w(p) - \alpha) \geq u(q) - \beta \geq u(p) \geq 0$$
for all $p \in E \backslash B$. Furthermore, for all $p \in B \cap L$ we have
$$w_\gamma(p) = u(q) + \gamma(w(p) - \alpha) \geq u(q) > u(p),$$
because u peaks at q, and so it follows that $w_\gamma \geq u$ on L. Since $\overline{B} \subseteq E$ and $w_\gamma \geq 0$ on $E \backslash B$, the minimum principle ensures that $w_\gamma \geq 0$ on E. Hence $w_\gamma \geq R_u^L$ on E, and so
$$u(q) > u(q) + \gamma(w(q) - \alpha) = w_\gamma(q) \geq R_u^L(q),$$
as required. $\square$

THEOREM 9.39. *Let F be a closed subset of $\mathbb{R}^{n+1}$, and let $q \in \partial F$. Then q is a thermal fine limit point of F if and only if there exist an open neighbourhood V of q, and a positive supertemperature w on $V \backslash F$, such that*

(9.13) $$\lim_{p \to q,\, p \in V \backslash F} w(p) = 0.$$

PROOF. We first suppose that q is a thermal fine limit point of F. The result is a local one, and so we can assume that $F \subseteq B(q, 1)$, because we could otherwise replace F with $F \cap \overline{B}(q, \tfrac{1}{2})$. We put $B = B(q, 1)$, and define u on B by putting $u(p) = 1 - |p - q|^2$. Then $u > 0$ and, by Example 9.37, $\Theta u < 0$ and u peaks at q. We consider the smoothed reduction of u over F relative to B. Since $q \in F^\Theta \cap B$ we have $\widehat{R}_u^F(q) = u(q)$, by Corollary 9.28. Therefore

(9.14) $$\liminf_{p \to q,\, p \in B \backslash F} \widehat{R}_u^F(p) \geq \widehat{R}_u^F(q) = u(q).$$

We put $w = u - \widehat{R}_u^F$ on $B \backslash F$. By Theorem 7.27(d), $\widehat{R}_u^F$ is a temperature on $B \backslash F$, and so w is a nonnegative supertemperature. If there was a point $p \in B \backslash F$ such that $w(p) = 0$, then we would have $w = 0$ on $\Lambda(p, B \backslash F)$ by the strong minimum principle, so that u would be a temperature there contrary to the inequality $\Theta u < 0$. Hence $w > 0$ on $B \backslash F$. Moreover,
$$\limsup_{p \to q,\, p \in B \backslash F} w(p) = u(q) - \liminf_{p \to q,\, p \in B \backslash F} \widehat{R}_u^F(p) \leq 0$$
by (9.14), so that (9.13) holds with $V = B$.

We now suppose, conversely, that F is thermally thin at q. We suppose also that there exist an open neighbourhood V of q and a positive supertemperature w on $V \backslash F$ such that (9.13) holds, and show that this leads to a contradiction.

There are two cases to consider. If q is not a limit point of F, then the set $V \backslash F$ is an open neighbourhood of q, so that w is defined at q and $w(q) = 0$ by (9.13). The strong minimum principle now shows that $w = 0$ on $\Lambda(q, V \backslash F)$, contrary to the hypothesis that $w > 0$.

If q is a limit point of F, then Corollary 9.16 shows that there exist an upper bounded *subtemperature* v on $\mathbb{R}^{n+1}$ and a ball $B = B(q, \rho)$ such that $v(p) < -1$ for all $p \in F \cap B \backslash \{q\}$ but $v(q) > 1$. We can suppose that $V = B$. We put $B_1 = B(q, \rho/2)$, and shall use the maximum principle to show that there is a positive number β such that $v - \beta w \leq 0$ on $B_1 \backslash F$. The set $U = \{p : v(p) < -1\}$ is open, because v is upper semicontinuous. On the compact set $\partial B_1 \backslash U$ we have $w > 0$, so that the lower semicontinuity of w ensures that w has a positive minimum over $\partial B_1 \backslash U$. Therefore the fact that v is upper bounded implies that we can find a positive number β such that $v - \beta w < 0$ on $\partial B_1 \backslash U$. Hence the upper semicontinuity of $v - \beta w$ implies that, for any point $p \in \partial B_1 \backslash U$ we have

$$\limsup_{p' \to p,\, p' \in B_1 \backslash F} (v(p') - \beta w(p')) \leq 0.$$

This inequality also holds whenever $p \in \partial(B_1 \backslash F) \cap U$, because then $v(p') < -1$. We now use the maximum principle of Theorem 7.9 (with $Z = \{q\}$), and with the fact that $v - \beta w$ is upper bounded on $B_1 \backslash F$, to deduce that $v - \beta w \leq 0$ on $B_1 \backslash F$. Hence

$$\limsup_{p \to q,\, p \in B \backslash F} v(p) \leq \beta \lim_{p \to q,\, p \in B \backslash F} w(p) = 0.$$

Since $v(p) < -1$ for all $p \in B \cap F \backslash \{q\}$, it follows that $\limsup_{p \to q} v(p) \leq 0$. Now Lemma 3.16 gives the contradiction

$$1 < v(q) = \limsup_{p \to q} v(p) \leq 0.$$

$\square$

THEOREM 9.40. *Let E be an open set, and let q be a finite point of ∂E. Then q is a thermal fine limit point of $\mathbb{R}^{n+1} \backslash E$ if and only if q is a regular point of $\partial_n E$.*

PROOF. We put $F = \mathbb{R}^{n+1} \backslash E$. If $q \in \partial_a E$, then there is an open half-ball $H(q, \delta) \subseteq E$. Since $H(q, \delta)$ is a thermal fine deleted neighbourhood of q by Example 9.21, we see that $q \notin F^{\Theta}$. We may therefore assume that $q \in \partial_n E$. Since $q \in \partial F$, it is a thermal fine limit point of F if and only if there exist an open neighbourhood V of q, and a positive supertemperature w on $V \backslash F = V \cap E$ such that

$$\lim_{p \to q,\, p \in V \cap E} w(p) = 0,$$

by Theorem 9.39. Thus q is a thermal fine limit point of F if and only if there is a barrier w at q for E. The result now follows from Theorem 8.46(a). $\square$

EXAMPLE 9.41. Any heat ball $\Omega(q; c)$ is a deleted thermal fine neighbourhood of its centre q, because we showed in Example 8.36 that q is an irregular point for the Dirichlet problem on $\Omega(q; c)$, and so it follows from Theorem 9.40 that q is not a thermal fine limit point of $\mathbb{R}^{n+1} \backslash \Omega(q; c)$.

EXAMPLE 9.42. A parabolic tusk Γ is not thermally thin at its vertex. To see this, let E be an open set and let $q = (y, s) \in \partial_n E$. If Γ has vertex q and $\Gamma \cap (\mathbb{R}^n \times]r, s[) \subseteq \mathbb{R}^{n+1} \backslash E$ for some $r < s$, then q is a regular point of $\partial_n E$ by Theorem 8.52, so that q is a thermal fine limit point of Γ by Theorem 9.40.

We now prove a uniqueness theorem on a heat ball which it is interesting to compare with Remark 3.24.

THEOREM 9.43. *Let w be a supertemperature on an open superset E of $\overline{\Omega}(p_0; c_0)$. Then there is a unique supertemperature v on E such that v is a temperature on $\Omega(p_0; c_0)$ and $v = w$ on $E \backslash (\Omega(p_0; c_0) \cup \{p_0\})$.*

PROOF. We put $\Omega = \Omega(p_0; c_0)$, and take an open superset D of $\overline{\Omega}$ such that $\overline{D} \subseteq E$. We choose a real number α such that $w - \alpha \geq 0$ on D. If we prove the result with w and E replaced by $w - \alpha$ and D, then the result will hold for w with E replaced by D, and the result as stated will follow. It therefore suffices to prove the result for a nonnegative supertemperature.

We therefore assume that $w \geq 0$, and we put $M = E \backslash \Omega$. On the set $E \backslash M = \Omega$, the supertemperature $\widehat{R}_w^M$ is a temperature, by Theorem 7.27(d). Furthermore, every point of $\partial\Omega \backslash \{p_0\}$ is a regular point of $\partial_n \Omega$, by Corollary 3.41, so that $\partial\Omega \backslash \{p_0\} \subseteq M^\Theta$, by Theorem 9.40. Since $E \backslash \overline{\Omega}$ is thermal fine open, Corollary 9.13 shows that $E \backslash \overline{\Omega} \subseteq M^\Theta$. Thus $E \backslash (\Omega \cup \{p_0\}) \subseteq M^\Theta$, and therefore $\widehat{R}_w^M = w$ on $E \backslash (\Omega \cup \{p_0\})$ by Corollary 9.28. Hence the function $u = \widehat{R}_w^M$ is a supertemperature such as is described in the theorem. If v is another such supertemperature, then because $v = w$ on $M \backslash \{p_0\}$ we have $v \geq \widehat{R}_v^{M \backslash \{p_0\}} = \widehat{R}_w^{M \backslash \{p_0\}} = \widehat{R}_w^M = u$ on E, by Theorem 7.27(e). Furthermore, whenever $0 < c \leq c_0$, Theorem 6.45 shows that

$$\mathcal{M}(v; p_0; c) = v(p_0) = \mathcal{M}(v; p_0; c_0) = \mathcal{M}(u; p_0; c_0) = u(p_0) = \mathcal{M}(u; p_0; c).$$

Since $v - u$ is nonnegative and continuous on Ω, it follows that $v = u$ on Ω, and hence everywhere on E. □

9.6. Thermal Fine Limits and Euclidean Limits

Let N be a deleted neighbourhood of a point q. A function defined on N with a thermal fine limit at q need not have a Euclidean limit there. For example, the characteristic function of the half-space $\mathbb{R}^n \times {]}0, +\infty{[}$ is a supertemperature on $\mathbb{R}^{n+1}$, and hence has a thermal fine limit at every point of the hyperplane $\mathbb{R}^n \times \{0\}$, although no Euclidean limit exists at such a point. However, Theorem 9.45 below shows that if f has a thermal fine limit at q, then there is a thermally thin set L such that f has a Euclidean limit at q on approach through $N \backslash L$.

LEMMA 9.44. *Let $\{L_j\}$ be a sequence of sets each of which is thermally thin at the point $q \in \mathbb{R}^{n+1}$. Then there exist a set L which is thermally thin at q, and a sequence $\{r_j\}$ of positive numbers, such that $L_j \cap B(q, r_j) \subseteq L$ for all j.*

PROOF. For each $j \in \mathbb{N}$, we choose a positive number r_j and a function v_j, as follows. If q is not a limit point of L_j, then we choose r_j such that $L_j \cap B(q, r_j) \subseteq \{q\}$ and choose $v_j = 0$. On the other hand, if q is a limit point of L_j, Theorem 9.15 shows that there exists a heat potential w_j on $\mathbb{R}^{n+1}$ such that

$$+\infty = \lim_{p \to q,\, p \in L_j} w_j(p) > w_j(q).$$

We multiply w_j by a positive constant, if necessary, to obtain a heat potential $v_j = \kappa_j w_j$ such that $v_j(q) < 2^{-j}$. We then choose r_j such that $v_j \geq 1$ on the set $B(q, r_j) \cap L_j \backslash \{q\}$.

We now put $L = \bigcup_{j=1}^\infty (L_j \cap B(q, r_j))$ and $v = (\sum_{j=1}^\infty v_j) \wedge 2$ on $\mathbb{R}^{n+1}$. Then v is a nonnegative supertemperature on $\mathbb{R}^{n+1}$, and $v(q) < 1$. If q is not a limit point of L, then it is not a limit point of L_j for any j, so that $L_j \cap B(q, r_j) \subseteq \{q\}$ for all j and the result is trivial. If q is a limit point of L, then $L \not\subseteq \{q\}$ so that for at least one k the inclusion $L_k \cap B(q, r_k) \subseteq \{q\}$ is not valid, which implies that q is

a limit point of L_k. If $p \in L\setminus\{q\}$, then $p \in L_k \cap B(q, r_k)$ for some such k, so that $v(p) \geq v_k(p) \wedge 2 \geq 1$. Hence
$$\liminf_{p \to q,\, p \in L} v(p) \geq 1 > v(q),$$
so that L is thermally thin at q, by Theorem 9.11. □

THEOREM 9.45. *Let f be a function on a thermal fine deleted neighbourhood N of the point $q \in \mathbb{R}^{n+1}$. Then f has a thermal fine limit l at q if and only if there is a set L which is thermally thin at q such that*
$$\lim_{p \to q,\, p \in N \setminus L} f(p) = l.$$

PROOF. We can suppose that l is finite, for otherwise we could replace f by $\tan^{-1} \circ f$.

If f has a thermal fine limit l at q, then for each positive integer j we can find a thermal fine neighbourhood V_j of q such that $|f(p) - l| < j^{-1}$ whenever $p \in V_j \setminus \{q\}$. Therefore the set $\{p \in N \setminus \{q\} : |f(p) - l| < j^{-1}\}$ is a thermal fine deleted neighbourhood of q, and so the set
$$L_j = \{p \in N \setminus \{q\} : |f(p) - l| \geq j^{-1}\}$$
is thermally thin at q. By Lemma 9.44, there exist a set L which is thermally thin at q, and a sequence $\{r_j\}$ of positive numbers, such that $L_j \cap B(q, r_j) \subseteq L$ for all j. The set $N \setminus L$ is thus a thermal fine deleted neighbourhood of q, and is a subset of $N \setminus (L_j \cap B(q, r_j))$ for every j. It follows that, if $k \in \mathbb{N}$ and p belongs to $(N \setminus L) \cap B(q, r_k)$, then $p \in N \setminus (L_k \cap B(q, r_k))$ and so $|f(p) - l| < k^{-1}$. Thus $f(p) \to l$ as $p \to q$ through $N \setminus L$.

Conversely, if there is a set L which is thermally thin at q such that $f(p) \to l$ as $p \to q$ through $N \setminus L$, then for each $\epsilon > 0$ we can find $r > 0$ such that $|f(p) - l| < \epsilon$ whenever $p \in N \cap B(q, r) \setminus (L \cup \{q\})$. For each r the set $N \cap B(q, r) \setminus (L \cup \{q\})$ is a thermal fine deleted neighbourhood of q, and so f has a thermal fine limit l at q. □

9.7. Thermal Thinness and the Quasi-Lindelöf Property

The main purpose of this section is to establish the quasi-Lindelöf property of the thermal fine topology. This is a weak form of the Lindelöf property of, for example, the Euclidean topology on $\mathbb{R}^{n+1}$. It says that any collection of thermal fine open sets has a countable subcollection whose union differs from the union of the whole collection by a semipolar subset of $\mathbb{R}^{n+1}$. To prove this result, we first need to show that $L \setminus L^\Theta$ is a semipolar subset of $\mathbb{R}^{n+1}$ for any set L, and that there exist heat potentials which characterize thermal thinness.

THEOREM 9.46. *For any subset L of $\mathbb{R}^{n+1}$, the set of thermal fine limit points L^Θ is a G_δ set, and the set of points of L where L is thermally thin is a semipolar subset of $\mathbb{R}^{n+1}$.*

PROOF. Let $\{B_j\}$ be a sequence of balls which forms a base for the Euclidean topology of $\mathbb{R}^{n+1}$. Reductions here are relative to $\mathbb{R}^{n+1}$. By Lemma 9.25, for any point $q \notin L^\Theta$ we have
$$\lim_{r \to 0+} \widehat{R}_1^{L \cap B(q, r)}(q) = 0,$$

so that we can find j such that $\widehat{R}_1^{L \cap B_j}(q) \leq 1/2$. On the other hand, if $q \in L^\Theta$, then for any j such that $q \in B_j$ we have $q \in (L \cap B_j)^\Theta$, so that $\widehat{R}_1^{L \cap B_j}(q) = 1$ by Theorem 9.20. Writing $A_j = \{q \in B_j : \widehat{R}_1^{L \cap B_j}(q) \leq 1/2\}$ for all j, we therefore obtain the identity

$$L^\Theta = \mathbb{R}^{n+1} \backslash \left(\bigcup_{j=1}^\infty A_j\right) = \bigcap_{j=1}^\infty (\mathbb{R}^{n+1} \backslash A_j).$$

Since smoothed reductions are lower semicontinuous, each set A_j is closed, and hence L^Θ is a G_δ set.

For the last part, Theorem 9.27 shows that $\widehat{R}_1^{L \cap B_j} = R_1^{L \cap B_j} = 1$ on $L \cap B_j$ except for a semipolar subset of $\mathbb{R}^{n+1}$. Thus $L \cap A_j$ is semipolar for all j, and hence the set $L \backslash L^\Theta = \bigcup_{j=1}^\infty (L \cap A_j)$ is also semipolar. □

COROLLARY 9.47. *For any open set E, the union of $\partial_a E$ with the set of all irregular points of $\partial_n E$ is a semipolar subset of $\mathbb{R}^{n+1}$. The set of finite regular points of $\partial_n E$ is a G_δ set.*

PROOF. We put $F = \mathbb{R}^{n+1} \backslash E$, and observe that every point of $\mathbb{R}^{n+1} \backslash \overline{E}$ is a thermal fine limit point of F, in view of Corollary 9.13. By Theorem 9.40, any finite point of ∂E belongs to F^Θ if and only if it is a regular point of $\partial_n E$. Also, if E is unbounded then the point at infinity is regular, by Theorem 8.46(b). Therefore the union in question is $F \backslash F^\Theta$, and hence is a semipolar subset of $\mathbb{R}^{n+1}$ by Theorem 9.46. Moreover, F^Θ is a G_δ set, by Theorem 9.46, and so the same is true of $F^\Theta \cap \partial E$, which is the set of finite regular points of $\partial_n E$, by Theorem 9.40. □

COROLLARY 9.48. *For any function u on an open set E in $\mathbb{R}^{n+1}$, the functions $q \mapsto \Theta - f\limsup_{p \to q} u(p)$ and $q \mapsto \Theta - f\liminf_{p \to q} u(p)$ are Borel measurable on E.*

PROOF. For any real number a, we have

$$\{q \in E : \Theta - f\limsup_{p \to q} u(p) > a\} = \bigcup_{k=1}^\infty \{p \in E : u(p) \geq a + k^{-1}\}^\Theta.$$

By Theorem 9.46, each set in this union is a G_δ set, and so the union is a Borel set. Thus the first function in question is Borel measurable, and the arbitrariness of u implies that the second one is too. □

We now establish the existence of heat potentials which characterize thermal thinness.

THEOREM 9.49. *Given an open set E, there is a bounded, continuous heat potential $u^\#$ on E such that*

$$L^\Theta \cap E = \{p \in E : \widehat{R}_{u^\#}^L(p) = u^\#(p)\}$$

for every subset L of E.

PROOF. Given any open circular cylinder $D = B \times \,]b,c[$ such that $\overline{D} \subseteq E$, we take a circular cylinder $\Delta = B \times \,]a,c[$ such that $c - b < c - a < 2(c-b)$ and $\overline{\Delta} \subseteq E$. We then take a continuous increasing function f on $\mathbb{R}$ such that $f(t) = 0$ if $t \leq a$, $0 < f(t) < 1$ if $a < t < b$, and $f(t) = 1$ if $t \geq b$, and define a supertemperature u by putting $u(x,t) = f(t)$ for all $(x,t) \in E$. Since Δ is open, we have $R_u^\Delta = \widehat{R}_u^\Delta$ on E, by Theorem 7.31(a). Moreover, because R_u^Δ is thereby lower semicontinuous, and $R_u^\Delta \leq u$ on E, if there is a point p such that $R_u^\Delta(p) = u(p)$, then the inequalities

$$R_u^\Delta(p) \leq \liminf_{q \to p} R_u^\Delta(q) \leq \limsup_{q \to p} R_u^\Delta(q) \leq \lim_{q \to p} u(q) = u(p)$$

imply that R_u^Δ is continuous at p. We know that R_u^Δ is continuous on $E \backslash \partial \Delta$, because $R_u^\Delta = u$ on Δ, and R_u^Δ is a temperature on $E \backslash \overline{\Delta}$ by Theorem 7.27(d). Furthermore, on $\overline{B} \times \{a\}$ we have $0 \leq R_u^\Delta \leq u = 0$, so that equality holds and hence R_u^Δ is continuous. Moreover, Example 9.42 implies that $\partial \Delta \backslash (\overline{B} \times \{a\}) \subseteq \Delta^\Theta$, so that $R_u^\Delta = u$ there by Corollary 9.28, and hence R_u^Δ is continuous there. Thus R_u^Δ is continuous on E, and is a heat potential by Theorem 7.28. Clearly $0 \leq R_u^\Delta \leq u \leq 1$ on E, and $R_u^\Delta = u = 1$ on D.

We now take a sequence $\{D_j\}$ of open circular cylinders, with closures in E, which forms a base for the Euclidean topology of E. For each j, we define Δ_j and u_j relative to D_j, as we defined Δ and u relative to D above. Then the sequence of sets $\{\Delta_j\}$ also forms a base for the Euclidean topology of E. We put

$$u^\# = \sum_{j=1}^\infty 2^{-j} R_{u_j}^{\Delta_j}$$

on E. Since $0 \leq R_{u_j}^{\Delta_j} \leq 1$ on E for all j, the series converges uniformly on E, and so $u^\#$ is a continuous heat potential such that $0 \leq u^\# \leq 1$ on E. If $L \subseteq E$ and $p \in L^\Theta \cap E$, then $\widehat{R}_{u^\#}^L(p) = u^\#(p)$ by Corollary 9.28.

On the other hand, if $p \in E \backslash L^\Theta$, we take a contracting subsequence $\{D_{j_k}\}$ of $\{D_j\}$ such that $\bigcap_{k=1}^\infty D_{j_k} = \{p\}$, and put

$$v_1 = \sum_{k=1}^\infty 2^{-j_k} R_{u_{j_k}}^{\Delta_{j_k}}, \qquad v_2 = u^\# - v_1$$

on E. As for $u^\#$, the functions v_1 and v_2 are continuous heat potentials whose values lie in $[0,1]$. We show that v_1 peaks at p, with a view to applying Theorem 9.38. Let U be an open ball with centre p such that $E \backslash U \neq \emptyset$. We choose a point q and a positive number c, such that $p \in \Omega^*(q;c)$ and $\overline{\Omega}^*(q;4c) \subseteq U$. By our choice of $\{\Delta_{j_k}\}$, there is an integer k_0 such that $\Delta_{j_k} \subseteq \Omega^*(q;c)$ for all $k \geq k_0$. The function $v = (\tau(c)^{-1} G(\cdot;q)) \wedge 1$ is a nonnegative supertemperature on $\mathbb{R}^{n+1}$ which takes the value 1 at every point of $\Omega^*(q;c)$, so that $v \geq R_{u_{j_k}}^{\Delta_{j_k}}$ for all $k \geq k_0$. Moreover, on $\partial \Omega^*(q;4c) \backslash \{q\}$ we have $G(\cdot;q) = \tau(4c) = 2^{-n}\tau(c)$, so that $v = 2^{-n}$. Since the restriction of v to $\mathbb{R}^{n+1} \backslash \overline{\Omega}^*(q;c)$ is a bounded temperature which tends to zero at infinity, it follows from the maximum principle of Theorem 8.2 that $v \leq 2^{-n}$ on $\mathbb{R}^{n+1} \backslash \overline{\Omega}^*(q;4c)$, so that $R_{u_{j_k}}^{\Delta_{j_k}} \leq 2^{-n}$ on $E \backslash \overline{\Omega}^*(q;4c)$. Therefore

$$\sup_{E \backslash U} v_1 \leq \sup_{E \backslash \overline{\Omega}^*(q;4c)} v_1 \leq \sum_{k=1}^{k_0-1} 2^{-j_k} + \sum_{k=k_0}^\infty 2^{-j_k-n} < \sum_{k=1}^\infty 2^{-j_k} = v_1(p).$$

Hence v_1 peaks at p. If w is a nonnegative supertemperature on E such that $w \geq v_1$ on L, then $w + v_2 \geq u^\#$ on L, so that $w + v_2 \geq \widehat{R}_{u^\#}^L$ on E, and hence $R_{v_1}^L + v_2 \geq \widehat{R}_{u^\#}^L$ on E. Thus $\widehat{R}_{u^\#}^L - v_2$ is a lower semicontinuous minorant of $R_{v_1}^L$, so that it is also

one of $\widehat{R}^L_{v_1}$. Since v_1 peaks at p, Theorem 9.38 shows that $\widehat{R}^L_{v_1}(p) < v_1(p)$. Hence
$$\widehat{R}^L_{u^\#}(p) \leq \widehat{R}^L_{v_1}(p) + v_2(p) < v_1(p) + v_2(p) = u^\#(p),$$
as required. □

We now use Theorems 9.46 and 9.49 to prove the *quasi-Lindelöf property* of the thermal fine topology.

THEOREM 9.50. *Let $\{D_\alpha : \alpha \in I\}$ be a collection of thermal fine open sets. Then there is a countable subcollection $\{D_\alpha : \alpha \in K\}$ such that the set*
$$\left(\bigcup_{\alpha \in I} D_\alpha\right) \setminus \left(\bigcup_{\alpha \in K} D_\alpha\right)$$
is a semipolar subset of $\mathbb{R}^{n+1}$.

PROOF. We put $F_\alpha = \mathbb{R}^{n+1} \setminus D_\alpha$ for all $\alpha \in I$, and note that the set in question can be written as $\left(\bigcap_{\alpha \in K} F_\alpha\right) \setminus \left(\bigcap_{\alpha \in I} F_\alpha\right)$. By Theorem 9.49, we can find a bounded continuous heat potential $u^\#$ on $\mathbb{R}^{n+1}$ such that
$$L^\Theta = \{p : \widehat{R}^L_{u^\#}(p) = u^\#(p)\}$$
for every subset L of $\mathbb{R}^{n+1}$. For each subset J of I, we put
$$w_J = \inf\{\widehat{R}^{F_\alpha}_{u^\#} : \alpha \in J\}.$$
By Lemma 7.12, there is a countable set $K \subseteq I$ such that $\widehat{w}_K = \widehat{w}_I$ on $\mathbb{R}^{n+1}$. We put $F = \bigcap_{\alpha \in K} F_\alpha$, and take any point $q \in F$. If $q \in F^\Theta$ and $\beta \in I$, then
$$\widehat{R}^{F_\beta}_{u^\#}(q) \geq \widehat{w}_I(q) = \widehat{w}_K(q) \geq \widehat{R}^F_{u^\#}(q) = u^\#(q).$$
Therefore equality holds throughout, so that $q \in F^\Theta_\beta$, and hence $q \in F_\beta$ because F_β is thermal fine closed. Thus $F^\Theta \subseteq \bigcap_{\beta \in I} F_\beta$, so that $F \setminus \left(\bigcap_{\beta \in I} F_\beta\right) \subseteq F \setminus F^\Theta$, which is a semipolar subset of $\mathbb{R}^{n+1}$ by Theorem 9.46. □

COROLLARY 9.51. *Any thermal fine open set can be written as the union of a (Euclidean) F_σ set and a semipolar subset of $\mathbb{R}^{n+1}$.*

PROOF. Let D be a thermal fine open set. By Lemma 9.6, for each point $p \in D$ we can find a compact set $K_p \subseteq D$ that is a thermal fine neighbourhood of p, and hence a thermal fine open set D_p such that $p \in D_p \subseteq K_p$. By Theorem 9.50, the collection $\{D_p : p \in D\}$ has a countable subcollection $\{D_p : p \in J\}$ such that $D \setminus \left(\bigcup_{p \in J} D_p\right)$ is a semipolar subset of $\mathbb{R}^{n+1}$. The set $K = \bigcup_{p \in J} K_p$ is an F_σ set such that $D \setminus K \subseteq D \setminus \left(\bigcup_{p \in J} D_p\right)$, so that $D \setminus K$ is a semipolar subset of $\mathbb{R}^{n+1}$. □

We conclude with an example which shows that the thermal fine topology does not possess the full Lindelöf property, and that, indeed, the word "semipolar" in the conclusion of Theorem 9.50 cannot be replaced by "polar".

EXAMPLE 9.52. It follows from Example 9.41 that, for any point $q \in \mathbb{R}^{n+1}$ and any positive number c, the set $\Omega(q;c) \cup \{q\}$ is thermal fine open. For any fixed $c > 0$, the union of the collection $\{\Omega(q;c) \cup \{q\} : q \in \mathbb{R}^n \times \{0\}\}$ is the strip $\mathbb{R}^n \times\,]-c, 0]$. Any countable subcollection can cover only countably many points of $\mathbb{R}^n \times \{0\}$, so that the set of points of $\mathbb{R}^n \times \{0\}$ that remain uncovered has infinite thermal capacity, by Theorem 7.55, and hence is not polar, by Theorem 7.46.

9.8. Notes and Comments

Most of the results of this chapter were taken from Doob [**14**], which contains many other results on the thermal fine topology and closely related topics, as does Bliedtner & Hansen [**7**]. For example, Doob shows that a set is a semipolar subset of E if and only if it is a cosemipolar subset of E.

Theorem 9.43 comes from Watson [**80**]. We do not have a characterization of the open sets that have a uniqueness property similar to that of the heat ball in Theorem 9.43. This remains an **open question**.

Generalizations of Example 9.42 to other parabolic regions, have been given by Hansen [**32**] and Netuka [**55**].

The Wiener test, for the regularity of normal boundary points relative to the Dirichlet problem, has been generalized by Brzezina [**9**] to a test for the thermal thinness of an arbitrary set at a point.

Bibliography

[1] U. G. Abdulla, *First boundary value problem for the diffusion equation. 1. Iterated logarithm test for the boundary regularity and solvability*, SIAM J. Math. Anal. **34** (2003), 1422-1434.

[2] U. G. Abdulla, *Multidimensional Kolmogorov-Petrovsky test for the boundary regularity and irregularity of solutions to the heat equation*, Boundary Value Probl. **2005** (2005), 181-199.

[3] D. H. Armitage & S. J. Gardiner, *Classical Potential Theory*, Springer-Verlag, London, 2001.

[4] D. G. Aronson, *Non-negative solutions of linear parabolic equations*, Ann. Scuola Norm. Sup. Pisa Cl. Sci. **22** (1968), 607-694.

[5] H. Bauer, *Harmonische Räume und ihre Potentialtheorie*, Lecture Notes in Mathematics 22, Springer-Verlag, Berlin, 1966.

[6] H. Bauer, *Heat balls and Fulks measures*, Ann. Acad. Sci. Fenn. Ser. A I Math. **10** (1985), 67-82.

[7] J. Bliedtner & W. Hansen, *Potential Theory. An analytic and probabilistic approach to balayage*, Springer-Verlag, Berlin, 1986.

[8] T. Bousch & Y. Heurteaux, *Caloric measure on domains bounded by Weierstrass-type graphs*, Ann. Acad. Sci. Fenn. Math. **25** (2000), 501-522.

[9] M. Brzezina, *On the base and essential base in parabolic potential theory*, Czech. Math. J. **40** (1990), 87-103.

[10] M. Brzezina, *Capacitary interpretation of the Fulks measure*, Exposition. Math. **11** (1993), 469-474.

[11] M. Brzezina, *A note on the convexity theorem for mean values of subtemperatures*, Ann. Acad. Sci. Fenn. Math. **21** (1996), 111-115.

[12] C. Constantinescu & A. Cornea, *Potential Theory on Harmonic Spaces*, Grundlehren der mathematischen Wissenschaften vol. 158, Springer-Verlag, New York, 1972.

[13] M. Dont, *On the continuity of heat potentials*, Časopis Pěst. Mat. **106** (1981), 156-167.

[14] J. L. Doob, *Classical Potential Theory and its Probabilistic Counterpart*, Grundlehren der mathematischen Wissenschaften vol. 262, Springer-Verlag, New York, 1984.

[15] E. G. Effros & J. L. Kazdan, *Application of Choquet simplexes to elliptic and parabolic boundary value problems*, J. Differential Equations **8** (1970), 95-134.

[16] E. G. Effros & J. L. Kazdan, *On the Dirichlet problem for the heat equation*, Indiana Univ. Math. J. **20** (1970-1), 683-693.

[17] L. C. Evans, *Partial Differential Equations*, Graduate Studies in Mathematics vol. 19, Amer. Math. Soc., Providence, RI, 1998.

[18] L. C. Evans & R. F. Gariepy, *Wiener's criterion for the heat equation*, Arch. Rational Mech. Anal. **78** (1982), 293-314.

[19] E. Fabes & S. Salsa, *Estimates of caloric measure and the initial-Dirichlet problem for the heat equation in Lipschitz cylinders*, Trans. Amer. Math. Soc. **279** (1983), 635-650.

[20] E. B. Fabes, N. Garofalo & S. Salsa, *Comparison theorems for temperatures in non-cylindrical domains*, Atti Accad. Naz. Lincei. Rend. Cl. Sci. Fis. Mat. Natur. **77** (1984), 1-12.

[21] T. M. Flett, *Temperatures, Bessel potentials and Lipschitz spaces*, Proc. London Math. Soc. **22** (1971), 385-451.

[22] A. Friedman, *On the uniqueness of the Cauchy problem for parabolic equations*, Amer. J. Math. **81** (1959), 503-511.

[23] A. Friedman, *Partial differential equations of parabolic type*, Prentice-Hall, Englewood Cliffs, 1964.

[24] W. Fulks, *A mean value theorem for the heat equation*, Proc. Amer. Math. Soc. **17** (1966), 6-11.

[25] R. Gariepy & W. P. Ziemer, *Removable sets for quasilinear parabolic equations*, J. London Math. Soc. **21** (1980), 311-318.

[26] N. Garofalo & E. Lanconelli, *Asymptotic behaviour of fundamental solutions and potential theory of parabolic operators with variable coefficients*, Math. Ann. **283** (1989), 211-239.

[27] R. Ya. Glagoleva, *An a priori estimate for the Hölder norm and the Harnack inequality for the solution of a linear parabolic differential equation of second order with discontinuous coefficients*, Mat. Sbornik **76** (1968), 167-185; Math. U.S.S.R. Sbornik **5** (1968), 159-176.

[28] R. Ya. Glagoleva, *Liouville theorems for the solution of a second-order linear parabolic equation with discontinuous coefficients*, Mat. Zametki **5** (1969), 599-606; Math. Notes **5** (1969), 359-363.

[29] A. L. Gusarov, *On a theorem of the Liouville type for parabolic equations*, Vestnik Moskov. Univ. Ser. I. Mat. Meh. **30** (1975), 62-70; Moskow Univ. Math. Bull. **30** (1975), 50-56.

[30] A. L. Gusarov, *On a sharp Liouville theorem for solutions of a parabolic equation on a characteristic*, Mat. Sbornik (N.S.) **97**(139) (1975), 379-94,463; Math. U.S.S.R. Sbornik **26** (1975), 349-364.

[31] J. Hadamard, *Extension à l'équation de la chaleur d'un théorème de A. Harnack*, Rend. Circ. Mat. Palermo **3** (1954), 337-346.

[32] W. Hansen, *Fegen und dünnheit mit anwendungen auf die Laplace- und wärmeleitungsgleichung*, Ann. Inst. Fourier (Grenoble) **21** (1971), 79-121.

[33] L. L. Helms, *Introduction to Potential Theory*, Wiley-Interscience, New York, 1969.

[34] Y. Heurteaux, *Mesure harmonique et l'equation de la chaleur*, Ark. Mat. **34** (1996), 119-139.

[35] S. Hofmann, J. L. Lewis & K. Nyström, *Caloric measure in parabolic flat domains*, Duke Math. J. **122** (2004), 281-346.

[36] K. M. Hui, *Another proof for the removable singularities of the heat equation*, Proc. Amer. Math. Soc. **138** (2010), 2397-2402.

[37] W. Kaiser & B. Müller, *Removable sets for the heat equation*, Vestnik Moscov. Univ. Ser. I Mat. Meh. **28** (1973), 26-32; Moscow Univ. Math. Bull. **28** (1973), 21-26.

[38] R. Kaufman & J.-M. Wu, *Singularity of parabolic measures*, Compositio Math. **40** (1980), 243-250.

[39] R. Kaufman & J.-M. Wu, *Parabolic potential theory*, J. Differential Equations **43** (1982), 204-234.

[40] R. Kaufman & J.-M. Wu, *Parabolic measure on domains of class $Lip\frac{1}{2}$*, Compositio Math. **66** (1988), 201-207.

[41] R. Kaufman & J.-M. Wu, *Dirichlet problem of heat equation for C^2 domains*, J. Differential Equations **80** (1989), 14-31.

[42] J. T. Kemper, *Temperatures in several variables: Kernel functions, representations, and parabolic boundary values*, Trans. Amer. Math. Soc. **167** (1972), 243-262.

[43] J. Král, *Hölder-continuous heat potentials*, Atti Accad. Naz. Lincei Rend. Cl. Sci. Fis. Mat. Natur. **51** (1971), 17-19.

[44] M. Krzyżański, *Sur les solutions non négatives de l'équation linéaire normale parabolique*, Rev. Roumaine Math. Pures Appl. **9** (1964), 393-408.

[45] L. P. Kuptsov, *Mean property for the heat-conduction equation*, Mat. Zametki **29** (1981), no. 2, 211-223; Math. Notes **29** (1981), no. 2, 110-116.

[46] E. Lanconelli, *Sul problema di Dirichlet per l'equazione del calore*, Ann. Mat. Pura Appl. **97** (1973), 83-114.

[47] E. Lanconelli & A. Pascucci, *Superparabolic functions related to second order hypoelliptic operators*, Potential Anal. **11** (1999), 303-323.

[48] E. M. Landis, *Necessary and sufficient conditions for the regularity of a boundary point for the Dirichlet problem for the heat equation*, Dokl. Akad. Nauk SSSR **185** (1969), 517-520; Soviet Math. Dokl. **10** (1969), 380-384.

[49] E. M. Landis, *Second Order Equations of Elliptic and Parabolic Type*, Translations of Math. Monographs vol. 171, Amer. Math. Soc., Providence, RI, 1998.

[50] J. L. Lewis & J. Silver, *Parabolic measure and the Dirichlet problem for the heat equation in two dimensions*, Indiana Univ. Math. J. **37** (1988), 801-839.

[51] F.-Y. Maeda, *Capacities on harmonic spaces with adjoint structure*, Potential Theory, Ed. J. Král et al, Plenum Press, New York, 1988, 231-236.

[52] J. Moser, *A Harnack inequality for parabolic differential equations*, Comm. Pure Appl. Math. **17** (1964), 101-134; Correction, ibidem **20** (1967), 231-236.

[53] J. Moser, *On a pointwise estimate for parabolic differential equations*, Comm. Pure Appl. Math. **24** (1971), 727-740.

[54] P. I. Mysovskikh, *Parabolic capacity of a set and some of its properties and bounds*, Differentsial'nye Uravneniya **23** (1987), 1567-1574; Differential Equations **23** (1987), 1081-1086.

[55] I. Netuka, *Thinness and the heat equation*, Časopis Pěst. Mat. **99** (1974), 293-299.

[56] L. Nirenberg, *A strong maximum principle for parabolic equations*, Comm. Pure Appl. Math. **6** (1953), 167-177.

[57] K. Nyström, *The Dirichlet problem for second order parabolic operators*, Indiana Univ. Math. J. **46** (1997), 183-245.

[58] I. Petrowsky, *Zur ersten randwertaufgabe der wärmeleitungsgleichung*, Compositio Math. **1** (1935), 383-419.

[59] B. Pini, *Sulla soluzione generalizzata di Wiener per il primo problema di valori al contorno nel caso parabolico*, Rend. Sem. Mat. Univ. Padova **23** (1954), 422-434.

[60] B. Pini, *Maggioranti e minoranti delle soluzione delle equazioni paraboliche*, Ann. Mat. Pura Appl. **37** (1954), 249-264.

[61] L. E. Slobodetskij, *On the Cauchy problem for nonhomogeneous parabolic systems*, Dokl. Akad. Nauk S.S.S.R. **101** (1955), 805-808.

[62] E. P. Smyrnélis, *Sur les moyennes des fonctions paraboliques*, Bull. Sci. Math. **93** (1969), 163-174.

[63] W. Sternberg, *Über die Gleichung der Wärmeleitung*, Math. Ann. **101** (1929), 394-398.

[64] N. Suzuki, *On the essential boundary and supports of harmonic measures for the heat equation*, Proc. Japan Acad. Ser. A Math. Sci. **56** (1980), 381-385.

[65] N. Suzuki & N. A. Watson, *A characterization of heat balls by a mean value property of temperatures*, Proc. Amer. Math. Soc. **129** (2001), 2709-2713.

[66] N. Suzuki & N. A. Watson, *Mean value densities for temperatures*, Colloq. Math. **98** (2003), 87-96.

[67] S. J. Taylor & N. A. Watson, *A Hausdorff measure classification of polar sets for the heat equation*, Math. Proc. Cambridge Phil. Soc. **97** (1985), 325-344.

[68] V. S. Umanskiĭ, *On removable sets of solutions of second-order parabolic equations*, Dokl. Akad. Nauk S.S.S.R. **294** (1987), 545-548; Soviet Math. Dokl. **35** (1987), 566-569.

[69] N. A. Watson, *A theory of subtemperatures in several variables*, Proc. London Math. Soc. **26** (1973), 385-417.

[70] N. A. Watson, *Classes of subtemperatures on infinite strips*, Proc. London Math. Soc. **27** (1973), 723-746.

[71] N. A. Watson, *On the definition of a subtemperature*, J. London Math. Soc. **7** (1973), 195-198.

[72] N. A. Watson, *Green functions, potentials, and the Dirichlet problem for the heat equation*, Proc. London Math. Soc. **33** (1976), 251-298; Corrigendum, ibidem **37** (1978), 32-34.

[73] N. A. Watson, *Thermal capacity*, Proc. London Math. Soc. **37** (1978), 342-362.

[74] N. A. Watson, *The rate of spatial decay of non-negative solutions of linear parabolic equations*, Arch. Rational Mech. Anal. **68** (1978), 121-124.

[75] N. A. Watson, *Uniqueness and representation theorems for the inhomogeneous heat equation*, J. Math. Anal. Appl. **67** (1979), 513-524.

[76] N. A. Watson, *Positive thermic majorization of temperatures on infinite strips*, J. Math. Anal. Appl. **68** (1979), 477-487.

[77] N. A. Watson, *The asymptotic behaviour of temperatures and subtemperatures*, Proc. London Math. Soc. **42** (1981), 501-532.

[78] N. A. Watson, *The weak maximum principle for parabolic differential inequalities*, Rend. Circ. Mat. Palermo **33** (1984), 421-425.

[79] N. A. Watson, *Parabolic Equations on an Infinite Strip*, Monographs and Textbooks in Pure and Applied Mathematics vol. 127, Marcel Dekker, New York, 1989.

[80] N. A. Watson, *A convexity theorem for local mean values of subtemperatures*, Bull. London Math. Soc. **22** (1990), 245-252.

[81] N. A. Watson, *Mean values and thermic majorization of subtemperatures*, Ann. Acad. Sci. Fenn. Ser. A I. Math. **16** (1991), 113-124.

[82] N. A. Watson, *Cofine boundary behaviour of temperatures*, Hiroshima Math. J. **22** (1992), 103-113.

[83] N. A. Watson, *Mean values of subtemperatures over level surfaces of Green functions*, Ark. Mat. **30** (1992), 165-185.

[84] N. A. Watson, *Generalizations of the spherical mean convexity theorem on subharmonic functions*, Ann. Acad. Sci. Fenn. Ser. A I. Math. **17** (1992), 241-255.

[85] N. A. Watson, *Nevanlinna's first fundamental theorem for supertemperatures*, Math. Scand. **73** (1993), 49-64.

[86] N. A. Watson, *Time-isolated singularities of temperatures*, J. Austral. Math. Soc. Ser. A **65** (1998), 416-429.

[87] N. A. Watson, *Elementary proofs of some basic subtemperature theorems*, Colloq. Math. **94** (2002), 111-140.

[88] N. A. Watson, *A generalized Nevanlinna theorem for supertemperatures*, Ann. Acad. Sci. Fenn. Math. **28** (2003), 35-54.

[89] N. A. Watson, *A unifying definition of a subtemperature*, New Zealand J. Math. **38** (2008), 197-223.

[90] N. A. Watson, *Caloric measure for arbitrary open sets*, J. Austral. Math. Soc. (to appear)

[91] D. V. Widder, *Positive temperatures on an infinite rod*, Trans. Amer. Math. Soc. **55** (1944), 85-95.

[92] J.-M. G. Wu, *On parabolic measures and subparabolic functions*, Trans. Amer. Math. Soc. **251** (1979), 171-185; Erratum, ibidem **259** (1980), 636.

[93] J.-M. Wu, *On heat capacity and parabolic measure*, Math. Proc. Cambridge Phil. Soc. **102** (1987), 163-172.

[94] J.-M. Wu, *An example on null sets of parabolic measures*, Proc. Amer. Math. Soc. **107** (1989), 949-961.

Index

$A(x_0, t_0; b, c)$, 5
$C(S)$, 17
$C^{2,1}(E)$, 1
$C_c(E)$, 140
$C_c^{2,1}(E)$, 140
G_E, 127
G_E^*, 132
$H(p_0, r)$, 196
$H^*(p_0, r)$, 197
L^Θ, 239
L_f, 66
L_f^E, 199
$L_u(\phi)$, 140
$M_b(v; t)$, 105
$Q(x_0 - x, t_0 - t)$, 4
$R(x_0; \rho; a, b)$, 64
R_u^L, 170
S_f, 66
S_f^E, 205
U_f, 66
U_f^E, 199
W, 2
$\Delta(x, t; c)$, 48
Λ-sequence, 161
$\Lambda(p_0; E)$, 25, 197
$\Lambda^*(q : E)$, 130
$\Omega(p_0; c)$, 2
$\Omega_m(x_0, t_0; c)$, 19
Φ_a, 119
Σ_b, 107
Θ, 1
Θ^*, 1
$\Theta_{x,t}$, 1
$\langle \cdot, \cdot \rangle$, 3
$\mathcal{C}(K)$, 175
$\mathcal{C}(S)$, 179
$\mathcal{C}^*$, 179
$\mathcal{C}^*(K)$, 176
$\mathcal{C}_+(S)$, 178
$\mathcal{C}_-(S)$, 178
$\mathcal{L}(u; x, t; c)$, 48
$\mathcal{M}(u; x_0, t_0; c)$, 4
$\mathcal{V}(u; x_0, t_0; c)$, 15
$\mathcal{V}_m(u; x_0, t_0; c)$, 20
$\mathfrak{L}_f$, 66
$\mathfrak{L}_f^E$, 199
$\mathfrak{U}_f$, 65
$\mathfrak{U}_f^E$, 199
μ_p^E-measurable, 208
∇_x, 3
$\nu = (\nu_x, \nu_t)$, 3
ω_n, 2
$\partial_n D$, 18
$\partial_n R$, 64
$\pi_D w$, 62
σ, 3
$\tau(c)$, 2
$\widehat{R}_u^L$, 170
$\widehat{u}$, 162

abnormal boundary point, 197
affine, 9
analytic set, 184
Arzelà-Ascoli Theorem, 29

barrier, 214
boundary
    abnormal, 197
    essential, 197
    lateral
        of a circular cylinder, 18
    normal
        of a circular cylinder, 18
        of a convex domain of revolution, 64
        of an arbitrary open set, 197
    semi-singular, 197
    singular, 197
boundary point
    abnormal, 197
    irregular, 214
    normal, 197
    regular, 214
    semi-singular, 197
    singular, 197

caloric measure
    for a circular cylinder, 45

for an arbitrary open set, 208
capacitable, 179
Cauchy problem, 35
Choquet's topological lemma, 162
circular cylinder, 18
  caloric measure for, 45
    invariance under parabolic dilation, 46
    translation invariance, 45
  initial surface, 18
  lateral boundary, 18
  normal boundary, 18
  Poisson integral, 45
class Φ_a, 119
class Σ_b, 107
coarser, 231
coheat potential, 134
completion, 208
concave, 9
convex, 9
  domain of revolution, 64
  function of ψ, 9
cotemperature, 1
cothermal
  capacitable, 179
  capacitary distribution, 176
  capacitary potential, 176
  capacity, 179
    of a compact set, 176
  inner capacity, 179
  outer capacity, 179

directional derivative, 39
Dirichlet problem
  classical solution, 198
  on a circular cylinder, 37
  on a convex domain of revolution, 64
distributional heat operator, 140
double layer heat potential, 39
  density, 39
downward-directed, 28

equicontinuous
  at a point, 29
  on a set, 29
  uniformly, 29

finer, 231
first boundary value problem, 37
first initial-boundary value problem, 37
fundamental convergence theorem, 244

Gauss-Weierstrass
  integral
    boundary behaviour, 91
    of a function, 88, 123, 125
    of a measure, 88, 99, 122
  kernel, 88
gradient, 3, 39
Green

  cofunction, 132
  formula, 3
  function, 127, 128
    for a convex domain of revolution, 128
    for a rectangle in $\mathbb{R}^2$, 129

Harnack
  inequality, 26, 51
  monotone convergence theorem, 25
Hausdorff maximality theorem, 59
heat
  annulus, 5
    centre, 5
    inner radius, 5
    outer radius, 5
  ball, 2
    centre, 2
    modified, 19
    radius, 2
  cylinder, 48
    centre, 48
    radius, 48
  equation, 1
    adjoint, 1
  operator, 1
    adjoint, 1
    distributional, 140
  potential, 134
    characterization in terms of thermic minorant, 148
    double layer, 39
    double layer density, 39
  sphere, 3
hyperplane mean, 105
hypertemperature, 199
hypotemperature, 57
  characterization, 61
  in terms of $\mathcal{M}$, 76
  in terms of $\mathcal{V}$, 77

initial value problem, 35
inner thermal capacity, 178
invariant
  translation, 1
  under parabolic dilation, 2
irregular, 214

lower class
  on a convex domain of revolution, 66
  on an arbitrary open set, 199
lower PWB solution
  on a convex domain of revolution, 66
  on an arbitrary open set, 199
lower semicontinuous, 53
  smoothing, 162

majorant, 82
maximum principle
  boundary

for hypotemperatures, 59, 161, 197
  for hypotemperatures on a circular
    cylinder, 60
  for hypotemperatures on a convex
    domain of revolution, 65
  for real continuous subtemperatures on
    a circular cylinder, 18
  strong, 58
mean value
  hyperplane, 105
  over heat ball, 15
  over heat spheres, 4
  over normal boundary of heat cylinder,
    48
  volume, 15
minimal, 101
minimum principle
  for temperatures on a circular cylinder,
    18
minorant, 82
modified heat ball, 19

natural order decomposition, 168
normal boundary point, 197

outer thermal capacity, 178

parabolic tusk, 223
parabolically continuous, 29
peaks at, 249
Poisson integral
  for circular cylinder, 45
polar set, 159
  Borel
    characterization in terms of heat
      potentials, 187
    characterization in terms of smoothed
      reductions, 170
problem
  Cauchy, 35
  Dirichlet
    on a circular cylinder, 37
  first boundary value, 37
  first initial-boundary, 37
  initial value, 35
PWB solution
  on a convex domain of revolution, 66
  on an arbitrary open set, 205

quasi-bounded, 126
quasi-everywhere, 159
quasi-Lindelöf property, 256
quasi-regular, 227

reduction, 170
  smoothed, 170
  strong subadditivity property, 174
regular
  boundary point, 214

set, 214
resolutive
  for a convex domain of revolution, 67
  for an arbitrary open set, 205
Riesz decomposition theorem, 146
Riesz measure, 142
  associated with a heat potential, 144

saturated family, 64
semi-singular boundary point, 197
semicontinuous
  lower, 53
  upper, 53
semigroup property, 92
  characterization, 112, 157
  of Gauss-Weierstrass integrals, 92
  of nonnegative temperatures, 98
semipolar, 240
singular boundary point, 197
smoothed reduction, 170
  strong subadditivity property, 174
Stone-Weierstrass theorem, 68
strong subadditivity property, 174
subtemperature, 55
  characterization
    in terms of $\mathcal{M}$, 79
    in terms of $\mathcal{V}$, 79
  characterization in terms of $\mathcal{V}_m$, 151, 152
  real continuous, 17, 77
  smooth, 7, 17, 77
    in terms of $\mathcal{M}$, 8
    in terms of $\mathcal{V}$, 16
    in terms of $\mathcal{V}_m$, 21
  monotone approximation by, 150
supertemperature, 56
  characterization, 166
  fundamental, 56
  monotone approximation by smooth
    ones, 150

temperature, 1
  characterization, 167
  characterization in terms of $\mathcal{L}$, 49
  characterization in terms of $\mathcal{M}$, 8, 49
  characterization in terms of $\mathcal{V}$, 16, 49
  characterization in terms of $\mathcal{V}_m$, 21, 49
  characterization in terms of the Poisson
    integral, 47
  fundamental, 2
  minimal, 101, 165
  on $\overline{D}\backslash\partial_n D$, 37
thermal
  capacitable, 179
  capacitary distribution, 175
  capacitary potential, 175
  capacity, 179
    of a compact set, 175
  fine topology, 231
  inner capacity, 178

outer capacity, 178
thermally thin, 233
thermic majorant, 82
  least, 82
thermic minorant, 82
  characterization in terms of heat
     potential, 148
  greatest, 82, 131

uniqueness principle
  for temperatures on a circular cylinder, 19
upper class
  on a convex domain of revolution, 65
  on an arbitrary open set, 199
upper finite, 56
upper PWB solution
  on a convex domain of revolution, 66
  on an arbitrary open set, 199
    boundary behaviour, 219
upper semicontinuous
  at a point, 54
  on a set, 53
upward-directed, 28

variable
  spatial, 1
  temporal, 1
Vitali-Carathéodory theorem, 108

weak compactness theorem, 98
weak convergence theorem, 98